Aviation Cybersecurity

Related titles on radar:

Advances in Bistatic Radar Willis and Griffiths
Airborne Early Warning System Concepts, 3rd Edition Long
Bistatic Radar, 2nd Edition Willis
Design of Multi-Frequency CW Radars Jankiraman
Digital Techniques for Wideband Receivers, 2nd Edition Tsui
Electronic Warfare Pocket Guide Adamy
Foliage Penetration Radar: Detection and characterisation of objects under trees Davis
Fundamentals of Ground Radar for Air Traffic Control Engineers and Technicians Bouwman
Fundamentals of Systems Engineering and Defense Systems Applications Jeffrey
Introduction to Electronic Warfare Modeling and Simulation Adamy
Introduction to Electronic Defense Systems Neri
Introduction to Sensors for Ranging and Imaging Brooker
Microwave Passive Direction Finding Lipsky
Microwave Receivers with Electronic Warfare Applications Tsui
Phased-Array Radar Design: Application of radar fundamentals Jeffrey
Pocket Radar Guide: Key facts, equations, and data Curry
Principles of Modern Radar, Volume 1: Basic principles Richards, Scheer and Holm
Principles of Modern Radar, Volume 2: Advanced techniques Melvin and Scheer
Principles of Modern Radar, Volume 3: Applications Scheer and Melvin
Principles of Waveform Diversity and Design Wicks *et al.*
Pulse Doppler Radar Alabaster
Radar Cross Section Measurements Knott
Radar Cross Section, 2nd Edition Knott *et al.*
Radar Design Principles: Signal processing and the environment, 2nd Edition Nathanson *et al.*
Radar Detection DiFranco and Rubin
Radar Essentials: A concise handbook for radar design and performance Curry
Radar Foundations for Imaging and Advanced Concepts Sullivan
Radar Principles for the Non-Specialist, 3rd Edition Toomay and Hannan
Test and Evaluation of Aircraft Avionics and Weapon Systems McShea
Understanding Radar Systems Kingsley and Quegan
Understanding Synthetic Aperture Radar Images Oliver and Quegan
Radar and Electronic Warfare Principles for the Non-specialist, 4th Edition Hannen
Inverse Synthetic Aperture Radar Imaging: Principles, algorithms and applications Chen and Martorella
Stimson's Introduction to Airborne Radar, 3rd Edition Baker, Griffiths and Adamy
Test and Evaluation of Aircraft Avionics and Weapon Systems, 2nd Edition McShea
Angle-of-Arrival Estimation Using Radar Interferometry: Methods and applications Holder
Biologically-Inspired Radar and Sonar: Lessons from nature Balleri, Griffiths and Baker
The Impact of Cognition on Radar Technology Farina, De Maio and Haykin
Novel Radar Techniques and Applications, Volume 1: Real aperture array radar, imaging radar, and passive and multistatic radar Klemm, Nickel, Gierull, Lombardo, Griffiths and Koch
Novel Radar Techniques and Applications, Volume 2: Waveform diversity and cognitive radar, and target tracking and data fusion Klemm, Nickel, Gierull, Lombardo, Griffiths and Koch
Radar and Communication Spectrum Sharing Blunt and Perrins
Systems Engineering for Ethical Autonomous Systems Gillespie
Shadowing Function from Randomly Rough Surfaces: Derivation and applications Bourlier and Li
Photo for Radar Networks and Electronic Warfare Systems Bogoni, Laghezza and Ghelfi
Multidimensional Radar Imaging Martorella
Radar Waveform Design Based on Optimization Theory Cui, De Maio, Farina and Li
Micro-Doppler Radar and Its Applications Fioranelli, Griffiths, Ritchie and Balleri
Maritime Surveillance with Synthetic Aperture Radar Di Martino and Antonio Iodice
Electronic Scanned Array Design Williams
Advanced Sparsity-Driven Models and Methods for Radar Applications Li
Deep Neural Network Design for Radar Applications Gurbuz
New Methodologies for Understanding Radar Data Mishra and Brüggenwirth
Radar Countermeasures for Unmanned Aerial Vehicles Clemente, Fioranelli, Colone and Li
Holographic Staring Radar Oswald and Baker

Aviation Cybersecurity

Foundations, principles, and applications

Edited by
Houbing Song, Kenneth Hopkinson, Tomaso De Cola,
Tom Alexandrovich and Dahai Liu

The Institution of Engineering and Technology

Published by SciTech Publishing, an imprint of The Institution of Engineering and Technology, London, United Kingdom

The Institution of Engineering and Technology is registered as a Charity in England & Wales (no. 211014) and Scotland (no. SC038698).

First published 2021

The Institution of Engineering and Technology
Michael Faraday House
Six Hills Way, Stevenage
Herts, SG1 2AY, United Kingdom

www.theiet.org

British Library Cataloguing in Publication Data
A catalogue record for this product is available from the British Library

ISBN 978-1-83953-321-1 (hardback)
ISBN 978-1-83953-322-8 (PDF)

Typeset in India by MPS Limited

Contents

About the editors

Houbing Song is the Director of the Security and Optimization for Networked Globe Laboratory (SONG Lab, http://songlab.us/) and an assistant professor of Electrical Engineering and Computer Science at the Embry-Riddle Aeronautical University, FL, USA. He is the editor of eight books, the author of more than 100 articles, and the inventor of two patents. His research interests include cyber-physical systems/internet of things, cybersecurity and privacy, edge computing, AI/machine learning, big data analytics, unmanned aircraft systems, connected vehicle, smart and connected health, and wireless communications and networking.

Kenneth Hopkinson is a professor and head of the Department of Electrical and Computer Engineering at the Air Force Institute of Technology, OH, USA. He is a senior member of the IEEE and ACM professional societies. His research interests include networking, security, cryptography, remote sensing, sensor fusion, critical infrastructure protection, and space applications.

Tomaso De Cola is a group leader at the German Aerospace Center (DLR), Institute of Communications and Navigation. He is co-author of more than 100 papers. His main research activity concerns: TCP/IP protocols, satellite networks, transport protocols for wireless links, interplanetary networks as well as delay tolerant networks. He is a member of the IEEE Communications Society, where he has served as chair of the Satellite and Space Communications (SSC) technical Committee (TC). He is also serving as associate editor for many IEEE journals and magazines. He is also appointed as area director of the Space Internetworking Services within the Consultative Commmittee for Space Data Systems (CCSDS).

Tom Alexandrovich is head of the Aviation Guidance Department at the Israeli National Cyber Directorate. The Cyber Unit at the Office of Israel's State Attorney is a new national unit, which was established in 2015, in view of the need recognized to coordinate efforts in dealing with crime and terrorism in cyberspace.

Dahai Liu is a professor of School of Graduate Studies at the College of Aviation at Embry-Riddle Aeronautical University, FL, USA. He has authored over 100 articles and a book. He is a referee for many journals and a member of the Institute of Industrial Engineering. In 2019, he was awarded Researcher of The Year for the College of Aviation.

Preface

Making cyberspace secure and trustworthy is a critical challenge confronting society. The fragility and vulnerability of cyberspace have exposed societies and individuals to untold risks with severe consequences. Achieving a more secure cyberspace demands overcoming major scientific challenges, and realizing privacy and trust in cyberspace requires reconciling technology with human and societal needs. New advances in technology for cyberspace, changes in society, and adoption in new domains will also necessitate a rethinking of the interplay between security, privacy, and trust in cyberspace.

In the aviation domain, the past decade has seen the rise of emergent technologies, including cyber, unmanned aircraft, and spectrum-dependent systems that generate economic and social benefits, but also challenge the cybersecurity of the Aviation Ecosystem. Cybersecurity exists within an ever-changing competition with those who intend to do harm to the system. As directed in the National Strategy for Aviation Security (NSAS) and the International Civil Aviation Organization (ICAO)'s Aviation Cybersecurity Strategy, these evolving "disruptive technologies," along with on-going traditional threats, require a holistic and adaptive approach to protect and defend aviation cyberspace against harmful actions by determined adversaries and to assess their effectiveness, towards ensuring the cybersecurity, and prosperity of the Aviation Ecosystem. There is an urgent need for sound mathematical and scientific foundations, principled design methodologies, and metrics for evaluating success or failure for securing aviation cyberspace.

This edited book, *Aviation Cybersecurity: Foundations, Principles, and Applications*, aims to present the scientific foundations and engineering principles needed to advance the field of aviation cybersecurity, with focus on fundamentally new, principled approaches to deter, protect, detect, and respond to actions that compromise or threaten to compromise the confidentiality, integrity, or availability of computer- and network-based aviation systems. Towards this goal, this book is organized into three parts: Foundations, Principles, and Applications.

Part 1 is composed of four chapters. This part presents various scientific foundations of aviation cybersecurity, including counter-unmanned aircraft systems (C-UAS) (Chapter 1), formal verification of software systems (Chapter 2), drone surveillance (Chapter 3), and L-band digital aeronautical communication system (LDACS) (Chapter 4). Both Chapters 1 and 3 address the cybersecurity risks posed by unmanned aircraft systems (UAS), also known as unmanned aerial vehicles (UAV) or drones, which can be used for malicious schemes by terrorists, criminal organizations (including transnational organizations), and lone actors with specific objectives.

Part 2 is composed of three chapters. This part presents various engineering principles of aviation cybersecurity, including Global navigation satellite systems (GNSS) (Chapter 5), cybersecurity risk management (Chapter 6), and physical layer security (Chapter 7).

Part 3 is composed of three chapters. This part presents different perspectives of aviation cybersecurity, spanning privacy (Chapter 8), workforce development (Chapter 9), and cyber-physical systems (Chapter 10).

This book is intended as an introductory aviation cybersecurity book which should be useful to practitioners in the aviation cybersecurity industry. In addition, this book is an ideal textbook for undergraduate courses and for first-year graduate courses in Cybersecurity, Computer Science, Computer Engineering, Electrical Engineering, and Aviation.

This book would not have been possible without the help of many people. First, we would like to thank all the contributors and reviewers of the book from all over the world. Second, we would like to thank our graduate research assistants, Yongxin Liu, Jian Wang, Faisal Alrefaei, Yupeng Yang, Jiahao Yu, Sang-A Lee, all in the Security and Optimization for Networked Globe Laboratory (SONG Lab, http://songlab.us/) and Modeling and Simulation Lab for Intelligent Transportation Systems at Embry-Riddle Aeronautical University, who provided indispensable support at all stages of the editorial process of the book. Third, we would like to acknowledge the US National Science Foundation under Grant No. 1956193 and the US Department of Transportation under Grant No. 69A3551747125 via the Center for Advanced Transportation Mobility (CATM). Fourth, we would like to thank Srinivasan N and Olivia Wilkins, both at IET, for guiding us through the publication process of the book.

Houbing Song
Kenneth Hopkinson
Tomaso De Cola
Tom Alexandrovich
Dahai Liu
October 2021

Chapter 1
Counter unmanned aerial vehicle for aviation security

Chengtao Xu, Yongxin Liu, Jian Wang, and Houbing Song[1]

1.1 Introduction

The unmanned aircraft system (UAS) is an aircraft that is operated without direct human intervention. The associated elements, including communication, control and sensing, are required for safe operation in the national airspace. In recent years, technology development in material and manufacturing have made unmanned aerial vehicle (UAV) more reliable. With the price dropping on amateur drones, the market of UAS has been expanding for a while. Advances in miniaturization technology for air and space systems accelerate such a trend in the civilian application of UAS. According to the aerospace forecast fiscal years 2019–2039 of Federal Aviation Administration (FAA), the model UAS is set to grow from the present 1.25 million to around 1.39 million units by 2023 and the non-model UAS is set to grow from the present 277,000 aircraft to over 835,000 aircraft by 2023 [1]. The positive usage of UAS has the potential to save lives, enabling more effective science and engineering research [2]. These uses may include experiments with small UAS, performing numerous functions, including aerial photography and personal recreational activity, package and medical supply delivery, and providing support for search and rescue missions.

However, the imminent threat brought by UAS can also lead to the less-secured airspace for those sensitive facilities, such as airports, nuclear power plants, and electrical transformers. UAS can be used for malicious schemes by terrorists, transnational organizations, and lone personnel with specific objectives. UAS-based threats stem from increasing reliance on computer and communication technologies, placing public safety, national security, and individual privacy at risk. A system with a more complex operation suffers more vulnerability from the threats of malicious UAS, such as air transportation. A severe delay in local airport caused by the unknown UAS may lead to disaster propagation of air flight delays and potential safety issues.

[1]Department of Electrical Engineering and Computer Science, Embry–Riddle Aeronautical University, Daytona Beach, FL, USA

There is an urgent need for innovative technologies to promote safe, privacy-respecting UAS operations toward aviation security for disabling, identifying, and mitigating UAS. A counter-UAS (C-UAS) system is defined as a system or device capable of lawfully and safely detecting, tracking, disrupting, or seizing control of an unmanned aircraft or UAS [3]. Typically, such a system comprises two subsystems: one for detection and the other for neutralization [4,5]. The ideal UAS detection subsystem will detect, identify one or multiple unmanned aircraft, or UAS, which has a small footprint, and support highly automated operations. Typically, the UAS negation system requires actively disrupting the communication link of UAS operation and successfully hacking into UAS through their wireless link or giving dummy commands on localization and control.

Due to the fact that UAS is usually controlled by a remote pilot or programmed to fly autonomously, protecting aviation facilities against UAS-based threats is significantly challenging. Different from conventional high altitude UAV detection with high-power radar devices, interruption brought by the ground object usually causes high-rate false alarms of the same detection system. Other methods of sensing the presence of UAS near ground have drawbacks as well, with respect to the type and strength of signal emitted from UAS. Therefore, a high accurate UAV intrusion detection system is essential. Especially, with the development technique of airborne networks of UAV and a higher rate of group UAV intruding events, it is necessary to develop a reliable technique to detect the presence of single or multiple UAVs and identify their features.

Over the last five years, many research efforts have been made to detect, track, identify, and mitigate UAS: detection technologies are based on acoustic [6], vision [7], passive radio frequency (RF) [8], radar [9], and data fusion [10]; and mitigation technologies include physical capture (containment netting) [11], jamming (RF command and control (C2) jamming and spoofing, or GPS jamming and spoofing) [12], and destruction (RF C2 intercept and control) [13,14]. However, these efforts are not mature because it is usually not scalable, transferable, modifiable, and affordable.

In this chapter, we provide a comprehensive discussion of existing literature in the area of UAS detection and negation, identify the challenges in countering adversary UAS, and evaluate the trends against UAS-based threats. The objective of this chapter is to present a systematic introduction of C-UAS technologies, thus fostering a research community committed to the safe integration of UAS into the airspace system.

The remainder of this chapter is structured as follows: The "Background" section presents various UAS-based threats and introduces restrictions that commonly affect UAS flights. "State-of-the-Art UAS Detection" and "State-of-the-art UAS detection" and "state-of-the-art mitigation technologies" are presented respectively. Challenges and unsolved problems of methods in UAS detection and mitigation are discussed in countering unauthorized or unsafe UAS. We also discuss the future trends of integrating such detection and mitigation based C-UAS into national airspace system (NAS) for protecting safety of air transportation. In the end, we would conclude this chapter with remarks.

1.2 Background

The UAS-based threats can be categorized into three categories: public safety, national security, and individual privacy. UAS-based public safety threats are due to UAS operation near other aircraft, especially near airports; over groups of people, public events, or stadiums full of people; near emergencies such as fires or hurricane recovery efforts; or under the influence of drugs or alcohol; UAS-based national security threats are due to UAS operation over designated national security facilities, such as military bases, national landmarks, and certain critical infrastructure, among others [15]; UAS-based privacy threats are due to operating UAS with their camera on when pointing inside a private residence [16].

1.2.1 Reported UAS sightings

The architecture of the UAS can be roughly sketched as consisting of a bottom sensor layer, a middle network layer, and a top application layer. One of the primary information-acquiring means at the bottom layer of the tags has found increasingly widespread applications in various business areas, with the expectation that the use of RFID tags will eventually replace the existing bar codes in all business areas.

1.2.2 Safety threats

The unmanned nature of UAS operations raises two unique safety concerns that are not present in manned aircraft operations: the pilot of the small UAS, who is remotely located from it during flight, may not have the ability to see flights in the air in time to prevent a mid-air collision, and the pilot of the small UAS could lose control of it due to a failure of the communications [17]. Safety risks related to the use of UAS include the potential for unintentional collisions between a small UAS and a manned aircraft or other airport objects, causing damage to property or persons [17]. Operating UAS around airplanes, helicopters, and airports is dangerous and illegal.

1.2.3 Security threats

UAS-related national security threats may include the following:

(1) Weaponized or Smuggling Payloads: Depending on power and payload size, UAS may be capable of transporting contraband, chemical, or other explosive/weaponized payloads.
(2) Prohibited Surveillance and Reconnaissance: UAS are capable of silently monitoring a large area from the sky for nefarious purposes.

(3) Intellectual Property Theft: UAS can be used to perform cyber-crimes involving the theft of trade secrets, technologies, or sensitive information.

1.2.4 Airspace restrictions applicable to UAS flights

In the United States, there are several types of airspace restrictions that commonly affect UAS flights [18]. The most stringent restrictions are security-sensitive airspace restrictions, which prohibit UAS operations from the ground up to 400 feet above ground level, and apply to all types and purposes of UAS flight operations; for stadiums and sporting events, restrictions are valid during a gathering of a crowd; Washington, DC, has the largest spatial restriction circle (30 miles). The FAA established a series of rules and regulations, which apply to UAS operations, based on the type of UAS flier: recreational fliers and modeler community-based organizations, certificated remote pilots including commercial operators, public safety and government, and educational users [1]. UAS that weigh more than 0.55 pounds must be registered with the FAA. In addition, flying UAS that are less than 55 pounds for work or business requires remote pilot certificates [1].

Although the FAA has established guidelines and regulations, reports of UAS sightings from pilots, citizens, and law enforcement have increased dramatically over the past five years. Therefore, there is an urgent need for promoting safe, secure, and privacy-respecting UAS operations. We envision that an integrated system capable of detecting and negating UAS will be essential to the safe integration of UAS into the airspace system. Such a system needs novel technologies in the following three main areas.

(1) Detection: The ideal UAS detection will detect, track, and identify an unmanned aircraft or UAS, have a small footprint, support highly automated operations and be robust for its deployment location.
(2) Mitigation: The ideal UAS mitigation system will lawfully and safely disable, disrupt, or seize control of the UAS, while ensuring low collateral damage and low cost per engagement.
(3) Machine learning system: It performs a central information-gathering role within such a system. It has the ability to continuously learn the model of each attack of unknown UAS, online and off-line processing of upcoming data, and analyze the pattern of fused signal from distributed sensors.

Such an C-UAS system is leveraging communications and signal processing to develop low footprint UAS detection solutions, in terms of size, weight, power, and manning, as well as varied and low collateral damage UAS mitigation techniques and effectors, toward safe integration of UAS into the airspace system.

1.3 UAS detection

Since 2014, five technologies have been proposed for UAS detection, including audio, vision, passive RF, radar, and data fusion. We summarized the evolution of UAS detection technologies in Figure 1.1, where different technologies are ranked based on the popularity for each year. From Figure 1.1, we know that data-fusion approaches have been the most popular technology for UAS detection while acoustic-based approaches have been the least used. In this section, we will introduce each UAS-detection technology and discuss its advantages and disadvantages.

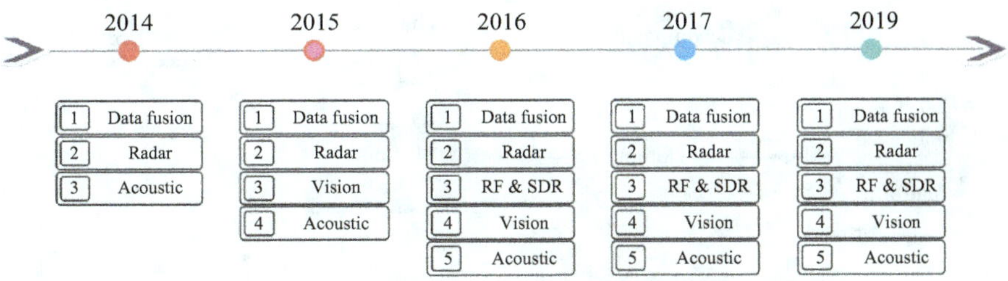

Figure 1.1 Evolution of UAS-detection technologies

Multiple types of signal sources could be used to obtain unique information to identify objects. We determined the potential detectable signal source as RF in the UAS communication link with remote controller, acoustic, vision, and radar reflection brought by the Doppler effects of swing spinning. Therefore, multiple sensors are desired to be integrated into one detector to combine the detection ability in different spectrums or evenly distributed near the protected field. The candidate sensor includes MIMO antenna (SDR based), microphone, infrared sensor, camera, and active radar. The signal decomposition method is used in the signal-processing module, in which the unique features from the obtained signal can be extracted for the following object identification. Figure 1.2 presents a diagram of the detection module in multi-sensor system for UAS detection.

The tracking module of the C-UAS system can be concluded as shown in Figure 1.3. The sensor-sensing information would also be acquired by the tracking function module. Using the localization algorithm to localize the signal source emitted from the intruding object, the module helps the directional antenna locate the signal source for the subsequent efforts on blocking UAS's communication link. Synchronization function is necessary to be applied in the scenario with detection sensor array to achieve higher accuracy of identifying multiple intruding UAS. Path estimation provides the information for the following evaluation on whether the intruding object is successfully neutralized.

With the generated estimated moving path of UAS, the directional antenna blocks the communication link between UAS and remote controller. Then the spoofing signal

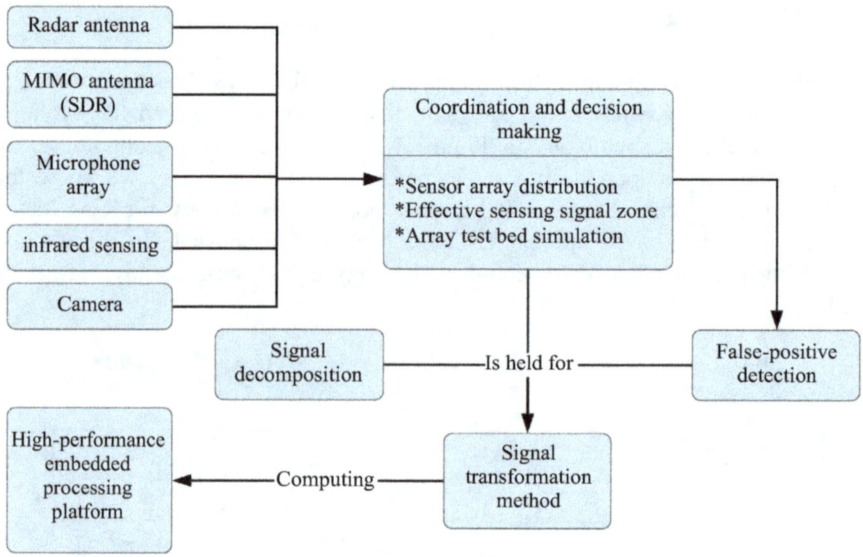

Figure 1.2 Detection module of C-UAS system

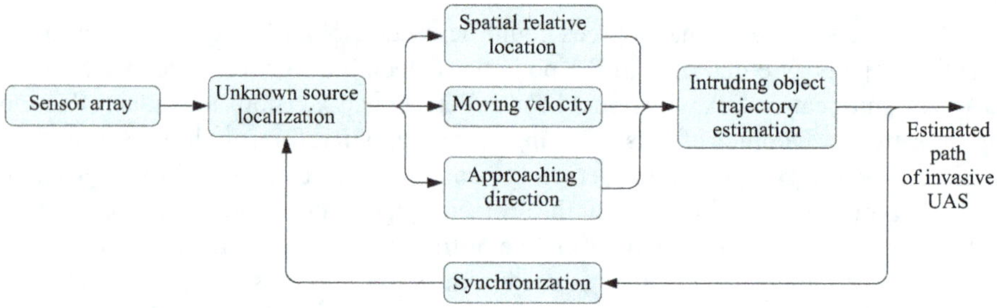

Figure 1.3 UAS tracking module of C-UAS system

has the potential to successfully pair with UAS, in which this protocol-aware spoofing is based on the identity in the identification module.

To verify the result of this protocol aware mitigation approach, the continuous UAS moving path could be compared between the estimated path from tacking and the command sent by the mitigation controller. The update of this comparison result happens within the evaluation module. It sends the result to machine learning (ML) and identification modules to update the model, which includes but not limited to reinforcement learning algorithm, neural network model, and the local signal pattern database. Figure 1.4 denotes a diagram of the proposed mitigation and evaluation module.

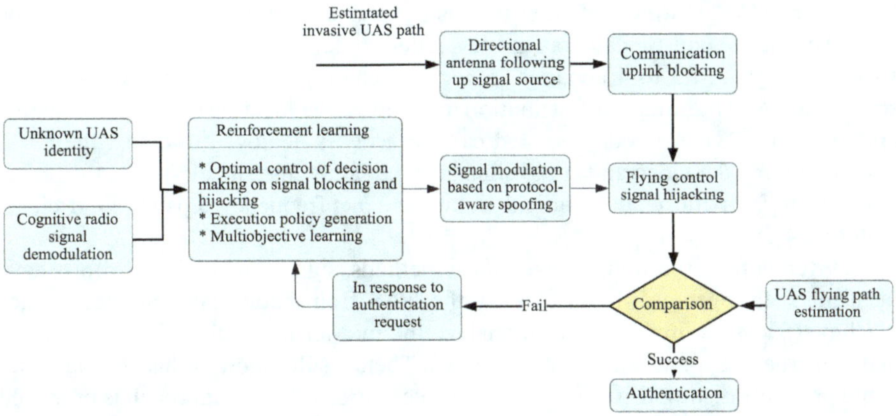

Figure 1.4 Mitigation and evaluation module of C-UAS system

1.3.1 Acoustic-based UAS detection

Acoustic-based UAS detection leverages acoustic sensors to capture the sound of UAS, and identify and track the UAS with an audio signal. Acoustic sensor arrays, which are deployed around the restricted areas, record the audio signal periodically and deliver the audio signal to the ground stations. The ground stations extract the features of the audio signal to determine whether the UAS are approaching based on its pattern. Conventionally, after receiving the audio signal of UAS, the power or frequency spectrum will be analyzed to identify the UAS. Vilımek and Buřita adopted the linear predictive coding to distinguish the sound of the UAS engine from the sound of car engine, but the performance is subject to weather conditions [19]. Kim *et al.* designed a real-time UAS sound detection and analysis system, which could acquire the real-time sound data from the sensor and recognize the UAS [6]. Jang *et al.* applied the Euclidean distance and scale-invariant feature transform (SIFT) to distinguish the UAV engine sound from the background sound and demonstrated their effectiveness, even though the power spectrum of the noise is larger than that of the UAS sound. However, in practice, their processing efficiency is poor [20].

Due to light weight, low-cost and easy assembly, acoustic sensors could be used to construct an acoustic acquiring array and deployed in the target area to locate and track the trajectories of the UAS. An approach is proposed to deploy an acoustic sensor array, which acquires the sound of engine in UAS. The acoustic sensor array consists of 24 custom-built microphones which locate and track the UAS collaboratively. They calibrate each sensor with a time delay of arrival (TDOA) and predict the UAS flight path with beamforming. They could track the flight path well, but this approach cannot work well in a large-scale space, and the accuracy of the sensor is dependent seriously on calibration [21]. Two arrays consist of four microphone sensors to improve the capability of the UAS localization. Due to the multi-path effect, they provided a Gauss prior probability density function to improve the TDOA estimation. Their arrays could be deployed in specific areas efficiently and achieve good performance to track the UAS. However, their system is not stable when it works for a long time [22]. Advanced acoustic cameras are leveraged to detect and track the UAS. To be specific, they used 2 to 4 acoustic cameras to capture the strength distribution of sound and fused the strength distribution to compute the location of the UAS indoors and outdoors [23]. An audio-assisted camera array is deployed to detect the UAS, which captured the video and audio signals at the same time and classified the object with a histogram of oriented gradient feature and mel frequency cepstral coefficients feature [24].

Different from the above conventional methods, significant research efforts have been made to leverage ML to classify the UAS from audio data. Support vector machine (SVM) is implemented to analyze the mid-term signal of the UAS engine and construct the signal fingerprint of UAS. Their results showed that the classifier could precisely distinguish the UAS in some scenarios [25]. An approach is proposed to transform the detection of the presence of UAS to a binary classification problem, and used Gaussian mixture model (GMM), convolutional neural network (CNN), and recurrent neural network to detect UAS. Their results showed that it could work well with the short input signal within 240 ms [26].

The current acoustic-based UAS detection technologies can recognize and locate the UAS precisely to meet the accuracy requirement of UAS detection. However, the nature of acoustic approaches limits the deployment and detection of UAS on a large scale. ML presents a tremendous opportunity of integrating acoustic sensing into acoustic-based UAS detection for improved UAS detection performance.

1.3.2 Passive RF-based detection

For UAS controlled by a remote pilot, it usually maintains at least one RF communication link to its controller to either receive control commands or deliver aerial data. The spectral patterns of such transmission are used as an important evidence for the detection and localization of UAS. Different passive RF technologies are shown in Figure 1.5. In most cases, software-defined radio (SDR) receivers are employed to intercept the RF channels.

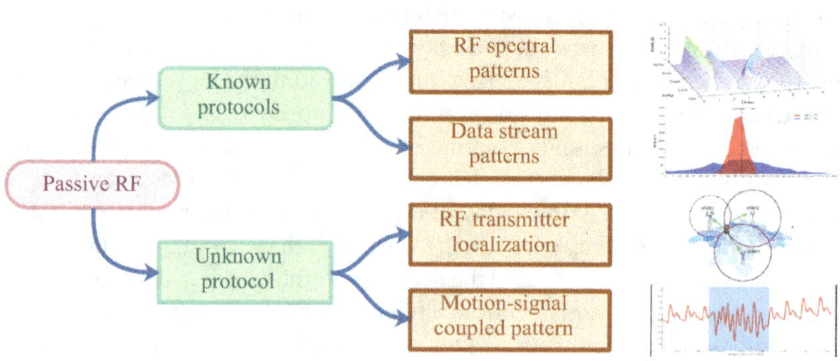

Figure 1.5 Categorization of passive RF approaches

To utilize the spectrum patterns of UAS, in [27], an artificial neural networks (ANN) detection algorithm for a UAS RF signal was proposed, which employs three signal features: improved slope, improved skewness, and improved kurtosis. It was shown that the proposed algorithm based on ANN outperforms other recognition technologies of the improved slope, skewness, and kurtosis of the signal spectrum.

Data traffic patterns are also an important feature to specify UAS. In [28], a UAS detection and identification system, which utilized commercial off-the-shelf hardware to passively listen to the wireless signal between UAS and controllers for packet transmission characteristics, was proposed. They mainly extracted the packet length distribution of UAS and evaluated the prototype system with three types of UAS. Their experiment results demonstrated the feasibility of using the data frame length to identify different UAS within 20 s. The increasing amount of commercial UAS, using WiFi as control and first-person-view (FPV) video streaming protocol, motivated the method in [8], a UAS detection approach based on WiFi fingerprint. The method identifies the presence of unauthorized UAS in a nearby area by monitoring the data traffic.

However, it is still challenging to develop a method that monitors and detects multiple UASs from the air space based on the RF data traffic. The advanced signal processing is desired to successfully extract each UAS's signal from mixed sensor data, which are located unevenly over the spectrum. Moreover, with the uncertain number of UAS operators, UASs, distance from UASs to the RF receiver, each UAS's transmission data link signal remains at different strength within the mixed signal, which makes the detection of UAS swarm even more difficult. Time-series analysis for such mixed signals of multiple UASs could usually be difficult. Therefore, a method such as Short-Time Fourier Transform (STFT) could be applied in revealing the time-frequency-energy (TFE) features of an intercept signal. The obtained TFE graph is also called a spectrogram, which can be calculated as follows:

$$SG(f,t) = |\,\mathrm{STFT}\,(f,t)|^2 \tag{1.1}$$

where

$$\mathrm{STFT}\,(f,t) = \sum_{n=-\infty}^{\infty} s(t)w(t-m)e^{-j2\pi kn/N} \tag{1.2}$$

Here, $s(t)$ represents the received quadcopter signal and $w(t)$ denotes the window function sampling signal. $N, n, and\ m$ denote the number of samples in the sampling window, sample number, position of the window, respectively.

Methods such as feature engineering algorithms provide convenience for post STFT operation processing because it reduces unnecessary noise from the environment and other wireless devices. Typically, feature reduction is what is desired in the situation of UAV detection. Principal components analysis, usually known as PCA, combines highly correlated high-dimensional variables into a less-dimensional variable with low linear relativity, called as principal components. By calculating the covariance matrix of a dataset, we get the (i,j) element, which is the covariance of i_{th} and j_{th} element covariance. In the n-dimensional covariance matrix, we have:

$$C = \begin{bmatrix} \text{cov}(x_1,x_1)\ \text{cov}(x_1,x_2)\cdots & \text{cov}(x_1,x_n) & \\ \text{cov}(x_2,x_1) & \ddots & \text{cov}(x_2,x_n) \\ \vdots & \ddots & \vdots \\ \text{cov}(x_n,x_1)\ \text{cov}(x_n,x_2)\cdots & \text{cov}(x_n,x_n) & \end{bmatrix} \tag{1.3}$$

For example, for dataset A with 1024×1024 dimension of data, an eigenvector with $1024 \times n$ dimension is needed to map A into the first main portion of dataset. Similarly, in the signal energy matrix, Gaussian noise potentially affects the following classification task done by the ML algorithm. A lower dimension of training dataset reduces the neural networking training time, which contributes to the real-time detection of UAV controller signal.

In [29], a transient signal emitted from the remote controller is applied in detecting the number of presented UASs. The results show that ML algorithms such as CNN, SVM, and KNN could successfully detect the number of UAVs in a less noisy RF environment. Feature reduction appears to be effective in enhancing the accuracy of ML classifier in recognizing the UASs' RF signal pattern. However, different strengths of the source signal emitted from UAS remain to be a big challenge for detecting multiple UASs in the air space.

Adaptive RF fingerprint decomposition ahead of spectrogram representation appears to solve the mixed multi-UAS signal problem. Some available signal separation method, such as blind signal separation (BSS), empirical mode decomposition (EMD), ensemble empirical mode decomposition method (EEMD) provides the potential solution toward effectively reducing noise portion in the extracted original signal. In detecting the UAS using dynamically changed communication link signal, EMD and EEMD have been applied in [30] to extract separated intrinsic mode function (IMF) to retrieve the original source RF signal. The results show that the dynamics of UAS uplink and downlink channels could barely affect the detection accuracy of the ML classifier in recognizing the video and control signal of UAS.

Data traffic-based methods or pure spectrum pattern methods depend highly on the telemetry protocol or RF front-ends of UAS. These methods may not be able to identify UAS operating with unknown telemetry protocols. Therefore, some research focuses on using the coupled kinetics motion patterns of flying UAS from their radio signal to specify their presence. In [31], Matthan, a cost-effective and passive RF-based UAS detection system was introduced. Their system detected the presence of UAS by identifying the unique signatures of its coupled vibration and shifting patterns in the transmitted wireless signals. The joint detector integrated evidence from both a frequency-based detector that indicated UAS body vibration and a wavelet-based detector that captured the sudden shifts of drone frame by computing wavelets at different scales from the temporal RF signal. A similar approach was also presented in [32].

Localization is also an important part of passive RFbased UAS detection. In [33], the authors divided the complete detection procedure for UAS detection into RF spectrum sensing and the direction of arrival (DoA) estimated. This work presented a useful experiment, which used a commercial off the shelf field programmable gate array (FPGA)-based SDR system for detecting and locating small UASs. Their results demonstrated that it is possible to develop a UAS detection system capable of detecting small UASs with error of 50–75 m using commercial FPGA-based SDR system. The SDR system with optimized clock synchronization would radically reduce the measurement error in distance. Implementing robust localization algorithms with acceptable accuracy on ubiquitous hardware is still an open problem in passive RF detection of UAS. How to deploy such passive RF-based systems on the ground stations or other UAS platforms is another open problem in the detection of UAS.

1.3.3 Vision-based UAS detection

Vision-based UAS detection technologies mainly focus on image processing. Videos and cameras are adopted to capture the images of trespassing UAS. The ground stations figure out the appearance of UAS from the videos and pictures with computational methods. Conventional methods mainly rely on the methods of image segmentation. The differential of UAS and environment in images determines whether the restricted areas have the UAS. A vision-based UAS detection approach is presented in [34], which could separate the UAS from background efficiently. Similar work was reported in [7] and [35]. The common challenges for their approaches are how to separate UAS from background images and how to distinguish UAS from flying birds.

In contrast, state-of-the-art image segmentation methods use neural networks to directly identify the appearance of UAS. An approach leverages the thermal camera to detect the UAS and neural network to identify the UAS [36]. An outstanding research is presented in [37]. A lightweight and fast algorithm could operate on embedded system (Nvidia Jetson TX1) and identify the UAS in movement.

A real-time vision-based UAS detection system is designed, which is based on two vision-processing platforms: FPGA-based platform, which can operate below 10 W (i.e., power saving), and graphics processing unit (GPU)-based platform, which is able to process more frames. However, for FPGA, it is impossible to change algorithms in real time [38]. Saqib *et al.* compared different CNNs' performance in detecting UAS and their results showed that the visual geometry group (VGG 16) network with Faster R-CNN achieved outstanding performance [39]. An approach is proposed to combine different pictures to generate synthetic images to extend the image dataset to train the CNN to enhance the performance of the UAS detection [40].

Birds are a serious factor, which downgrades the identification of UAS from the images. Significant research efforts have been made to use CNNs to enhance the identification of UAS. A survey about the challenges of detection of UAS and birds is presented in [41]. The survey concluded that the neural network algorithms are promising in identifying UAS and birds. It compared policy-based approaches and neural network-based algorithms for recognizing birds and UAS using datasets of videos and pictures. The results showed that the neural network-based approaches can outperform the policy-based approaches over 100 times in terms of accuracy and efficiency. A UAS detection framework was presented in [42], which was based on video streams and classified the objects into different types with CNN. The work mainly focused on distinguishing the birds and UAS in different scenarios.

At the same time, some attempts were made to apply infrared cameras to identify the UAS. Infrared sensors are leveraged to detect small variations of UAS in heat to identify the UAS. The drawback of this approach is that the heat from batteries has significant effects on result detection [43]. Different from other research on classifying the frame of images, dynamic vision sensors are applied to capture the rotating frequency of the propeller to distinguish the UAS from birds efficiently [44].

Currently, the vision-based approaches can be implemented in some specific scenarios to recognize the features of UAS from the environment. The evolution of deep neural networks stimulated image processing, which could have multiple positive effects on the UAS detection in vision field. The real-time attempts showed that the vision approaches have the potential for efficiency. However, how to implement the recognized algorithms in multiple and variable environments is challenging. The novel approaches are supposed to be robust, adjustable, and precise. The vision-based approaches need to be robust to the quick variation of the environment. The image distortion caused by weather change could be mitigated by the multiple-level image processors, which capture the features of UAS in the different spectrum. The mobility of UAS poses a challenge to the vision-based approaches, i.e., the images are supposed to be captured and recognized in different levels of mobility of UAS. The bio-inspired robots are limited in detection accuracy, which poses the risk of distinguishing between UAS and birds mistakenly. The enhanced accuracy of the distinguishing can improve the efficiency of UAS detection and mitigation greatly.

1.4 Radar-based UAS detection

Radar has several advantages in detecting airborne objects compared with other sensors in terms of day and night operating capability, weather independence, and ability to measure range and velocity simultaneously. However, regular radar systems focus on detecting air targets of medium and large size with radar cross-section larger than 1 m or 2 m, which makes it infeasible to detect small-size and low-speed UAS in [9] and [45]. It is difficult to detect UAS because of its slow moving speed, and the Doppler effects of the structure of aircraft are typically weak. Therefore, efforts are needed to either develop a new radar sensing technique or increase the detective resolution of conventional systems. In this section, we will discuss three categories of radar-based UAS detection technologies: active detection, passive detection, and posterior signal processing.

1.4.1 Active detection

Typically, there are two ways to increase the resolution of conventional radar detection systems for UAS surveillance: utilizing higher frequency carriers and using multiple input–multiple output (MIMO) beamforming radio front-ends, as shown in Figure 1.6.

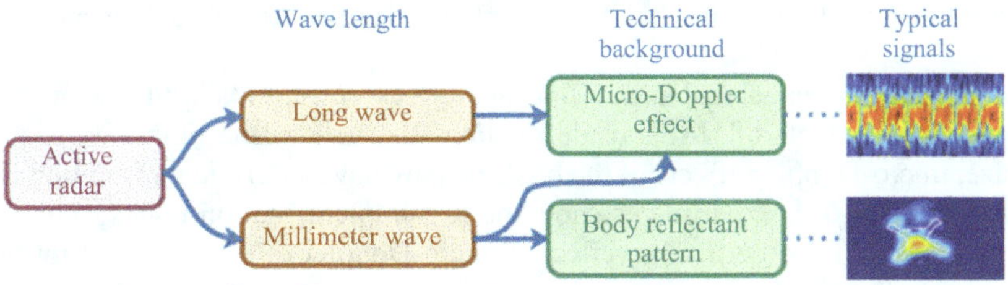

Figure 1.6 Categorization of active radar-based approaches

To utilize shorter wavelength, in [46] and [47], X-band and W-band frequency-modulated continuous wave (FMCW) radars were designed for UAS detection. Their solutions use the bistatic antenna and convert received signals into digital quadrature stream for posterior processing. The feasibility of using ultra wideband signals with a 24-GHz carrier was demonstrated in [48]. The selection of carrier frequency for UAS detection radar should be higher than 6 GHz (K-band) in [49] and [50].

Other approaches utilize multiple antennas to form MIMO front-ends. The benefit of such an approach lies in its applicability to radar system with lower carrier frequencies. The work in [51] demonstrated the detection of a small hexacopter using 32 by 8 element L-band receiver array, which achieved good detection sensitivity against micro-UAS. Similar research was presented in [52]: a ubiquitous FMCW radar system working at 8.75 GHz (X-band) with PC-based signal processor is presented. The results indicated that it can detect a micro-UAS at a range of 2 km with an excellent range-speed association. In [53], a Ka-band radar system that uses 16 transmit and 16 receive antennas to form 256 virtual antenna elements is presented. Their experiment proves that even in a non-stationary clutter environment, the UAS could be detected at a range of about 150 m. Similar work was presented in [54]. Another important characteristic of MIMO systems is that they generate a large quantity of data for further processing in [55]. The authors use the concept of data cube and classifier to assure the presence and location of an incoming UAS. A simplified approach, multiple input–single output, was utilized for UAS detection in [56].

Noise radar is considered an efficient way to detect a slow moving UAS, and its benefit is that UAS can be detected by using simple antenna components and lower carrier frequency. In [57] and [58], the feasibility of using random sequence radar for UAS detection was demonstrated in sub-X band, and their results indicate that such radar can be the future of cost-efficient UAS detection solutions.

The advances of intelligent computation enable another radar, the SDR-based multi-mode radar [59]. Such radar is small-size and highly configurable. However, the operational performance of SDR relies highly on the back-end processor. In [60], two different implementations of FMCW radar and continuous wave noise radar are presented to test their feasibility for UAS detection. And their findings indicate that the analog implementation has a higher updating rate and signal–noise ratio (SNR). The apparent drawback of active radars is that they need specially designed transmitters, which might not be easy to deploy and are vulnerable to anti-radioactive attacks.

Traditional radar signal processing also gives constraints while dealing with micro-doppler effects of UAS's blade rotation. Because of the limitation of physical size, micro-Doppler effects of the blade usually have a deficient resolution in the radar graph. In [61], experiments show the real implementation issues of applying radar to detect micro-UAS. The reflected micro-Doppler effects graph of intruding UAS shares the similar features and energy density compared to the wings vibration of birds. Low resolutions of STFT algorithm in processing the short radar signals cause the mode mixing of the representing reflection features of these two objects.

1.4.2 Passive radar

Passive radars do not require specially designed transmitters. Existing radioactive sources such as cellular signals can be leveraged to illuminate the space. In this section, we classify passive radars into two categories: single station passive radar and distributed synthetic passive radar, as shown in Figure 1.7.

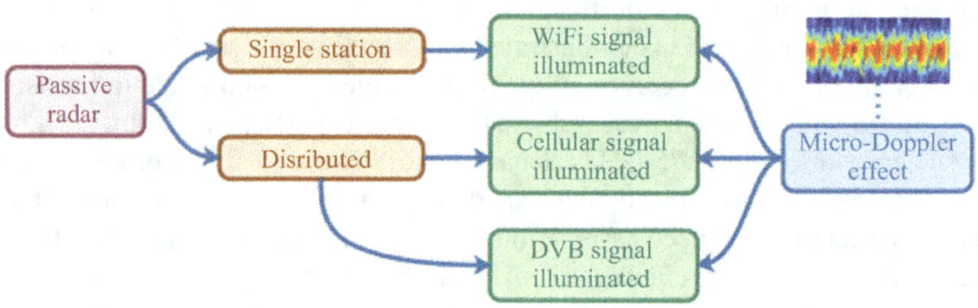

Figure 1.7 Categorization of passive radar-based approaches

1.4.2.1 Single station passive radar

This type of passive radar exploits only one illumination source. The variation of received signals can be analyzed to specify the appearance of UAS. In [62], a WiFi-based passive radar was presented for the detection and two-dimensional (2-D) localization of small aircraft. Obviously, this is the most direct adaptation of active radars.

1.4.3 Distributed synthetic passive radar

Distributed station leverages the existing telecommunication infrastructures as illumination sources to enhance the UAS detection. Unlike active radar, distributed passive radar can be securely hidden in the location without obviously exposing it in the attack zone. For line of sight detection, active radar usually shows more advances in its low false alarm rate, but it is not energy efficient and could easily become an object of malicious attack. There are mainly two approaches: cellular system-based solutions and digital video broadcasting (DVB) system-based solutions.

(1) Cellular system-based passive radars: An approach is proposed to enhance the detection system, which could locate and track the UAS by using a reflected signal of the global system for mobile communications (GSM) [63]. An approach is proposed to receive the 3G cellular reflecting signal for tracking UAS. Doppler features of 3G cellular signal are leveraged to monitor the target area and track the trajectory of UAS. The results showed that UAS could be tracked obviously in the waterfall graph. The drawback of this approach is that it needs a reference receiver to calibrate the received signal. The accuracy of detection is dependent on the calibration accuracy seriously [64]. A 5G mm-wave radar deployment infrastructure is constructed to detect amateur UAS. The deployed radars capture the signal of amateur UAS and upload it to cloud, which could analyze whether there exist hazards. This approach is applicable in security guard of the city which cope with the challenges of resource management, non-line-of-sight radar operation, noise mitigation, and big data management [65]. Similar implementation was presented in [66], where a passive radar array system was proposed to receive and process the orthogonal frequency division multiplexing (OFDM) echoes of UAS, which was originally transmitted by the nearby base stations.
(2) DVB system-based passive radars: The pervasive digital television signals are considered as an efficient illumination source for passive drone detection radars. In [67–69], passive drone detection radars were designed and tested. Similar to active radar approaches, the micro-Doppler effect could be employed. In [70], UAS classification tests were conducted through propeller-driven micro Doppler signature and ML. Their experiment reveals that the micro-Doppler signature of a plastic propeller is much less visible than a carbon fiber propeller. An apparent drawback for passive radar is that a large amount of post-processing efforts or multiple receivers are needed to achieve acceptable detection accuracy.

1.5 Data-fusion-based UAS detection

Data fusion, which is the process of integrating multiple data sources to produce more consistent, accurate, and useful information than that provided by any individual data source, can generate fused data that are more informative and synthetic than the original inputs.

The data-fusion approaches can leverage the advantages of each method to acquire a combined result that is more robust, accurate, and efficient than the single approaches. For the UAS detection, data fusion can be used to improve the performance of the UAS detection system to overcome the disadvantages of the single approach, which exist in some specific scenarios. Based on the above discussion, the disadvantages and the advantages of each approach have been presented. The first general problem of the single approach is the limitation of the detection range and accuracy. The second general problem of the single approach is the vulnerability to detection scenarios. The third general problem of the single approach is the efficiency of the computation.

Based on the above general problems, the research on the data-fusion-based UAS detection can be classified into three categories: (1) multiple-sensor data fusion; (2) multiple-type sensor data fusion; and (3) multiple-sensing algorithm fusion.

(1) Multiple-Sensor Data Fusion: Each type of sensor has its own advantages and disadvantages in the UAS detection scenarios. The general problem of the single approach is the limited detection range. The more straight and efficient approach to improve the performance of the single approach is designing a specific type of sensors to avoid the drawbacks of nature materials.

    A classical example is the acoustic sensor array. Distributed acoustic sensors are deployed in the detection areas. Each sensor can record the audio and deliver the record to the ground stations to make a combined evaluation of the environment in the sound spectrum [71]. The researchers extracted the phrase difference in the sound to locate the UAS. In the signal processing, the researchers need to adjust the weight of each sensor to obtain a higher accuracy of the location. Apart from the accuracy of detection, the data fusion can improve the functions of UAS detection. The RF-based detection can acquire the RF signal with omnidirectional antennas. The researchers could adjust the detection phase of the signals in the receiving process with multiple omnidirectional antennas. Thereafter, the system with omnidirectional antennas could acquire signals in the specific direction [72]. This function could enable the ground stations to track trajectories of UAS and determine whether the UAS has the malicious intentions. The combined function also could enlarge the range of UAS detection, which cannot be gained by the single antenna. The multiple sensor approaches combine multiple sensors with the same type to obtain better accuracy or additional functions to improve the performance of the UAS detection. The combination of the multiple sensors extends the capacity of the single sensor and maximizes the range of detection geographically.

(2) Multiple-Type Sensor Data Fusion: In some scenarios, the improvement on the amount cannot mitigate the disadvantages of the single sensors. Different UAS detection approaches are tested in [35]. The acoustic sensors are sensitive to humidity, temperature, and vibration in the environment. The cameras are invalid when sunlight projects on the lens directly. And the RF antennas are hard to recognize the target signal when it is buried in the white Gaussian noise environment. More specifications can be found in [35]. They concluded that the fusion of acoustic and radar could give more precise detection than other approaches. The conventional single sensors cannot achieve outstanding performance on a variable environment and meet multiple requirements of UAS detection ranges.

    Concurrently, the cost of developing high-quality functional sensors is prohibitively high. The researchers resort to the combination of different types of sensors. Based on the disadvantages and the advantages of each type of sensor, the researchers could combine different types of sensors to achieve accuracy and long-distance detection. Long- and short-range detection technologies are combined, where the passive RF receivers detect UAS's telemetry signals while video and acoustic sensors are used to increase the detection accuracy in the near field. Different range sensor systems including acoustic, optic, and radars are integrated for UAS detection in a target area. They deployed a 120-node acoustic array, which uses acoustic camera to locate and track the UAS, and 16 high-resolution optical cameras to detect the UAS in the middle distance. In the long distance, they adopted MIMO radar to operate three different band radars to achieve remote detection [73]. The resulting combination overcomes the drawbacks of each type of sensor on the UAS detection and maximizes the advantages of each type of sensor. Simultaneously, the combination reduced the cost of deploying sensors on a large scale. In [73], the deployment of cameras can meet the middle distance detection requirement and reduce the cost on the deployment of acoustic sensors and radars.

The combination of different types of sensors could achieve an outstanding performance for the restricted areas. However, the deployment and the configuration of this approach require much more specific and professional technologies and technicians with related backgrounds in maintenance. The different types of sensor combinations are a promising approach to maximize the capacity of sensors at the physical levels. In the future, the investigation should be focused on exploring more combinations and characteristics of different sensors for more robust, affordable solutions.

(3) Multiple-Sensing Algorithm Fusion: The conventional approaches of data fusion fuse multiple data acquired from the sensors. Many data fusion approaches had maximized the capacity of sensors greatly. However, for UAS detection, the efficiency and the accuracy still cannot meet the requirement. The novel approaches are needed to combine the multiple sensing algorithms to achieve the efficiency and the accuracy required for UAS detection. The sensing algorithms could be triggered according to the status of detection. The activated sensors deliver the information to the ground stations. Thereafter, the ground stations set the status of the detection, and the relevant algorithms will be swapped into the processing to extract the features of the target information. The UAS detection system will generate the threat outcome according to the result of the sensing algorithms.

In this part, the sensing algorithms receive the data delivered from the sensors and extract the target features according to the types of sensors. The UAS detection system can adjust the sensing algorithm accuracy to meet the requirement of detection once the abnormal signals are detected. In [71], the researchers leverage the unsupervised approaches to extract the features of signal from various acoustic sensors under different scenarios (bird, airplanes, thunderstorm, rain, wind, and UAS). Based on the recognition of scenarios, the system, proposed in this research, triggers SVM and K nearest neighbor, separately, to detect the amateur drones in the restricted areas. To achieve the tracking efficiency, the authors, in [74–77], implemented multiple sensing algorithms on the passive mm-wave radar system to achieve different accuracies of tracking according to the requirement of the recognition. The combination of the multiple sensing algorithms could achieve advantages of the efficiency, accuracy, less overhead of system, etc., according to the requirement of the UAS detection. However, how to make a reasonable arrangement for the sensing algorithms still needs more efforts. And the reasonable schedules of sensing algorithms can be a stimulation for the UAS detection system in the future.

Other data-fusion schemes can be based on different platform integration. The researchers deploy multiple sensors into different platforms to leverage the mobility of different platforms, thus maximizing the capacities of the sensors. The authors deployed the cameras on the surveillance UAS to make sure the amateur drones entering the restricted areas after the deployed acoustic sensors, in the sensing areas, send alarms to the ground stations [78]. In this research, they could recognize different targets such as birds.

The data-fusion approaches combine the advantages of each approach in the detection. The attempts show that the data-fusion approaches have obvious advantages compared with single methods. According to the characteristics of each type of approach, the detection deployment could contain multiple schemes in different areas, which is apart from the center-restricted areas in distance differently. Thereafter, how to implement the data-fusion algorithms to achieve the consistency of the detection system on the results will be the next challenge. Another challenge of the data-fusion approaches is how to balance the weight of each approach in the final decision to achieve optimal detection results.

The distinction between "capture and retrieve" and "disable and drop" is important. Most malicious UASs are captured by the defenders with physical capture, directional electromagnetic pulse (EMP), RF jamming, and hacking. However, only the technologies of RF jamming and hacking can realize the function of retrieve. The retrieve function is supposed to be robust, accurate, and efficient. The defenders are supposed to be confident that their systems have high probabilities of retrieving the malicious UAS again with the protection of the property and the public. For the "disable and drop," only the physical capture methods just drop the UAS from the flight. The technologies of directional EMP, RF jamming, and hacking have the capacity of both disabling and dropping. The directional EMP, RF jamming, and hacking can disable the UAS sensors, circuit, control system, and communication devices to disable the control from remote attackers. However, these technologies can go deeper, such as damaging control circuits and control algorithms, which can drop the UAS from the flight directly.

1.6 UAS mitigation

The technologies of detection and mitigation are still immature. The research on UAS mitigation is limited. Arteche *et al.* [79] developed an architecture of UAS defense system. In this architecture, they specified the effective engagement range, including initial target range, detection range, and neutralization range, which is dominant for response. And their report showed that when the range is over 4000 feet, the hardware-based reaction and neutralization could operate efficiently. Based on the architecture, the approaches could be classified into three main categories. The first one is the physical capture, which focuses on capturing UAS with physical methods. The second is to leverage the noise generator to jam the systems or sensors, thus rendering the UAS inoperable by the UAS controller. The third is to exploit vulnerabilities of the system or sensors to acquire priority of control.

1.6.1 Physical capture

1.6.1.1 Net capture

Net capture is a physical method to negate the UAS. The defenders adopt guns or some specific weapons to trigger the net to catch UAS. The net is stretched when the net is shot and closed to disable the mobility of the drone when the net touches the drone. Kilian *et al.* [11] invented a deployable net capture system that could be installed in the airplane or authenticated UAS. When the unauthorized or unsafe UAS is located, the system could capture the unauthorized or unsafe UAS. Practical approaches to neutralize UAS are attracting the attentions of the military. In [80], a spin-launched UAS projectile was developed. This projectile aims to launch a net to capture a flying UAS. The net is stored in the warhead of a projectile, which allows soldiers to shoot it with regular guns.

1.6.1.2 Directional electromagnetic pulse

Electromagnetic pulses have been mainly used to counter illegal electronic facilities in the car, which could restart or disable the operation of the control system. Based on the function of electromagnetic pulse, Gomozov *et al.* [13] focused on the functional neutralization of on-board radio electronic system on UAS, and they adopted spatio-temporal pulse of the wavelength 0.25 cm to neutralize the UAS and their results showed that their approach could provide an aimed impact on UAS in the range from 0.5 to 1 km and no harm to biological protection.

The physical capture mainly focuses on disabling the mobility of drone and control system. The physical capturing approaches have advantages of easy manipulation, light weight, quick assembling, etc. Once the drone is captured by the physical capturing approaches, the drone will experience damages at different levels. The physical capturing approaches are efficient and low cost, but not friendly to pilots.

1.6.2 Jamming

Jamming is the most popular method used in neutralizing UAS entering restricted areas. The defender's leverage noise signal to interfere the operation of the UAS sensors or systems for neutralization. In this section, we classify jamming into three main categories. Among these attack methods, the main targets are UAS sensors and systems. Zhou *et al.* [81] proposed an approach to leverage a team of UAS to form an air defense radar network, which could jam the targets' sensors. This approach could detect and negate the unauthenticated UAS and their experimental results showed that they could track and jam the phantom developed by DJI to leave the restricted areas, and proofed that the N UAS, in a team, could negate at most 1D targets. Li *et al.* [82] used the direct track deception and fusion to invade the control priority of navigation system and trajectory control system. Based on the GPS deception jamming theory, they leveraged the trajectory cheating to lead the unauthenticated UAS to fly out from the restricted areas. Their results showed that both the direct track and the fusion track could make UAS drift off the restricted areas. Pärlin *et al.* [12] proposed an approach to use the SDR to realize a protocol-aware UAS jamming system. They used an SDR to achieve the remote controller's signal and recognized the communication protocol with analysis. The SDR generates a command to control the UAS to fly away from restricted areas. They compared three different approaches (tone, sweep, and protocol-aware) to evaluate the performance of their approach, which showed that protocol-aware is more efficient than tone and sweep jamming.

- Tone: a narrow band signal at the center of a single channel.
- Sweep: a linear chirp swept across the entire 2.4 GHz ISM band.
- Protocol-aware: a signal imitating either Futaba advanced spectrum spread technology (FASST) and advanced continuous channel-shifting technology (ACCST).

Li *et al.* [83] proposed an approach to leverage the UAS with a jammer to neutralize other UAS eavesdroppers. They used the mobility of UAS to get close to malicious UAS and the jammer installed on the UAs could impact the malicious UAS's trajectories. Sliti *et al.* [84] presented different attacks, which focus on the UAS network for neutralization.

- Jamming attack: transmitting a jamming signal to disrupt communications between a drone and the pilot, forcing the drone to return to "home" location, i.e., from where it took off.
- Black hole attack: a type of denial of service attack, which discards the incoming or outgoing traffic of communication.
- Replay attack: a network attack that vises to maliciously repeat a valid communication so that the communication of the UAS could be analyzed and invaded.

Curpen *et al.* [85] focus on neutralizing the UAS, which is based on long-term evolution (LTE) network. The spectrum analysis in two different cell networks showed that the efficient jamming range for LTE UAS is approximately 60 m. Mototolea *et al.* [86] leveraged the SDR to analyze and hijack the small UAS. In this work, the decoding protocol of DSM2 could get the fingerprinting and pairing process. Willner [87] invented a system that could neutralize remotely explosive UASs in a combat zone. This system could be equipped on the ground or installed on the authenticated UAS, which transmits the jamming signal once the target is locked. Bhattacharya and Başar [88] developed a game theoretic approach to optimize the jamming method on expelling the UAS attacker.

Jamming approaches can provide friendly and zerodamage schemes to neutralize the drone entering the restricted areas. The jamming approaches can provide the neutralizing effects on the drones in different levels (from hardware to software). The jamming can be deployed on a large scale and take effect for a long time. However, the current jamming cannot make a directional effect, which can be controlled by the defenders. The effects of jamming are omnidirectional, which could affect the devices in the restricted areas, and energy-consuming. The jamming needs a long time to take effect when the drone receives enough jamming signals. In the future, jamming are supposed to be controlled and directional with quick reactions.

1.6.3 Vulnerabilities

There are three main methods to exploit the UAS vulnerabilities. Most vulnerabilities exploitation work focuses on GPS control using sensors and communication protocol. The defenders leverage the spoofing methods to GPS and control using sensors, and adopt modification and invading to control using sensors and communication protocol. Rodday *et al.* [89] demonstrated an approach to exploit the identified vulnerabilities of the UAS control systems and performed Man-in-the-Middle attack to inject the control commands to interact with the UAS. Dey *et al.* [90] presented a cracking software development kit (SDK), reversing engineering, and GPS spoofing to hijack the UAS. They compared the DJI and Parrot drone performances under the exploitation attack. The results showed that the DJI is more secure than Parrot, which means that the DJI is hard to rush into the restricted area. Chen *et al.* [91] analyzed the popular altitude estimation algorithms utilized in the navigation system of UAS and proposed several effective attacks (described in "Vulnerabilities" section) to the vulnerabilities.

1.6.3.1 KF-based sensor fusion

- Maximum false data injection: Modifying the calculation functions of GPS and barometer measurements.

1.6.3.2 Altitude estimation based on accurate sensor noise models

- Blocking GPS: Disable the GPS readings.
- Modifying barometer readings: Manipulating the barometer and inject bad data.

1.6.3.3 First-order low-pass filter

- Inject the barometer readings and influence the accuracy of estimation.

Esteves *et al.* [92] demonstrated a simulation on a locked target to gain access to the internal sensors for neutralizing the UAS from restricted areas. Marty [93] presented an approach to hijack the MAVLink protocol on the ArduPilot Mega 2.5 autopilot. Katewa *et al.* [94] proposed a probabilistic attack model to neutralize the UAS, which executed denial of service attack against a subset of sensors based on Bernoulli process. They also described the vulnerabilities on the sensors of the UAS and strategies to negate the UAS using jamming sensors or access control systems. Huang and Wang [95] proposed a spoofing attack based on the physical layer to utilize the angle of arrival, distance-based path loss, and the Rician ◇ k factor to recognize the UAS and source where the signal comes from.

Penetrating vulnerabilities of the system have affected the processing of security in computer field for a long time. The evolution of the amateur drones enables the drones to have their own operation system (OS), which gives the defenders a great chance to exploit the system's vulnerabilities. The integration of embedded system and sensors extend the "blind spot" of unauthorized intruding drones.

For the deployment of detection technologies, the deployment decides the capacity of each type of approach significantly. According to the nature of sensors, the deployment on ground-based stations, UAS, or manned aircraft are varying. The acoustic sensors are sensitive to sound, which needs the environment noise to be stable and quiet. This means the acoustic sensors are not suitable for mobile platforms. The passive RF-based detection has a specific requirement of the antenna distance between each other, which is important to achieve an accurate result for passive RF signal detection. The passive RF-based detection needs the platform to have a powerful computation capacity, which is only for ground-based stations and manned aircraft. The current UAS cannot provide suitable computation and power supply. The vision-based detection can be deployed on ground-based stations, UAS, and manned aircraft. The key sensors of vision-based detection are mainly cameras. Meanwhile, various light weight, low-energy consumption cameras are suitable for the deployment of vision-based detection. The current radar-based detection has the disadvantages of heavy-weight, which is a challenge for the mobile UAS platform. The payload and power supply of the UAS are limited, which cannot satisfy the requirement of radar. Most cases of the deployment of radar-based UAS detection are ground-based stations and manned aircraft. The fanciest approaches are the data-fusion-based detection methods, which are just constrained by computation capacity. Concurrently, the data-fusion-based detection approaches are mainly implemented in the ground-based stations and manned aircraft. Apart from the above, the combination of different platforms also has promising potential to improve the capacity of detection for big properties. The different deployment on the ground-based stations, UAS, and manned aircraft could achieve scalable detection to malicious UAS.

1.7 Conclusion and remarks

In addition to recreational use, UASs, also known as UAVs or drones, are used across our world to support firefighting and search and rescue operations, to monitor and assess critical infrastructure, to provide disaster relief by transporting emergency medical supplies to remote locations, and to aid efforts in securing our borders. However, UAS can also be used for malicious schemes by terrorists, criminal organizations, and lone actors with specific objectives. To promote safe, secure, and privacy-respecting UAS operations, there is an urgent need for innovative technologies for detecting and mitigating UAS. Over the last five years, significant research efforts have been made to counter UAS: detection technologies are based on acoustic, vision, passive RF, or data fusion; and mitigation technologies include physical capture or jamming. In this article, we provided a comprehensive survey of existing literature in the area of UAS detection and mitigation, identified the challenges in countering unauthorized or unsafe UAS, and evaluated the trends of detection and mitigation for protecting against UAS-based threats. We envision that an integrated system capable of detecting and mitigating UAS will be essential to the safe integration of UAS into the airspace system.

References

[1] Federal Aviation Administration, "FAA Aerospace Forecast 2019-39," Apr. 2019. Accessed: Sep. 26, 2021. [Online]. Available: https://www.faa.gov/newsroom/federal-aviation-administration-faaaerospace-forecast-fiscal-years-fy-2019-2039.

[2] H. Song, R. Srinivasan, T. Sookoor, and S. Jeschke. *Smart Cities: Foundations, Principles, and Applications*. John Wiley & Sons, 2017.

[3] Department of Homeland Security & Department of Defense & Federal Aviation Administration, "Federal Contract Opportunity for Air Domain Awareness and Protection in the NAS 2019–2020 Equipment Demonstration and Evaluation," 2019–2020. [Online]. Available: https://govtribe.com/opportunity/federal-contract-opportunity/air-domain-awareness-and-protection-in-the-nas-70rsat20rfi000001#

[4] H. Song, G. Fink, and S. Jeschke. *Security and Privacy in Cyber-Physical Systems: Foundations, Principles, and Applications*. Wiley-IEEE Press, 2017.

[5] X. Yue, Y. Liu, J. Wang, H. Song, and H. Cao, "Software defined radio and wireless acoustic networking for amateur drone surveillance." *IEEE Communications Magazine*, vol. 56, no. 4 (2018): 90–97.

[6] J. Kim, C. Park, J. Ahn, Y. Ko, J. Park, and J. C. Gallagher, "Real-time UAV sound detection and analysis system." In 2017 IEEE Sensors Applications Symposium (SAS), pp. 1–5. IEEE, 2017.

[7] F. Christnacher, S. Hengy, M. Laurenzis, *et al.*, "Optical and acoustical UAV detection." In Electro-Optical Remote Sensing X, vol. 9988, p. 99880B. International Society for Optics and Photonics, 2016.

[8] I. Bisio, C. Garibotto, F. Lavagetto, A. Sciarrone, and S. Zappatore, "Unauthorized amateur UAV detection based on WiFi statistical fingerprint analysis." *IEEE Communications Magazine*, vol. 56, no. 4 (2018): 106–111.

[9] M. E. Rovkin, V. A. Khlusov, N. D. Malyutin, *et al.*, "Radar detection of small-size UAVs." In 2018 Ural Symposium on Biomedical Engineering, Radioelectronics and Information Technology (USBEREIT), pp. 371–374. IEEE, 2018.

[10] B.-H. Sheu, C.-C. Chiu, W.-T. Lu, C.-I. Huang, and W.-P. Chen, "Development of UAV tracing and coordinate detection method using a dual-axis rotary platform for an anti-UAV system." *Applied Sciences*, vol. 9, no. 13 (2019): 2583.

[11] J. C. Kilian, B. J. Wegener, E. Wharton, and D. R. Gavelek, "Counter-unmanned aerial vehicle system and method," Jul. 21, 2015, U.S. Patent 9,085,362, issued July 21, 2015.

[12] K. Pärlin, M. M. Alam, and Y. Le Moullec, "Jamming of UAV remote control systems using software defined radio." In 2018 International Conference on Military Communications and Information Systems (ICMCIS), pp. 1–6. IEEE, 2018.

[13] A. V. Gomozov, D. V. Gretskih, V. A. Katrich, and M. V. Nesterenko, "Functional neutralization of small-size UAVs by focused electromagnetic radiation." In 2017 XXIInd International Seminar/Workshop on Direct and Inverse Problems of Electromagnetic and Acoustic Wave Theory (DIPED), pp. 187–189. IEEE, 2017.

[14] H. Lee, S. Eom, J. Park, and I. Lee, "UAV-aided secure communications with cooperative jamming." *IEEE Transactions on Vehicular Technology*, vol. 67, no. 10 (2018): 9385–9392.

[15] Federal Aviation Administration, "Security Sensitive Airspace Restrictions," 2019, Accessed: Apr. 29, 2019. [Online]. Available: https://www.faa.gov/uas/recreationalfliers/where-can-i-fly/airspace-restrictions/security-sensitive/.

[16] U.S. Department of Homeland Security, "Countering unmanned aircraft systems," Accessed: Apr. 29, 2019. [Online]. Available: https://www.dhs.gov/publication/countering-unmanned-aircraft-systems.

[17] U.S. Government Accountability Office, "Small unmanned aircraft systems: FAA should improve its management of safety risks," 2018.

[18] E. B. Carr, "Unmanned aerial vehicles: Examining the safety, security, privacy and regulatory issues of integration into US airspace." National Centre for Policy Analysis (NCPA). Accessed: Sept. 23, 2013. [Online] 2014. Available: http://www.ncpathinktank.org/pdfs/sp-Drones-long-paper.pdf.

[19] J. Vilímek and L. Buřita. "Ways for copter drone acustic detection." In 2017 International Conference on Military Technologies (ICMT), pp. 349–353. IEEE, 2017.

[20] B. Jang, Y. Seo, B. On, and S. Im, "Euclidean distance based algorithm for UAV acoustic detection." In 2018 International Conference on Electronics, Information, and Communication (ICEIC), pp. 1–2. IEEE, 2018.

[21] E. E. Case, A. M. Zelnio, and B. D. Rigling, "Low-cost acoustic array for small UAV detection and tracking." In 2008 IEEE National Aerospace and Electronics Conference, pp. 110–113. IEEE, 2008.

[22] X. Chang, C. Yang, J. Wu, X. Shi, and Z. Shi, "A surveillance system for drone localization and tracking using acoustic arrays." In 2018 IEEE 10th Sensor Array and Multichannel Signal Processing Workshop (SAM), pp. 573–577. IEEE, 2018.

[23] J. Busset, F. Perrodin, P. Wellig, *et al.*, "Detection and tracking of drones using advanced acoustic cameras." In Unmanned/Unattended Sensors and Sensor Networks XI; and Advanced Free-Space Optical Communication Techniques and Applications, vol. 9647, p. 96470F. International Society for Optics and Photonics, 2015.

[24] H. Liu, Z. Wei, Y. Chen, J. Pan, L. Lin, and Y. Ren, "Drone detection based on an audio-assisted camera array." In 2017 IEEE Third International Conference on Multimedia Big Data (BigMM), pp. 402–406. IEEE, 2017.

[25] A. Bernardini, F. Mangiatordi, E. Pallotti, and L. Capodiferro, "Drone detection by acoustic signature identification." *Electronic Imaging*, 2017, no. 10 (2017): 60–64.

[26] S. Jeon, J.-W. Shin, Y.-J. Lee, W.-H. Kim, Y.-H. Kwon, and H.-Y. Yang, "Empirical study of drone sound detection in real-life environment with deep neural networks." In 2017 25th European Signal Processing Conference (EUSIPCO), pp. 1858–1862. IEEE, 2017.

[27] H. Zhang, C. Cao, L. Xu, and T. A. Gulliver, "A UAV detection algorithm based on an artificial neural network." *IEEE Access*, vol. 6 (2018): 24720–24728.

[28] P. Kosolyudhthasarn, V. Visoottiviseth, D. Fall, and S. Kashihara, "Drone detection and identification by using packet length signature." In 2018 15th International Joint Conference on Computer Science and Software Engineering (JCSSE), pp. 1–6. IEEE, 2018.

[29] C. Xu, B. Chen, Y. Liu, F. He, and H. Song, "RF fingerprint measurement for detecting multiple amateur drones based on stft and feature reduction." In 2020 Integrated Communications Navigation and Surveillance Conference (ICNS), pp. 4G1-1. IEEE, 2020.

[30] C. Xu, F. He, B. Chen, Y. Jiang, and H. Song, "Adaptive RF Fingerprint Decomposition in Micro UAV Detection based on Machine Learning." In ICASSP 2021-2021 IEEE International Conference on Acoustics, Speech and Signal Processing (ICASSP), pp. 7968–7972. IEEE, 2021.

[31] P. Nguyen, H. Truong, M. Ravindranathan, A. Nguyen, R. Han, and T. Vu, "Matthan: Drone presence detection by identifying physical signatures in the drone's RF communication." In Proceedings of the 15th annual international conference on mobile systems, applications, and services, pp. 211–224. 2017.

[32] P. Nguyen, M. Ravindranatha, A. Nguyen, R. Han, and T. Vu, "Investigating cost-effective RF-based detection of drones." In Proceedings of the 2nd workshop on micro aerial vehicle networks, systems, and applications for civilian use, pp. 17–22. 2016.

[33] S. Basak and B. Scheers, "Passive radio system for real-time drone detection and DoA estimation." In 2018 International Conference on Military Communications and Information Systems (ICMCIS), pp. 1–6. IEEE, 2018.

[34] Q. Dong and Q. Zou, "Visual UAV detection method with online feature classification." In 2017 IEEE 2nd Information Technology, Networking, Electronic and Automation Control Conference (ITNEC), pp. 429–432. IEEE, 2017.

[35] S. Hengy, M. Laurenzis, S. Schertzer, *et al.*, "Multimodal UAV detection: study of various intrusion scenarios." In Electro-Optical Remote Sensing XI, vol. 10434, p. 104340P. International Society for Optics and Photonics, 2017.

[36] V. M. Sineglazov, "Multi-functional integrated complex of detection and identification of UAVs." In 2015 IEEE International Conference

Actual Problems of Unmanned Aerial Vehicles Developments (APUAVD), pp. 320–323. IEEE, 2015.

[37] C. Briese, A. Seel, and F. Andert, "Vision-based detection of non-cooperative UAVs using frame differencing and temporal filter." In 2018 International Conference on Unmanned Aircraft Systems (ICUAS), pp. 606–613. IEEE, 2018.

[38] T. Perschke, K. Moren, and T. Müller, "Real-time detection of drones at large distances with 25 megapixel cameras." In Emerging Imaging and Sensing Technologies for Security and Defence III; and Unmanned Sensors, Systems, and Countermeasures, vol. 10799, p. 107990N. International Society for Optics and Photonics, 2018.

[39] M. Saqib, S. D. Khan, N. Sharma, and M. Blumenstein, "A study on detecting drones using deep convolutional neural networks." In 2017 14th IEEE International Conference on Advanced Video and Signal Based Surveillance (AVSS), pp. 1–5. IEEE, 2017.

[40] C. Aker and S. Kalkan, "Using deep networks for drone detection." In 2017 14th IEEE International Conference on Advanced Video and Signal Based Surveillance (AVSS), pp. 1–6. IEEE, 2017.

[41] A. Coluccia, A. Fascista, A. Schumann, *et al.*, "Drone-vs-bird detection challenge at IEEE AVSS2019." In 2019 16th IEEE International Conference on Advanced Video and Signal Based Surveillance (AVSS), pp. 1–7. IEEE, 2019.

[42] A. Schumann, L. Sommer, J. Klatte, T. Schuchert, and J. Beyerer, "Deep cross-domain flying object classification for robust UAV detection." In 2017 14th IEEE International Conference on Advanced Video and Signal Based Surveillance (AVSS), pp. 1–6. IEEE, 2017.

[43] P. Andraši, T. Radišić, M. Muštra, and J. Ivošević, "Nighttime detection of UAVs using thermal infrared camera." *Transportation Research Procedia*, vol. 28 (2017): 183–190.

[44] Q. Dong and Q. Zou, "Visual UAV detection method with online feature classification." In 2017 IEEE 2nd Information Technology, Networking, Electronic and Automation Control Conference (ITNEC), pp. 429–432. IEEE, 2017.

[45] J. Ochodnický, Z. Matousek, M. Babjak, and J. Kurty, "Drone detection by Ku-band battlefield radar." In 2017 International Conference on Military Technologies (ICMT), pp. 613–616. IEEE, 2017.

[46] S. Park and S.-O. Park, "Configuration of an X-band FMCW radar targeted for drone detection." In 2017 International Symposium on Antennas and Propagation (ISAP), pp. 1–2. IEEE, 2017.

[47] M. Caris, W. Johannes, S. Sieger, V. Port, and S. Stanko, "Detection of small UAS with W-band radar." In 2017 18th International Radar Symposium (IRS), pp. 1–6. IEEE, 2017.

[48] R. Nakamura and H. Hadama, "Characteristics of ultra-wideband radar echoes from a drone." *IEICE Communications Express*, vol. 6, no. 9 (2017): 530–534.

[49] M. Krátký and L. Fuxa, "Mini UAVs detection by radar." In International Conference on Military Technologies (ICMT) 2015, pp. 1–5. IEEE, 2015.

[50] D. Shin, D. Jung, D. Kim, J. Ham and S. Park, "A distributed FMCW radar system based on fiber-optic links for small drone detection." *IEEE Transactions on Instrumentation and Measurement*, vol. 66, no. 2 (2016): 340–347.

[51] M. Jahangir, C. J. Baker, and G. A. Oswald, "Doppler characteristics of micro-drones with L-band multibeam staring radar." In 2017 IEEE Radar Conference (RadarConf), pp. 1052–1057. IEEE, 2017.

[52] Á. D. de Quevedo, F. I. Urzaiz, J. G. Menoyo, and A. A. López, "Drone detection with X-band ubiquitous radar." In 2018 19th International Radar Symposium (IRS), pp. 1–10. IEEE, 2018.

[53] J. Klare, O. Biallawons, and D. Cerutti-Maori, "UAV detection with MIMO radar." In 2017 18th International Radar Symposium (IRS), pp. 1–8. IEEE, 2017.

[54] F. Hoffmann, M. Ritchie, F. Fioranelli, A. Charlish, and H. Griffiths, "Micro-Doppler based detection and tracking of UAVs with multistatic radar." In 2016 IEEE radar conference (RadarConf), pp. 1–6. IEEE, 2016.

[55] M. Jian, Z. Lu, and V. C. Chen, "Drone detection and tracking based on phase-interferometric Doppler radar." In 2018 IEEE Radar Conference (RadarConf18), pp. 1146–1149. IEEE, 2018.

[56] G. Sacco, E. Pittella, S. Pisa, and E. Piuzzi, "A MISO radar system for drone localization." In 2018 5th IEEE International Workshop on Metrology for AeroSpace (MetroAeroSpace), pp. 549–553. IEEE, 2018.

[57] M. Zywek, G. Krawczyk, and M. Malanowski, "Experimental results of drone detection using noise radar." In 2018 19th International Radar Symposium (IRS), pp. 1–10. IEEE, 2018.

[58] S. J. Lee, J. H. Jung, and B. Park, "Possibility verification of drone detection radar based on pseudo random binary sequence." In 2016 International SoC Design Conference (ISOCC), pp. 291–292. IEEE, 2016.

[59] Y.-K. Kwag, I.-S. Woo, H.-Y. Kwak, and Y.-H. Jung, "Multi-mode SDR radar platform for small air-vehicle Drone detection." In 2016 CIE International Conference on Radar (RADAR), pp. 1–4. IEEE, 2016.

[60] K. Stasiak, M. Ciesielski, A. Kurowska, and W. Przybysz, "A study on using different kinds of continuous-wave radars operating in C-band for drone detection." In 2018 22nd International Microwave and Radar Conference (MIKON), pp. 521–526. IEEE, 2018.

[61] H. Sun, B.-S. Oh, X. Guo, and Z. Lin, "Improving the Doppler resolution of ground-based surveillance radar for drone detection." *IEEE Transactions on Aerospace and Electronic Systems*, vol. 55, no. 6 (2019): 3667–3673.

[62] T. Martelli, F. Murgia, F. Colone, C. Bongioanni, and P. Lombardo, "Detection and 3D localization of ultralight aircrafts and drones with a WiFi-based passive radar." In International Conference on Radar Systems (Radar 2017), pp. 1–6. IET, 2017.

[63] B. Knoedler, R. Zemmari, and W. Koch, "On the detection of small UAV using a GSM passive coherent location system." In 2016 17th International Radar Symposium (IRS), pp. 1–4. IEEE, 2016.

[64] A. D. Chadwick, "Micro-drone detection using software-defined 3G passive radar." In International Conference on Radar Systems (Radar 2017), pp. 1-6, IET, 2017.

[65] D. Solomitckii, M. Gapeyenko, V. Semkin, S. Andreev, and Y. Koucheryavy, "Technologies for efficient amateur drone detection in 5G millimeter-wave cellular infrastructure." *IEEE Communications Magazine*, vol. 56, no. 1 (2018): 43–50.

[66] X. Yang, K. Huo, W. Jiang, J. Zhao, and Z. Qiu, "A passive radar system for detecting UAV based on the OFDM communication signal." In 2016 Progress in Electromagnetic Research Symposium (PIERS), pp. 2757–2762. IEEE, 2016.

[67] Y. Liu, X. Wan, H. Tang, J. Yi, Y. Cheng, and X. Zhang, "Digital television based passive bistatic radar system for drone detection." In 2017 IEEE Radar Conference (RadarConf), pp. 1493–1497. IEEE, 2017.

[68] G. Fang, J. Yi, X. Wan, Y. Liu, and H. Ke, "Experimental research of multistatic passive radar with a single antenna for drone detection." *IEEE Access*, vol. 6 (2018): 33542–33551.

[69] C. Schüpbach, C. Patry, F. Maasdorp, U. Böniger, and P. Wellig, "Micro-UAV detection using DAB-based passive radar." In 2017 IEEE radar conference (RadarConf), pp. 1037–1040. IEEE, 2017.

[70] H. Fu, S. Abeywickrama, L. Zhang, and C. Yuen, "Low-complexity portable passive drone surveillance via SDR-based signal processing." *IEEE Communications Magazine*, vol. 56, no. 4 (2018): 112–118.

[71] Z. Uddin, M. Altaf, M. Bilal, L. Nkenyereye, and A. K. Bashir, "Amateur Drones Detection: A machine learning approach utilizing the acoustic signals in the presence of strong interference." *Computer Communications*, vol. 154 (2020): 236–245.

[72] B. Nuss, L. Sit, M. Fennel, J. Mayer, T. Mahler, and T. Zwick, "MIMO OFDM radar system for drone detection." In 2017 18th International Radar Symposium (IRS), pp. 1–9. IEEE, 2017.

[73] U. Böniger, B. Ott, P. Wellig, *et al.*, "Detection of mini-UAVs in the presence of strong topographic relief: a multisensor perspective." In Target and Background Signatures II, vol. 9997, p. 999702. International Society for Optics and Photonics, 2016.

[74] V. K. Klochko, V. V. Strotov, and S. A. Smirnov, "Multiple objects detection and tracking in passive scanning millimeter-wave imaging systems." In MillimetreWave and Terahertz Sensors and Technology XII, vol. 11164, p. 111640E. International Society for Optics and Photonics, 2019.

[75] M. Martinez, "UAS detection, classification, and tracking in urban terrain." In 2019 IEEE Radar Conference (RadarConf), pp. 1–6. IEEE, 2019.

[76] B. M. Sazdić-Jotić, D. R. Obradović, D. M. Bujaković, and B. P. Bondžulić, "Feature extraction for drone classification." In 2019 14th International Conference on Advanced Technologies, Systems and Services in Telecommunications (TELSIKS), pp. 376–379. IEEE, 2019.

[77] T. B. Sarikaya, D. Yumus, M. Efe, G. Soysal, and T. Kirubarajan, "Track based UAV classification using surveillance radars." In 2019 22th International Conference on Information Fusion (FUSION), pp. 1–6. IEEE, 2019.

[78] J. Wang, X. Yue, Y. Liu, *et al.*, "Integrating ground surveillance with aerial surveillance for enhanced amateur drone detection." In Disruptive Technologies in Information Sciences, vol. 10652, p. 106520E. International Society for Optics and Photonics, 2018.

[79] D. Arteche, K. Chivers, B. Howard, *et al.*, "Drone defense system architecture for US navy strategic facilities." Naval Postgraduate School Monterey United States, 2017.

[80] B. Tomasz, F. Richard, and T. LaMar, "Scalable effect net warhead." Feb. 5, 2019, U.S. Patent 10 197 365.

[81] C. Zhou, Y. Liu, and Y. Song, "Detection and tracking of a UAV via hough transform." In 2016 CIE International Conference on Radar (RADAR), pp. 1–4. IEEE, 2016.

[82] C. Li and X. Wang, "Jamming research of the UAV GPS/INS integrated navigation system based on trajectory cheating." In 2016 9th International Congress on Image and Signal Processing, BioMedical Engineering and Informatics (CISP-BMEI), pp. 1113–1117. IEEE, 2016.

[83] A. Li, Q. Wu, and R. Zhang, "UAV-enabled cooperative jamming for improving secrecy of ground wiretap channel." *IEEE Wireless Communications Letters*, vol. 8, no. 1 (2018): 181–184.

[84] M. Sliti, W. Abdallah, and N. Boudriga, "Jamming attack detection in optical UAV networks." In 2018 20th international conference on transparent optical networks (ICTON), pp. 1–5. IEEE, 2018.

[85] R. Curpen, T. Bălan, I. A. Micloş, and I. Comănici, "Assessment of signal jamming efficiency against LTE UAVs." In 2018 International Conference on Communications (COMM), pp. 367–370. IEEE, 2018.

[86] D. Mototolea and C. Stolk, "Software defined radio for analyzing drone communication protocols." In 2018 International Conference on Communications (COMM), pp. 485–490. IEEE, 2018.

[87] B. J. Willner, "Methods and apparatuses for detecting and neutralizing remotely activated explosives." U.S. Patent 8,240,238, issued August 14, 2012.

[88] S. Bhattacharya, and T. Başar, "Game-theoretic analysis of an aerial jamming attack on a UAV communication network." In Proceedings of the 2010 American Control Conference, pp. 818–823. IEEE, 2010.

[89] N. M. Rodday, R. de O. Schmidt, and A. Pras, "Exploring security vulnerabilities of unmanned aerial vehicles." In NOMS 2016-2016 IEEE/IFIP Network Operations and Management Symposium, pp. 993–994. IEEE, 2016.

[90] V. Dey, V. Pudi, A. Chattopadhyay, and Y. Elovici, "Security vulnerabilities of unmanned aerial vehicles and countermeasures: An experimental study." In 2018 31st International Conference on VLSI Design and 2018 17th International Conference on Embedded Systems (VLSID), pp. 398–403. IEEE, 2018.

[91] W. Chen, Y. Dong, and Z. Duan, "Attacking altitude estimation in drone navigation." In IEEE INFOCOM 2018-IEEE Conference on Computer Communications Workshops (INFOCOM WKSHPS), pp. 888–893. IEEE, 2018.

[92] J. L. Esteves, E. Cottais, and C. Kasmi, "Unlocking the access to the effects induced by IEMI on a civilian UAV." In 2018 International Symposium on Electromagnetic Compatibility (EMC EUROPE), pp. 48–52. IEEE, 2018.

[93] J. A. Marty, "Vulnerability analysis of the mavlink protocol for command and control of unmanned aircraft." Air Force Institute of Technology. Theses and Dissertations, 613, 2014.

[94] V. Katewa, R. Anguluri, A. Ganlath, and F. Pasqualetti, "Secure reference-tracking with resource-constrained UAVs." In 2017 IEEE Conference on Control Technology and Applications (CCTA), pp. 1319–1325. IEEE, 2017.

[95] K. Huang and H. Wang, "Combating the control signal spoofing attack in UAV systems." *IEEE Transactions on Vehicular Technology*, vol. 67, no. 8 (2018): 7769–7773.

Chapter 2

Formal verification of safety-critical software using SPARK

Laura Humphrey[1], Ryan Baity[2], and Kenneth Hopkinson[2]

The increasing size, complexity, and interconnectedness of software-intensive systems are raising concerns about how to affordably ensure such systems are free of cybersecurity vulnerabilities and other faults, because even the smallest of faults can have severe safety and financial repercussions. Unfortunately, typical test-based approaches only achieve partial coverage of all possible software behaviors, leaving room for faults to go undetected. A possible solution is to supplement testing with the use of formal methods, i.e., mathematically based tools and techniques that verify software and other design artifacts through formal proof and rigorous analysis. In this chapter, we show how SPARK, a programming language and associated formal verification toolset, can be used to help prevent cybersecurity vulnerabilities and other software faults. We also give an overview of industrial efforts that have successfully used SPARK for this purpose, paying particular attention to how SPARK has been used in conjunction with other formal methods and more traditional test- and review-based approaches to increase software quality and reliability.

2.1 Introduction

The increasing size and complexity of software running on safety-critical systems such as those developed in the aerospace industry has raised concerns about how to continue to ensure software quality and reliability, especially in a cost-effective way [1,2]. At the same time, more and more devices are becoming interconnected, opening new avenues for cybersecurity attacks. While any type of software fault is of concern in a safety-critical system, cybersecurity-related faults are particularly pernicious; whereas a general fault might never encounter the circumstances necessary to trigger it, a cybersecurity-related fault will be actively sought out and exploited by attackers.

Even the smallest software faults can have severe safety and financial repercussions. For example, in 1990, AT&T lost more than $60 million in unconnected

[1]Air Force Research Laboratory, Wright-Patterson Air Force Base, Dayton, OH, USA
[2]Air Force Institute of Technology, Wright-Patterson Air Force Base, Dayton, OH, USA

calls because of a misplaced break statement in a network switch software patch that remained undetected even after rigorous testing [3]. The 2014 Heartbleed OpenSSL vulnerability, a simple fault consisting of a single unchecked memory copy, allowed attackers to remotely read protected memory from an estimated 24–55% of HTTPS sites [4]. In 2017, at least 465,000 St. Jude's Medical RF-enabled cardiac devices were recalled because of possible cybersecurity vulnerabilities [5]. In the same year, the WannaCry ransomware and NotPetya malware attacks, which made use of the EternalBlue exploit to target a vulnerability in Microsoft's implementation of the Server Message Block (SMB) protocol, caused an estimated $4 billion and $10 billion in damages, respectively [6,7]. Recently, concerns have been raised about the possible consequences of cybersecurity vulnerabilities in automobiles and aircraft, especially as driverless cars and unmanned air vehicles begin to see real-world use [8].

Unfortunately, it is extremely difficult to ensure that software is free of faults. One of the primary mechanisms for finding software faults is testing, but testing can only achieve partial coverage of all possible software behaviors, a problem that is exacerbated as the size and complexity of software increases [9]. Because testing only achieves partial coverage, it can only reveal the presence of faults; it cannot prove their absence. But given that even a single cybersecurity-related fault will be sought out and exploited if found, we would very much like to do our best to prove their absence!

A possible solution is the use of *formal methods*, i.e., mathematically based tools and techniques for design and verification of software and hardware, to provide evidence of software quality and reliability. Formal methods have their roots in formal logic, discrete mathematics, and computer-readable languages. Using these as a foundation, they aim to provide mathematical guarantees about system behavior through formal proof and analysis rather than testing, similar to the way in which one can prove that Pythagorean theorem is correct using geometric axioms rather than testing it against the set of all possible right triangles. Formal methods consist of two major activities, modeling and analysis, often done with the assistance of automated or semi-automated tools. They are analogous to mathematically based approaches used in traditional engineering disciplines, e.g., statics for civil engineering, dynamics for mechanical engineering, and stoichiometry for chemical engineering. Formal methods can be used to analyze a range of system artifacts, including high- and low-level requirements, architectures, source code, object code, and discrete logic hardware, for a variety of properties such as traceability, completeness, consistency, compliance, and robustness. Formal methods are an accepted means to satisfy verification objectives in certification standards across a variety of domains, for instance, in railway (EN 50128) [10] and industrial (IEC 61508) [11] processes and for avionics (DO-333 supplement to DO-178C) [12–14]. Certification standards for other domains such as the automotive (ISO 26262) [15] and nuclear (IEC 60880) and space (ECSS-QST-80C) [16] domains also recognize some uses of formal methods as a verification technique.

Given the variety of artifacts, properties, and underlying mathematical frameworks involved, formal methods are somewhat broad. Reference [17] gives an extensive overview of formal methods that discusses different categories of approaches according to the level of rigor; variations in the extent to which formal methods can be applied in terms of lifecycle stage, proportion of the system covered, and types of properties analyzed; the value of formalizing specifications and the importance of validating them; the benefits and fallibilities of formal methods; automated and semi-automated systems and tools; and a review of industrial applications. It also includes a quick introduction to mathematical logic, including propositional calculus, predicate calculus, first-order theories, modal logics, etc. The DO-333 [12] supplement to DO-178C [18] gives a more recent overview of formal methods, and [13] and [14] each provide an overview of DO-333 itself and illustrate the application of various tools and approaches from formal methods on concrete examples relevant to the aerospace domain.

In this chapter, we will discuss the use of formal methods to address cybersecurity and other important software concerns, in particular, using the SPARK formal verification language and toolset. In Section 2.2, we give some background on SPARK. In Section 2.3, we show how SPARK can be used to find, fix, and then prove that source code is free of cybersecurity vulnerabilities and other faults through an example. In Section 2.4, we give an overview of aerospace and cybersecurity-related projects that have or are currently using SPARK for similar purposes. We close in Section 2.4 with a summary and thoughts on future directions for formal methods.

2.2 Background on SPARK

SPARK is both a programming language and an associated toolset for formal verification [19]. The SPARK language is based on the Ada programming language [20]. Ada is a statically typed, imperative, object-oriented language with a strong type system, and the principles underlying its design are intended to encourage reliability, maintainability, and efficiency. These characteristics are inherited by SPARK, which both removes a few features from Ada that make verification difficult and adds a small number of features that facilitate verification. Several good resources are available for learning SPARK and Ada [19–22]. In what follows, we assume that the reader has some familiarity with SPARK/Ada or is otherwise able to deduce the meaning of basic SPARK/Ada language features.

In terms of formal verification, SPARK performs two forms of sound static analysis on source code. The first is flow analysis, which checks the initialization of variables and compliance with user-specified data flow contracts. Flow analysis can also identify unused assignments and unmodified variables, which often indicate extraneous/unnecessary code. The second is proof, which checks for both functional correctness, i.e., compliance with user-specified behavioral contracts, and the absence of runtime errors. It should be noted that these checks and contracts can be optionally compiled as executable assertions that are evaluated at runtime, e.g., to debug code during development or to verify code through testing if static analysis cannot be fully completed.

There are different levels to which one can use SPARK to verify a program [23]. Generally, the level of verification performed during development is incremental, going from lowest to highest. Colloquially, the levels are referred to in order as "stone," "bronze," "silver," "gold," and "platinum." Verification at the stone level is achieved when code is accepted by SPARK, because the SPARK language has stricter legality rules than Ada. Bronze is achieved when flow analysis returns with no error. Silver ensures that no runtime errors, e.g., division by zero or numeric overflow/underflow, will be encountered when executing the program. Gold consists of verifying compliance with key user contracts/specifications; however, at this level, the specifications only partially describe the desired behavior of the code. Platinum consists of verifying compliance of the code against a complete set of specifications. We briefly note that these levels have a rough correspondence with levels of assurance in certification standards, e.g., "software level" in DO-178C and "software integrity level" in IEC 61508, EN 50128, *et al.* As discussed in [23], silver is the baseline level for all software levels except Level E, i.e., software whose anomalous behavior would have no effect on aircraft operational capability or pilot workload. Software at Level A or Level B may aim for gold or platinum depending on whether key properties or full functional correctness can be more easily verified to a sufficient level by other means. Targeting platinum becomes less likely at Level C or Level D because verification by testing can be more easily argued to be sufficient. For software at Level E, silver might still be considered but could be weakened to bronze if other means were used to build sufficient confidence that the software is free of runtime errors or such errors were sufficiently mitigated.

At a high level, SPARK performs formal verification by translating SPARK programs, including checks and contracts to be verified, to the Why3 platform for deductive program verification [24]. Why3 then uses a weakest-precondition calculus to generate verification conditions (VCs), i.e., logical formulas whose validity would imply soundness of the code with respect to its contracts. Why3 then uses multiple provers, including but not limited to satisfiability modulo theory (SMT) solvers Alt-Ergo [25], CVC4 [26], and Z3 [27], to attempt to prove the validity of the VCs given the code. We briefly note that though SPARK aims to perform fully automated verification through this process, it is often necessary for the user to provide additional assertions, e.g., loop invariants and type invariants, to create additional VCs that help guide the provers toward proof of the original checks and contracts to be verified.

Listing 2.1 Example data dependency and data flow contracts

```
procedure Append_To_Stream(Message : in String;
                           Status  : out Boolean)
with   Global  => (In_out => Message_Stream),
       Depends => (Message_Stream => (Message_Stream, Message)
                   Status =>         (Message_Stream, Message));
```

As previously mentioned, SPARK performs two types of analysis: flow analysis and proof. The proof is covered in the next section, so let us briefly consider flow analysis. There are two types of contracts relevant to flow analysis: data dependency contracts and flow dependency contracts. Data dependency contracts describe what global data a subprogram depends on and whether the data are read, written, or both. Flow dependency contracts describe dependencies between a subprogram's inputs and outputs, including global data. Consider the example specification shown in Listing 2.1 for a hypothetical procedure `Append_To_Stream` that takes a `Message` string as input, appends it to a global `Message_Stream`, and reports whether the operation was successful through a boolean output `Status`. The data dependency contract specified with aspect `Global` indicates that `Message_Stream` has mode `In_Out`, meaning that it is both an input to the procedure that must be initialized before the procedure is called and also an output that the procedure will modify. The flow dependency contract specified with aspect `Depends` indicates that the value of `Message_Stream` after the procedure is called depends both on the current value of `Message_Stream` and the value of `Message`, and the value of `Status` depends on both of these as well. SPARK flow analysis will verify whether or not these contracts hold given an implementation (not shown).

2.3 An example of SPARK for cybersecurity

SPARK is able to verify that code is free of different types of cybersecurity vulnerabilities, including many of those identified by the MITRE Corporation. MITRE maintains two databases relevant to cybersecurity. One is the Common Weakness Enumeration (CWE) [28], a database that categorizes common types of software and hardware weaknesses, giving each a unique CWE number. The other is the Common Vulnerabilities and Exposures (CVE) [29], which tracks publicly known cybersecurity vulnerabilities and their root causes in terms of CWEs. For example, the Heartbleed OpenSSL vulnerability is CVE-2014-0160 and is caused by CWE-126 (buffer over-read), CWE-125 (out-of-bounds read), CWE-130 (improper handling of length parameter inconsistency), and CWE-843 (access of resource using incompatible type). The CWE also contains a running list of the 25 most dangerous software weaknesses based on CVEs over the last two years. The current list contains many CWEs that map to common runtime errors, e.g. out-of-bounds reads and writes, improper restriction of operations within the bounds of a memory buffer, and integer overflow or wraparound. The aforementioned CWEs are easily eliminated by proving SPARK code to the silver level. In fact, [30] shows how SPARK can be used to address many CWEs, some through fundamental stone-level language features and others by proving code to the silver level or beyond.

To demonstrate SPARK's ability to address cybersecurity concerns, let us consider how SPARK can be used to prove that code is both free from common types of cybersecurity vulnerabilities and functionally correct to the gold level through an example involving linear interpolation. Let us start with a brief review. Suppose we have a mathematical function $y = f(x)$ to be estimated as a piecewise linear function over a sequence of points (x_i, y_i) for $i = \{1, \ldots, n\}$. Suppose the points are ordered such that $x_i < x_{i+1}$ for $i = \{1, \ldots, n-1\}$. Then using linear interpolation, the value of $f(a)$ is estimated as follows:

- If $a < x_1$, then $f(a) = f(x_1) = y_1$
- If $a > x_n$, then $f(a) = f(x_n) = y_n$

- If $a = x_i$ for some i, then $f(a) = f(x_i) = y_i$
- Otherwise, $f(a) = y_i + (a - x_i)\dfrac{x_{i+1} - x_i}{y_{i+1} - y_i}$ for $x_i < a < x_{i+1}$

Listing 2.2 Candidate package specification interpolation.ads

```
package Interpolation with SPARK_Mode is

   subtype Arg is Integer;
   subtype Value is Integer;
   type Point is record
     X : Arg;
     Y : Value;
   end record;

   type Index is new Integer range 1 .. 100;
   type Func is array (Index range <>) of Point with
     Predicate => Func'First = 1;

   function Increasing (F : Func) return Boolean is
     (for all I in F'Range =>
       (for all J in F'Range =>
         (if I < J then F(I).Y <= F(J).Y)));

   subtype Monotonic_Incr_Func is Func with
     Predicate => Increasing (Monotonic_Incr_Func);

   function Eval (F : Monotonic_Incr_Func; A : Arg) return Value;

end Interpolation;
```

Let us now consider a candidate implementation in SPARK, in particular, for linear interpolation over monotonically increasing functions, i.e., where additionally $y_i \leq y_{i+1}$ for $i = \{1, \ldots, n-1\}$. A SPARK package consists of both a specification and a body. Listing 2.2 shows a candidate specification file `interpol.ads` for a package `Interpolation` that specifies relevant types, subtypes, procedures, and functions. We first define integer subtypes `Arg` and `Value` to represent x and y values, respectively. We then define a record type `Point` with fields `X` and `Y` of types `Arg` and `Value` to represent (x, y) point values. Next, we define an integer-derived type `Index` that can range from 1 to 100. Then, we define an array type `Func` with elements of type `Point` indexed by type `Index` to represent sequences of (x, y) point values representing a piecewise linear function. Note that type `Func` includes a type predicate specifying that the first index of all variables of this type should be 1; if not, the result would be a runtime error. We then define a function `Increasing` that operates on arrays of type `Func` and returns true if and only if the `Y` values of points stored in the array are in monotonically increasing order. Note that the logic for this function is defined directly in the specification using an expression function. Next, we define a subtype `Monotonic_Incr_Func` of type `Func` with a type predicate specifying that all variables of this type should satisfy the function `Increasing`. Finally, we define a function `Eval` that takes a `Monotonic_Incr_Func` and an `Arg` as arguments and returns a `Value`; this is the function that will perform linear interpolation. The candidate body for this function is given in file `interpol.adb`, shown in Listing 2.3. Note that integer arithmetic is used to perform linear interpolation.

Listing 2.4 shows a simple main package for exercising the interpolation function `Eval` through procedure `Test_Interpolation`. This package sets up a nominal test case with a constant `F` of type `Func` containing seven points. Note that the y-values of `F` satisfy `Increasing`, so that when we call `Eval` on `F`, `F` can be converted to subtype `Monotonic_Incr_Func` without violating its type predicate and raising a runtime error. This test case returns the expected result, printing monotonically increasing y-values for x-values ranging from −20 to 10.

Listing 2.3 Candidate package body interpolation.adb

```
package body Interpolation with SPARK_Mode is

  function Eval (F : Monotonic_Incr_Func; A : Arg) return Value is
  begin
    if A <= F(1).X then
      return F(1).Y;
    end if;

    for K in F'Range loop
      if A = F(K).X then
        return F(K).Y;
      elsif A > F(K).X and A < F(K+1).X then
        declare
          DX : constant Integer := F(K+1).X - F(K).X;
          DY : constant Integer := F(K+1).Y - F(K).Y;
        begin
          return F(K).Y + (A - F(K).X) * DY / DX;
        end;
      end if;
    end loop;

    return F(F'Last).Y;
  end Eval;

end Interpolation;
```

Listing 2.4 Initial main package test_interpolation.adb

```
with Ada.Text_IO; use Ada.Text_IO;
with Interpolation; use Interpolation;

procedure Test_Interpolation is
  F : constant Func :=
    ((-10, -10), (-1, -3), (0, 6), (5, 12), (12, 12), (18, 12), (20, 15));
begin
  for A in -10 .. 20 loop
    Put_Line ("X =" & A'Image & " Y =" & Eval(F, A)'Image);
  end loop;
end Test_Interpolation;
```

We can now use SPARK to analyze the code. SPARK can run in conjunction with another tool called CodePeer to report possible CWEs, and with that option on, SPARK reports five errors:

1. Line 5: array index check might fail if array empty [CWE-118]
2. Line 12: array index check might fail at F(K+1) [CWE-119]
3. Line 14: overflow check might fail at subtraction [CWE-190]
4. Line 15: overflow check might fail at subtraction [CWE-190]
5. Line 17: overflow check might fail at multiplication [CWE-190]

where CWE-118 is a "range error," CWE-119 is an "improper restriction of operations within the bounds of a memory buffer," and CWE-190 is an "integer overflow or wraparound." We note that CWE-119 and CWE-190 are both currently on the CWE list of the top 25 most dangerous weaknesses.

These reports help us find and correct errors in our code. By inspecting the body in Listing 2.3, we realize that the first error will occur if `F` contains no elements. This can be prevented by adding a precondition to `Eval` specifying that the length of `F` must be greater than `0`. The second error will occur if the loop variable `K` is equal to the length of `F`, which would cause `F(K+1)` to be out of bounds. This can be corrected by making the loop variable `K` range from `F'First .. F'Last-1` instead of over `F'Range`. The third, fourth, and fifth errors would occur if the corresponding arithmetic operations were to overflow. This can be corrected by restricting the range of values allowed for types `Arg` and `Value`, e.g., from −10,000 to 10,000 or other values appropriate to the application domain. With these changes, SPARK returns no errors.

Suppose we would then like to check functional correctness of `Eval` to the gold level. We would start by writing an appropriate specification, e.g.

```
function Eval (F : Monotonic_Incr_Func; A : Arg) return Value with
  Pre  => (F'Length > 0),
  Post => (if A <= F(1).X then Eval'Result = F(1).Y
           elsif A >= F(F'Last).X then Eval'Result = F(F'Last).Y
           else (for some K in 1 .. F'Last - 1 =>
                    A in F(K).X .. F(K+1).X and then
                    Eval'Result in F(K).Y .. F(K+1).Y));
```

SPARK analysis then returns the message `medium: postcondition might fail, cannot prove Eval'Result = F(1).Y`. This tells us that SPARK is unable to prove the postcondition. There could be several reasons for this. First, the function specification might be incorrect, i.e., it might not express what we intend. Second, the function implementation in the body might be incorrect. Third, the specification and implementation might be correct, but SPARK is unable to complete a proof of correctness. In such cases, it is useful to re-examine the code.

Upon re-examination of the specification in Listing 2.2, we notice that though subtype `Monotonic_Incr_Func` with type predicate `Increasing` specifies that y-values should be stored in monotonically increasing order, no ordering is specified for x-values. Indeed, we can modify the test case of Listing 2.4 to set

```
F : constant Func := ((-10,0), (10, 0), (20,0), (0,5));
```

to create a robustness test case. The values in this array do not describe a monotonically increasing function, yet `Eval` accepts it as an argument and returns an erroneous result in which all y-values are 0.

Listing 2.5 Corrected and extended specification interpol.ads

```
package Interpolation with SPARK_Mode is

   subtype Arg is Integer range -10_000 .. 10_000;
   subtype Value is Integer range -10_000 .. 10_000;
   type Point is record
      X : Arg;
      Y : Value;
   end record;

   type Index is new Integer range 1 .. 100;
   type Func is array (Index range <>) of Point with
     Predicate => Func'First = 1 and then
                  (for all I in Func'Range =>
                     (for all J in Func'Range =>
                        (if I < J then Func(I).X < Func(J).X)));

   function Monotonic_Increasing (F : Func) return Boolean is
     (for all I in F'Range =>
        (for all J in F'Range =>
           (if I < J then F(I).Y <= F(J).Y)));

   subtype Monotonic_Incr_Func is Func with
     Predicate => Monotonic_Increasing (Monotonic_Incr_Func);

   function Eval (F : Monotonic_Incr_Func; A : Arg) return Value with
     Pre  => (F'Length > 0),
     Post => (if A <= F(1).X then Eval'Result = F(1).Y
              elsif A >= F(F'Last).X then Eval'Result = F(F'Last).Y
              else (for some K in 1 .. F'Last - 1 =>
                      A in F(K).X .. F(K+1).X and then
                      Eval'Result in F(K).Y .. F(K+1).Y));

end Interpolation;
```

There are several ways to address this. Listing 2.5 shows a corrected specification file with several changes. These include the changes necessary to address runtime errors related to the CWEs discussed earlier, the gold-level postcondition for function `Eval`, and a modified type predicate on `Func` that specifies that all x-values should be stored in increasing order. We also rename function `Increasing` to `Monotonic_Increasing` to help emphasize the difference between this function, which specifies that y-values must be stored in non-decreasing order, and the predicate on type `Func`, which specifies that x-values must be stored in strictly increasing order. With these changes, the `Test_Interpolation` procedure now raises a runtime error for the input in the robustness test case because of the new type predicate on `Func`. However, SPARK analysis still reports that the postcondition might fail.

Let us now go back and examine the body of `Eval` in Listing 2.3. We make several changes that allow us to prove the postcondition of `Eval`, which are shown in Listing 2.6. At first glance, the original body can be improved for efficiency by

Listing 2.6 Corrected interpol.adb

```
package body Interpolation with SPARK_Mode is

  function Eval (F : Monotonic_Incr_Func; A : Arg) return Value is
  begin

    if A <= F(1).X then
      return F(1).Y;
    elsif A >= F(F'Last).X then
      return F(F'Last).Y;
    end if;

    for K in F'First .. F'Last - 1 loop
      pragma Loop_Invariant(A >= F(K).X);
      if A = F(K).X then
        return F(K).Y;
      elsif A > F(K).X and A < F(K+1).X then
        declare
          DX : constant Integer := F(K+1).X - F(K).X;
          DY : constant Integer := F(K+1).Y - F(K).Y;
        begin
          pragma Assert(F(K).Y + (A - F(K).X) * DY / DX <= F(K+1).Y);
          pragma Assert(F(K).Y + (A - F(K).X) * DY / DX >= F(K).Y);
          return F(K).Y + (A - F(K).X) * DY / DX;
        end;
      end if;
    end loop;

    raise Program_Error;
  end Eval;

end Interpolation;
```

checking for the case that `A >= F(F'Last).X` before looping over the points comprising `F`. We would then also want to add a `raise Program_Error` statement after the loop, because going past the end of the loop without returning a value should not be possible. As part of proving the absence of runtime errors, SPARK will attempt to prove that this statement is not reachable. However, with only these changes, SPARK still reports that the postcondition might fail and that the `Program_Error` might be reachable. Note that loops are challenging for automated provers, which attempt to prove properties about loops inductively; so, often a loop invariant is needed to help the provers develop an inductive proof. Here, we add a loop invariant expressing that `A >= F(K).X` in every iteration of the loop. Intuitively, this helps because it establishes that the value of `A` could not have been between `F(K).X` and `F(K+1).X` for values of `K` in earlier iterations of the loop, i.e., we did not somehow erroneously "skip over" the points that `A` lies between on a previous iteration. SPARK is able to prove that this loop invariant is true and then that the `Program_Error` is not reachable. However, SPARK is still not able to prove the postcondition. We then speculate the provers do not have enough information about the return result within the loop, so we add an assertion that the value computed by the expression used in the return statement does in fact lie between `F(K).Y` and `F(K+1).Y`. SPARK is now able to prove the postcondition.

Through the process of analyzing code with SPARK, reviewing and correcting the code in response to reported errors, and providing additional assertions such as assert statements and loop invariants when SPARK is unable to find proof, we are able to prove that our code is free of runtime errors and satisfy its user-specified functional contract. We note that finding the right assertions can be challenging. Several resources walk through other examples [31–33] or offer advice on writing good assertions [19,21], including the SPARK User's Guide.

2.4 Aerospace and cybersecurity projects utilizing SPARK

Many projects have or are currently using SPARK to help provide evidence of software security and correctness, often in conjunction with other formal methods and more traditional review-based and test-based approaches. Reference [34] gives a historical overview of several such projects. Here, we give a more detailed description of some of those projects along with other SPARK-related projects that have an aerospace or cybersecurity focus.

One of the first major applications of SPARK was to the Ship Helicopter Operating Limits Information System (SHOLIS), a safety-critical system that aids the safe operation of helicopters on Naval vessels [35]. The work was carried out under the 1991 version of UK Interim Defence Standards 00-55 and 00-56, which required the use of formal methods for safety-critical applications. In addition to SPARK, the effort used the Z notation or Z [36] to write parts of the specification. Z is based on set theory and a first-order predicate calculus and is therefore amenable to proof, either manually or with tool assistance. In this effort, the software requirement specification was written in a combination of English and Z. The software design specification, which refines the software requirement specification by adding implementation details, was written in a combination of English, Z, and SPARK. The code was written in SPARK. The effort therefore consisted of two categories of proof activity: Z proof and SPARK proof. The Z proof activities were mainly manual and included checking the consistency of global variables and constants, the existence of valid initial states, and that operator preconditions calculated from the Z specifications [37] matched the specifier's expectations. Based on a software safety analysis, some software components were classified as safety-critical. For every safety-critical subprogram, SPARK pre- and postconditions were written based on the Z specifications, and data flow and information flow contracts were written to check for separation between safety-critical and non-safety-critical code. SPARK tools were then used to statically prove these contracts and the absence of runtime errors. Informal feedback found that the process of proving Z specifications was more efficient at finding faults than testing, and the use of formal specification led to simpler code that was easier to understand and maintain. The effort also found that SPARK proof was favorable in comparison with unit testing, particularly for proving the absence of runtime errors.

SPARK was also used in the development of the Lockheed C130J "Hercules" [38,39], in particular, to implement the core of the mission computer, which performs the majority of the aircraft's mission critical functions. Formal modeling methods were used to develop the requirements, including the use of Parnas Tables [40] to write specifications relating inputs to required outputs. The requirements were then used to write data and information flow contracts and postconditions for SPARK subprograms. Prior to modified condition/decision coverage (MC/DC) testing, only basic flow analysis with SPARK was performed. Lockheed found that coding proceeded at near-normal Ada rates, yet there was an 80% savings in the expected budget allocated to MC/DC testing, in part because the code had an unusually low fault density of less than one-tenth the expected industry norm for safety critical software. SPARK was later used to perform a more thorough analysis [34].

In terms of cybersecurity, SPARK was used in the development of the Certification Authority for the Multos smart card [41]. The system's security-enforcing kernel was written in SPARK, the infrastructure in Ada, and the graphical user interface in C++. Other parts of the system were built using commercial off-the-shelf (COTS) components to save time and money. The effort used a rigorous process to develop the requirements, including a Formal Security Policy Model (FSPM), which was formalized in Z. The FSPM was manually reviewed for correctness, and a typechecker was used to check its consistency. A formal top-level specification (FTLS) was created to describe the functionality of the system. The FTLS was also formalized in Z, a typechecker was used to check that it was well formed, and some special features allowed it to be checked against the FSPM. A high-level design was developed to describe the system's internal structure and explain how the components work together. This was particularly important for establishing confidentiality and integrity of the system through separation between COTS components and the security-enforcing kernel. A detailed design refined the requirements, assigning functionality to specific software modules. The process structure was modeled by mapping sets of Z operations in the FTLS to actions in the communicating sequential processes (CSP) language [42]. The CSP model was used to check that the overall system was deadlock-free and that there was no concurrent processing of security-critical functions. Rigorous coding standards were used for all codes. Manual reviews were carried out to check for compliance with requirements, conformance to standards, and internal consistency. SPARK was used to carry out some proofs of the absence of runtime errors and data flow errors, though it was not used to prove functional correctness. Instead, manual reviews and tests were used to check correctness. In the first year of use, reported faults were far better than the industry average for new systems.

SPARK was also used in Tokeneer [43,44], an NSA-funded demonstrator of high-security software engineering approaches that was later made open source [45]. The Tokeneer system consists of a secure enclave containing workstations that should only be accessible to users who can be biometrically authenticated. Furthermore, the level of user access allowed depends on factors such as the user's allowed roles, security clearance, time of day, etc. The Tokeneer ID Station (TIS) project re-developed a core component of the Tokeneer system. Similar to the efforts on SHOLIS and the Multos smart card CA, a formal specification describing system states and operations was written in Z. To validate this specification, the existence of a valid initial state was proven and operator preconditions were calculated and checked against the specifier's expectations. A typechecker was also used to check for consistency of types in all expressions. Security properties were also expressed in Z, and the formal specification was manually proven to exhibit the security properties. A formal design describing the system in terms of concrete states and operations in Z was developed to refine the formal specification in areas where more details were needed to bring the specifications closer to an implementation. For operations where the refinement was non-trivial, proofs that the formal design correctly refined the formal specification were carried out. A rigorous design process was then used to develop a software architecture, map Z specifications to software packages, and put bounds on values in the software implementation, e.g., state components represented by unbounded integers in the formal design. The system was implemented in SPARK. Information flow contracts, data flow contracts, and gold-level functional specifications expressing the security properties in the form of pre- and postconditions were written and proven with SPARK. Platinum-level functional specifications were not written because of budget constraints and the fact that manual review of the code against the formal design was relatively straightforward. SPARK was also used to check for the absence of certain types of runtime errors. System testing was used to further test the code against the formal design. The effort found that the amount of time spent on system testing was significantly less than would normally be expected and that formal specification was useful for finding errors early in the design process. Independent system reliability testing found zero defects, and for a period of time after delivery, the number of defects found remained at zero. However, Reference [46] discusses two defects in the SPARK code and other issues that were found later, in part because of new developments that allow SPARK to more efficiently check for certain runtime errors that were too time-consuming to check for in the original effort. This includes checks for arithmetic overflow, which was the cause of one of the defects not found in the original effort.

Other cybersecurity-related efforts include the Muen x86/64 separation kernel [47], which is written in SPARK. A separation kernel aims to ensure cybersecurity properties such as confidentiality, integrity, and availability by providing an execution environment in which processes only have access to specified resources, only communicate with each other according to a specified policy, and are otherwise isolated. For Muen, SPARK analysis was used to prove the full absence of runtime errors and some functional specifications expressed as postconditions. Similarly, an effort to develop a verified SPARK kernel for the Genode operating system framework is underway [48]. SPARK was also used to create SPARKSkein [49], a reference implementation of the Skein cryptographic hash algorithm, one of the hash functions considered for the SHA-3 standard. SPARKSkein was implemented based on the Skein specification and existing C reference implementation, and SPARK was used to check for runtime errors, which uncovered a subtle corner-case error that persists in the C reference implementation. Reference test cases were also used to check the code, which uncovered an additional specification-related error because of a typo in a constant. A more extensive effort along these lines is SPARKNaCl, a SPARK implementation of the NaCl cryptographic library for which SPARK has been used to prove the absence of runtime errors and certain functional correctness properties [50]. As a final example, the RecordFlux framework [51] provides a domain-specific language

(DSL) for formally modeling binary protocol message formats, including invariant relations and dependencies between message fields, and a method to automatically generate SPARK implementations of message parsers from message models. SPARK can then be used to prove the absence of runtime errors in the message parsers and certain gold-level functional specifications, e.g., that optional message fields can be accessed if and only if all requirements defined in the specification are met. Demonstrations of RecordFlux include Ethernet frames, with a model that covers both the IEEE 802.3 and Ethernet II standards; Fizz, a TLS 1.3 implementation developed and used by Facebook; and a correct implementation of the TLS Heartbeat extension that was responsible for the Heartbleed vulnerability.

2.5 Summary

In this chapter, we have given a brief overview of formal methods and how they can be used to improve software quality and reliability. In particular, we have focused on how the SPARK language and formal verification toolset can be used to prove that code is free of certain cybersecurity faults and is functionally correct. In giving an overview of projects that have used SPARK for aerospace and cyber systems, we have also emphasized the need for complementary tools and approaches such as various levels of requirements formalization, refinement, and analysis; manual reviews; and integration and system testing.

It is interesting to think about what the future holds for formal methods. A recent survey of formal methods experts was performed as part of the 25th anniversary of the International Conference on Formal Methods for Industrial Critical Systems [52]. This survey found that experts believe that formal methods can improve system quality, safety, security; ease certification and maintenance; and potentially improve system performance, lower costs, and speed up development. Most believe that formal methods are essential for cybersecurity, and they anticipate the increased use of formal methods in new areas such as artificial intelligence, machine learning, biology, chemistry, finance, and law.

Along those lines, a few efforts have or are currently using formal methods to improve the safety and security of increasingly autonomous aerospace systems. For example, the Secure Mathematically-Assured Composition of Control Models project performed as part of the DARPA High-Assurance Cyber Military Systems program developed new tools for building software for unmanned air vehicles (UAVs) that are provably secure against many classes of cyber attacks [53,54]. The effort brought together four main concepts based on formal methods: modeling and analyzing the system architecture for key security and safety properties, synthesizing software components using languages that guarantee important security properties, using the seL4 formally verified microkernel [55] to guarantee enforcement of communication and separation constraints specified in the architecture, and automatically building the final system from the verified architecture model and component specifications. The prototyped technologies were first applied to a research-grade quadcopter, then to portions of the Vehicle Management System of the Boeing Unmanned Little Bird helicopter, where they prevented two cyberattacks that were previously successful on the original system. In terms of SPARK, SPARK is currently being used to re-implement and verify gold-level properties of services in the Unmanned Systems Autonomy Services (UxAS) [56,57], a framework for mission-level autonomy for teams of unmanned vehicles, especially UAVs. SPARK is also now a target language for the Salty synthesis tool [58], which synthesizes system controller logic directly from specifications, e.g., for teams of UAVs performing missions in dynamic environments. The use of SPARK with Salty quickly identified a previously undetected issue related to certain types of specifications that could lead to unexpected runtime errors [59]. Given the demand for increased automation onboard aerospace systems and the verification challenges that brings, we agree with the experts that formal methods will see increased use in areas such as these.

Acknowledgements

The authors would like to thank Yannick Moy of AdaCore for outlining the interpolation example and providing feedback. The authors would also like to thank M. Anthony Aiello of AdaCore and Roderick Chapman of Protean Code Limited. Along with Yannick, they provided additional insight and feedback on other aerospace and cybersecurity efforts that have used SPARK.

This work was partially funded under AFOSR contract #20RQCOR096. It was cleared for public release on December 31, 2020, case #AFRL-2020-0644. DISTRIBUTION STATEMENT A. Approved for public release: distribution unlimited.

References

[1] Kalra N and Paddock SM. Driving to safety: How many miles of driving would it take to demonstrate autonomous vehicle reliability? Santa Monica, CA: RAND Corporation; 2016. RR-1478-RC.

[2] Hansson J, Helton S and Feiler P. ROI analysis of the System Architecture Virtual Integration initiative. Software Engineering Institute, Carnegie Mellon University; 2018. CMU/SEI-2018-TR-002.

[3] Burke D. All circuits are busy now: The 1990 AT&T long distance network collapse. California Polytechnic State University; 1995. CSC440-01.

[4] Durumeric Z, Li F and Kasten J, *et al.* The matter of Heartbleed. In: ACM Internet Measurement Conference (IMC); 2014. p. 475–488.

[5] Alexander B, Haseeb S and Baranchuk A. Are implanted electronic devices hackable? *Trends in Cardiovascular Medicine*. 2019;29(8):476–480.

[6] What is WannaCry ransomware? [monograph on the Internet]. Kaspersky Lab; 2020 [cited 2020 Nov 19]. Available from: https://usa.kaspersky.com/resource-center/threats/ransomware-wannacry.

[7] Greenberg A. The untold story of NotPetya, the most devastating cyberattack in history [monograph on the Internet]. Wired; 2018 [cited 2020 Nov 19].

Available from: https://www.wired.com/story/notpetya-cyberattack-ukraine-russia-code-crashed-the-world/.

[8] Yağdereli E, Gemci C and Aktaş AZ. A study on cyber-security of autonomous and unmanned vehicles. *The Journal of Defense Modeling and Simulation*. 2015;12(4):369–381.

[9] Moy Y and Aiello MA. When testing is not enough [monograph on the Internet]. AdaCore; 2020 [cited 2020 Nov 19]. Available from: https://www.adacore. com/papers/when-testing-is-not-enough.

[10] Basile D, ter Beek MH, Fantechi A, *et al.* On the industrial uptake of formal methods in the railway domain. In: International Conference on Integrated Formal Methods (iFM). Springer; 2018. p. 20–29.

[11] Wijbrans K, Buve F, Rijkers R, *et al.* Software engineering with formal methods: Experiences with the development of a storm surge barrier control system. In: International Symposium on Formal Methods (FM). Springer; 2008. p. 419–424.

[12] RTCA Special Committee 205 (SC-205), EUROCAE Working Group 71 (WG-71). Formal methods supplement to DO-178C and DO-278A. RTCA, Inc.; 2011. DO-333.

[13] Cao Z, Lv W, Huang Y, *et al.* Formal analysis and verification of airborne software based on DO-333. *Electronics*. 2020;9(2):327.

[14] Cofer D and Miller S. DO-333 certification case studies. In: NASA Formal Methods Symposium (NFM). Springer; 2014. p. 1–15.

[15] Bahig G and El-Kadi A. Formal verification of automotive design in compliance with ISO 26262 design verification guidelines. *IEEE Access*. 2017;5: 4505–4516.

[16] Ledinot E, Blanquart JP, Astruc JM, *et al.* Joint use of static and dynamic software verification techniques: A cross-domain view in safety critical system industries. In: Embedded Real Time Software and Systems (ERTS); 2014. p. 1–10.

[17] Rushby J. Formal methods and the certification of critical systems. Menlo Park, CA: Computer Science Laboratory, SRI International; 1993. SRI-CSL-93-7.

[18] RTCA Special Committee 205 (SC-205), EUROCAE Working Group 71 (WG-71). Software considerations in airborne systems and equipment certification. RTCA, Inc.; 2011. DO-178C.

[19] McCormick JW and Chapin PC. *Building High Integrity Applications with SPARK*. Cambridge University Press; 2015.

[20] Barnes J. Programming in Ada 2012. Cambridge University Press; 2014.

[21] Introduction to SPARK [monograph on the Internet]; 2020 [cited 2020 Nov 6]. Available from: https://learn.adacore.com/courses/intro-to-spark/index.html.

[22] Introduction to Ada [monograph on the Internet]; 2020 [cited 2020 Nov 6]. Available from: https://learn.adacore.com/courses/intro-to-ada/index.html.

[23] Moy Y. Climbing the software assurance ladder – Practical formal verification for reliable software. Electronic Communications of the EASST. 2019;76.

[24] Filliâtre JC and Paskevich A. Why3 – Where programs meet provers. In: European Symposium on Programming (ESOP). Springer; 2013. p. 125–128.

[25] Conchon S, Coquereau A, Iguernlala M, *et al.* Alt-Ergo 2.2. In: International Workshop on Satisfiability Modulo Theories (SMT); 2018. p. 1–11.

[26] Barrett C, Conway CL, Deters M, *et al.* CVC4. In: International Conference on Computer Aided Verification (CAV). Springer; 2011. p. 171–177.

[27] De Moura L and Bjørner N. Z3: An efficient SMT solver. In: International Conference on Tools and Algorithms for the Construction and Analysis of Systems (TACAS). Springer; 2008. p. 337–340.

[28] Common Weakness Enumeration (CWE®) [database on the Internet]. 2020 Aug [cited 2020 Nov 6]. Available from: https://cwe.mitre.org.

[29] Common Vulnerabilities and Exposures (CVE®) [database on the Internet]. 2020 Nov [cited 2020 Nov 6]. Available from: https://cve.mitre.org.

[30] Chapman R and Moy Y. AdaCore Technologies for Cyber Security [monograph on the Internet]; 2018 [cited 2020 Nov 6]. [93]. Available from: https://www.adacore.com/books/adacore-tech-for-cyber-security.

[31] Dross C and Moy Y. Auto-active proof of red-black trees in SPARK. In: NASA Formal Methods Symposium (NFM). Springer; 2017. p. 68–83.

[32] Baity R, Humphrey L and Hopkinson K. Formal verification of a merge sort algorithm in SPARK. In: AIAA SciTech; 2020. p. 1–8.

[33] Dross C and Kanig J. Recursive data structures in SPARK. In: International Conference on Computer Aided Verification (CAV). Springer; 2020. p. 178–189.

[34] Chapman R and Schanda F. Are we there yet? 20 years of industrial theorem proving with SPARK. In: International Conference on Interactive Theorem Proving (ITP). Springer; 2014. p. 17–26.

[35] King S, Hammond J, Chapman R, *et al.* Is proof more cost-effective than testing? *IEEE Transactions on Software Engineering*. 2000;26(8):675–686.

[36] Woodcock J and Davies J. *Using Z: Specification, Refinement, and Proof.* Prentice Hall International; 1996.

[37] Garlan D. Preconditions for understanding (formal specification). In: Proceedings of the Sixth International Workshop on Software Specification and Design. IEEE Computer Society; 1991. p. 242–243.

[38] Chapman R. Industrial experience with SPARK. *ACM SIGAda Ada Letters*. 2000;20(4):64–68.

[39] Croxford M and Sutton J. Breaking through the V and V bottleneck. In: International Eurospace-Ada-Europe Symposium. Springer; 1995. p. 344–354.

[40] Parnas DL, Madey J and Iglewski M. Precise documentation of well-structured programs. *IEEE Transactions on Software Engineering*. 1994;20(12): 948–976.

[41] Hall A and Chapman R. Correctness by construction: Developing a commercial secure system. *IEEE Software*. 2002;19(1):18–25.

[42] Hinchey MG and Jarvis SA. *Concurrent Systems: Formal Development in CSP*. McGraw-Hill, Inc.; 1995.

[43] Barnes J, Chapman R, Johnson R, *et al.* Engineering the Tokeneer enclave protection software. In: International Symposium on Secure Software Engineering (ISSSE). IEEE; 2006. p. 1–9.

[44] Copper D and Barnes J. Tokeneer ID station EAL5 demonstrator: Summary report. Altran Praxis Limited; 2008. S.P1229.81.1.

[45] Tokeneer [database on the Internet]. AdaCore. 2020 [cited 2020 Nov 19]. Available from: https://www.adacore.com/tokeneer.

[46] Moy Y and Wallenburg A. Tokeneer: Beyond formal program verification. Embedded Real Time Software and Systems (ERTS). 2010;24.

[47] Buerki R and Rueegsegger AK. Muen – an x86/64 separation kernel for high assurance. University of Applied Sciences Rapperswil (HSR); 2013.

[48] Stein M. Spunky, a Genode kernel in Ada/SPARK. *Ada User Journal*. 2020;41(2):99–102.

[49] Chapman R, Botcazou E and Wallenburg A. SPARKSkein: A formal and fast reference implementation of Skein. In: Brazilian Symposium on Formal Methods. Springer; 2011. p. 16–27.

[50] Chapman R and Moy Y. SPARK 2014 re-implementation of the TweetNaCl crypto library [repository on the Internet]. GitHub; 2020 [cited 2020 Nov 19]. Available from: https://github.com/rod-chapman/SPARKNaCl.

[51] Reiher T, Senier A, Castrillon J, *et al.* RecordFlux: Formal message specification and generation of verifiable binary parsers. In: International Conference on Formal Aspects of Component Software (FACS). Springer; 2019. p. 170–190.

[52] Garavel H, ter Beek MH and van de Pol J. The 2020 expert survey on formal methods. In: International Conference on Formal Methods for Industrial Critical Systems (FMICS). Springer; 2020. p. 3–69.

[53] Cofer D, Backes J, Gacek A, *et al.* Secure mathematically-assured composition of control models. Air Force Research Laboratory, Information Directorate; 2017. AFRL-RI-RS-TR-2017-176.

[54] Cofer D, Gacek A, Backes J, *et al.* A formal approach to constructing secure air vehicle software. *Computer*. 2018;51(11):14–23.

[55] Klein G, Elphinstone K, Heiser G, *et al.* seL4: Formal verification of an OS kernel. In: ACM SIGOPS 22nd Symposium on Operating Systems Principles (SOSP); 2009. p. 207–220.

[56] Rasmussen S, Kingston D and Humphrey L. A brief introduction to Unmanned Systems Autonomy Services (UxAS). In: 2018 International Conference on Unmanned Aircraft Systems (ICUAS). IEEE; 2018. p. 257–268.

[57] Aiello MA, Dross C, Rogers P, *et al.* Practical application of SPARK to OpenUxAS. In: International Symposium on Formal Methods (FM). Springer; 2019. p. 751–761.

[58] Elliott T, Alshiekh M, Humphrey LR, *et al.* Salty–A domain specific language for GR(1) specifications and designs. In: International Conference on Robotics and Automation (ICRA). IEEE; 2019. p. 4545–4551.

[59] Humphrey LR, Hamil J and Huguet J. End-to-end verification of initial and transition properties of GR(1) designs in SPARK. In: International Conference on Software Engineering and Formal Methods (SEFM). Springer; 2020. p. 60–76.

Chapter 3

Drone surveillance system—RF/WiFi-based drone detection, localization, and tracking: a survey

Igor Bisio[1], Chiara Garibotto[1], Halar Haleem[1], Fabio Lavagetto[1], and Andrea Sciarrone[1]

Unmanned Air Vehicles (UAVs), also known as drones, have become dominant recently in all fields including military, commercial, and other multiapplications, such as agriculture, search and rescue, infrastructures monitoring, and surveillance, especially for safety and security purposes. On one hand, deployment of UAVs has facilitated end-users and, on the other hand, this enormous increase in the usage of UAVs has raised many concerns with regard to security, harm, threats, and privacy issues by different agencies, governments, private organizations even by people in their residencies, which has stimulated and motivated many researchers toward the designing of anti-drone systems in order to be secure from such risks. The main objective of this chapter is to provide a comprehensive survey concerning the work carried out in the designing of an anti-drone system for detection, localization, and tracking using Radio Frequency (RF-) and WiFi-based techniques. In the category of drone detection, drone identification, and classification are also being explored. In more detail, this chapter will highlight the key factors of the surveyed techniques and the characteristics of such systems to deploy them, practically.

Indeed, current studies provide systematic ways to understand the procedures, methods, tools, and techniques, as well as experimental setups, which have been carried out for the implementation of drone-surveillance systems. A comparison of the performance, found in the literature, of the designed systems in this field is presented and analyzed. Finally, open challenges, possible future research trends, and direction are highlighted in the context of drone detection, localization, and tracking systems. The presented survey chapter is aimed at providing a guide for the research community as it could be an ignition to work further with specific direction so making these crucial systems versatile and ready to be deployed practically to meet their potential market demand.

[1]Electrical, Electronics, Telecommunication Engineering and Naval Architecture Department (DITEN), University of Genoa, Italy

3.1 Introduction

Unmanned Aerial Vehicle's, also called drones, are the miniature drones that fly in the absence of a pilot and are operated by either the user from any remote location, called *tele-operated* mode, or autonomously operated, called *autonomous* mode. These miniature air vehicles have not only succeeded in capturing the attention of different people, including investors and hobbyists, but also have been given serious research concentration over the past few years [1,2]. UAVs were mainly deployed by the military in hostile territory for different custom operations with different missions, including to safeguard the pilot lives [2].

The enhancement and advancement in technology in the designing of drones have put a significant impact on incorporating the diversified features, for example, their smaller size and light weight. This has caused an exponential growth in the importance and deployment of drones in many commercial and civilian domains during the last couple of years, and end users are being facilitated in different fields with numerous applications in agriculture, airports, photography, security & surveillance, forest fire detection, infrastructure inspection, agriculture, airports, transportation, weather forecasting, smart policing, package delivery, and various other applications [2–5]. Drones are also being preferred to be deployed as base stations in order to facilitate the rural areas in replacement of ground base stations, implementation of UAVs in the Internet of Things (IoT) scenarios and also their deployment for building the cellular networks in the regions or countries where implementing the towers is quite expensive [6]. Technological development in the designing of drones with diversified features has enormously raised the demand and deployment of drones. More specifically, the total number of aircraft's, both manned and unmanned for the Department of Defence (DoD), of the United States only are predicted to increase to 27,000 in quantity equipped with advanced feature and attack capabilities as reported in [4].

Every technological development in a particular field has its own consequences in terms of the ways the technology is being utilized. On one hand, diversified features of UAVs/drones have achieved great importance and deployment by various organizations in military and commercial domains. On the other hand, the misuse of the same technology has dramatically raised concerns in terms of public privacy invasion, attacks, security, harms, threats, etc. This has been reported in the literature by many organizations as well as individuals for such negative consequences. For example, during the initial months of 2019, a continuous nuisance to airports of different cities of few countries (including the United States, England, Ireland, and Dubai) has been observed because of drones, as drones hitting aircrafts may cause catastrophic damage or even death, which has been reported in [7]. Further, a woman in Seattle reported to the police of a drone flying at the 26th-floor apartment window when she was partially dressed as given in [8], which reflects the privacy invasion attack on individuals. Not only this but drones are also being used in violating the privacy of the public and sensitive areas such as nuclear power plants in France, as highlighted in [9], and also an incident of a drone crashing in Bugey, France, as reported in [10]. One of the events involved the crashing of a drone on the lawn of the White House, mentioned in [11]. Also, the utilization of drones in the dropping of weapons and smuggling of drugs has been highlighted in [11]. These discussed scenarios have made the implementation of drone surveillance system a fundamental need for public and national sensitive areas. Considering these ground realities, survey work is prepared concerning drone surveillance systems in the context of detection, localization, and tracking of intruders so that existing designed solutions could be explored in this field and further research direction and trends may be identified and notified in general for further implementation. Three subsystems of drone surveillance system explored are:

- Drone Detection: Drone detection is the basic and primary consideration of the surveillance system and it is mainly concerned with drone detection and identification of its type.
- Drone Localization: In the category of drone localization, drone direction and its position are key points and lead the system for accurate drone localization.
- Drone Tracking: Detection and localization of the drones are not enough for anti-drone surveillance systems; tracking is also a crucial part of this system. This part of the system helps in monitoring the activities of any suspicious detected drone.

Keeping the above in consideration, this chapter aims to provide a survey of research work carried out so far using RF/WiFi-based techniques toward the realization of drone detection, localization, and tracking systems. Based on this research theme, two-fold objectives have been set, in particular, as:

i. To investigate the counter-drone systems designed so far based on RF/WiFi techniques in the context of detection, localization, and tracking of intruders.
ii. To explore and identify the different methodologies, tools, and techniques implemented so far, and hence mapping of the future research direction based on this critical review toward the designing of a drone surveillance system.

It is expected that the outcome of this survey work helps in various aspects, such as scrupulous understanding of the drone surveillance system, state-of-the-art-based identification of design challenges and further research trends. These would help the researcher community of this field to get benefited toward the designing and practical deployment of drone surveillance systems in terms of detection, localization, and tracking.

3.2 Drone surveillance system and RF/WiFi approach

3.2.1 Counter drone system

The drone surveillance system, also known as anti-drone system or counter-drone system, is responsible for the monitoring of drones and navigates continuously for the presence of any drone, observes the activities and behavior of the intruder drones, and thus takes the actions accordingly. The drone surveillance system is implemented in different ways depending upon the nature of the surveillance station, which could be a fixed ground station, fixed aerial station, or even another drone at some higher altitudes. This basic scenario of the drone surveillance system is depicted in Figure 3.1.

One can view a drone surveillance system from different aspects as Ding *et al.* [12] have classified the drone surveillance systems into four different classes based on the anti-drone technologies used in the military: Warning, Spoofing, Jamming, and Mitigation. Drones are detected and abducted in the warning and spoofing classes, respectively. In jamming, control and navigation information are destructed, whereas in mitigation, drones are crashed in emergencies through different means. The following are typical features or key challenges in designing the complete counter-drone system in all aspects:

- Detection of Drones
  - Recognition and classification;
  - Capturing the drone communication;
  - Detection of autonomous/ground-controlled drone;
  - Decision for authorized/unauthorized drone;
  - Detection of multiple drones;
  - Detection of the drone controller (If present);
- Drone Localization
  - Localization of the drone controller (If present);
- Drone Tracking
- Drone Interdiction.

Sometimes for surveillance systems, it is paramount to acquire more precise information about the drones causing threats which may include the identification and localization of the drone controller from where it is being operated. However, it is not necessary that drone is tele-operated so it could be in autonomous mode as well. Therefore, the feature of detecting the autonomous/ground-controlled drone

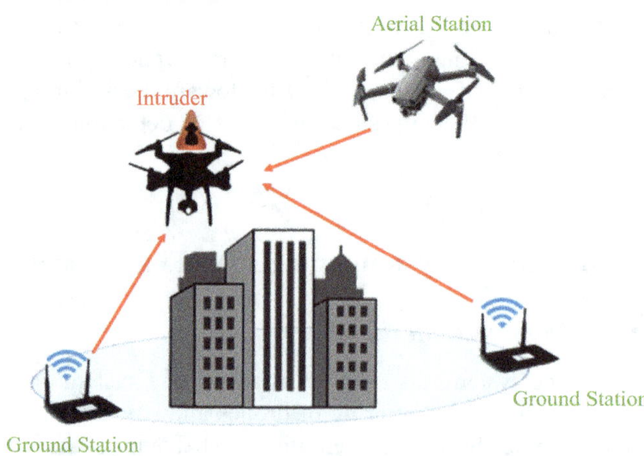

Figure 3.1 Drone surveillance system

is also paramount in the counter-drone systems. In certain scenarios, it is also crucial to not only capture the communication between the drone and its controller but also to decode it so that possible threats may be identified in advance. The advancement of the counter-drone technology also lies in its ability to detect multiple drones since it is not necessary to have the presence of a single drone in all cases and scenarios. In this survey, it is tried to search, and review all the relevant studies related to drone surveillance system features in the context of detection, localization, and tracking of intruders.

3.2.2 *Significance of RF/WiFi techniques*

Several techniques are available and have been used so far in order to realize the drone surveillance system, for example, image-based drone surveillance system (e.g., [13–15]), acoustic-based drone surveillance system (e.g., [16–18]), and radar-based drone surveillance system (e.g., [19,20]). Apart from these techniques, RF/WiFi-based techniques are also explored and being utilized during the last few years. All of these techniques have been reviewed and compared in the context of their pros and cons by many authors while presenting their primary research work on a specifically selected technique in the field of counter-drone systems technology, for example, [21–24]. The taxonomy of these techniques at their functional level is also summarized in Table 3.1 for a quick understanding of readers. A comparison of different surveillance technologies has been provided, which could be beneficial in deciding to select a particular technique. Further, it must be noted that entries, including numeric data, provided in Table 3.1, are obtained from the literature. For example, if we took range values, they would be dependant upon and vary for the defined environment, relative algorithm, type of drone used, and implementation of specific hardware.

Currently, RF-based detection technique has been given importance because of its versatile nature in terms of different factors such as passive detection approach, energy efficiency, increase in range using high-gain antenna along with efficient RF receivers [24,25]. Also, RF-based technique is effective and robust [26] because it works all the time and in all weathers [27]. This technique takes advantage of RF signals reflected or transmitted from the drone for detection purposes. The advantage of using WiFi-based technique is its unlicensed frequencies 2.45 GHz and 5 GHz from the RF spectrum. However, the RF spectrum covers a range of frequencies from very low frequencies to extremely high frequencies, because low-frequency signals exhibit low signal attenuation, cover larger distances, and vice versa. This characteristic is used in different drone detection systems, where low-frequency signals transmitted from drones are captured, processed, and used as a reference for its detection. WiFi frequencies are easily captured, but the difficulties are imposed on other RF frequencies, especially when interferences from other RF-operated devices are available in the vicinity. WiFi has a shorter and limited range unlike to rest of RF frequencies used in the detection system. Normally, RF fingerprinting and WiFi fingerprinting terminologies are a bit confusing because analysis has been carried out on RF signals in both these techniques and they both utilize RF receivers to capture the signals.

Table 3.1 Taxonomy of counter-drone detection techniques

Detection techniques	Target principle	Drone signature	Pros	Cons	Challenges	Detection range
Acoustic	Hissing high-frequency signals	Time–Frequency signals	NLoS detection if a sound is audible. Low cost depending upon microphone arrays.	Not useful in noisy and urban areas. Requires database development for drone acoustic signatures. Background noises and windy conditions affect the detection performance.	Accurate detection in the presence of different ambient noises.	25–30 ft
Video	Image & Motion detection	Image features	Usage of optical sensors having low cost. Accurate detection with human eye.	Line of Sight is mandatory. Difficult to distinguish between drone & similar object. Higher cost of high-definition cameras. Affected from dust, fog, & clouds.	Indistinguishable drones from similar small objects. Detection in the presence of obstacles.	Less than 350 ft
Radar	EM waves backscattering & Micro-Doppler effects	Micro-Doppler	NLoS detection. Can work in the presence of fog, dust, clouds.	Drone's smaller Radar Cross Section (RCS), which limits the detection. Detection range is affected because of significant pathloss of mmWave. Difficult to measure drone signature when radial velocity is low.	Low RCS. Low altitude objects detection.	More than 2k ft

(Continues)

Table 3.1 (*Continued*)

Detection techniques	Target principle	Drone signature	Pros	Cons	Challenges	Detection range
RF	Control & video signals over RF	RF fingerprint	NLoS detection. low-cost RF sensors. can work in both passive and active approaches. Higher detection ranges.	Requires database development for drone's RF signatures for detection & classification. Autonomously flying drone's detection is hard.	Transmitter & receiver sensitivity affects in detection range. Multipath fading. High-altitude & NLoS detection.	Less than 1,400 ft
WiFi	Control & video signals over WiFi	WiFi fingerprint	NLoS detection. low-cost devices. No need of active detection approach.	Can only detect the drones, which are WiFi operated. Detection range is limited to WiFi channel.	Detection in intermittent & fake Tx.	WiFi operational range

However, the main difference lie in the layers from which the fingerprint data are extracted. RF-based technique usually deals with the extraction of features at the physical layer while WiFi-based technique deals with it at the data link layer.

3.3 Drone detection system

A fundamental and necessary requirement of a drone surveillance system is its ability to detect and identify any intruder or malicious drone. In the context of the detection of drones, various solutions have been proposed in the literature while exploring this field. As detection itself is not enough, the concept of reliable detection because of the presence of similar objects in the nearby areas has made this task challenging. Therefore, several researchers have put their efforts into designing drone detection systems, which are not only able to detect the presence of drones in the surroundings but also able to classify the drones and also decide on which is an authorized or unauthorized drone. Apart from detection techniques, detection methods, in general, are employed, which describe the ways and procedures for incorporating these techniques. In this context, two main methods are incorporated for drone detection in the category of RF technique, which are as follows:

- Active approach
- Passive approach

In the active method, a signal is continually transmitted and the receiver waits for any reflected wireless signal, which is further processed for detection purposes based upon unique drone signatures. On the contrary, the passive method uses communication between the drone and its controller for detection purposes rather than transmitting the signal continuously. Because passive detection method has multiple advantages when compared with the active one, such as lower cost of hardware and software and no requirement of transmitter circuitry, easy deployment, portability, simple operation, and maintenance. Under the Umbrella of active and passive approaches, three main methods are adopted to detect the realistic presence of any drone as given in [26], which are:

- By capturing drone propeller signals.
- By eavesdropping on the communication signals between the drone and its controller.
- By capturing the signals generated due to the drone vibration.

The classification of work done on the drone detection systems using RF/WiFi techniques concerning drone detection and classification is shown in Figure 3.2.

3.3.1 Methodology

This section deals with the approaches and procedures employed in the detection of intruders. In general, detection is related to confirm the presence of drones versus nothing, while identification is the ability of a system to identify specific drones.

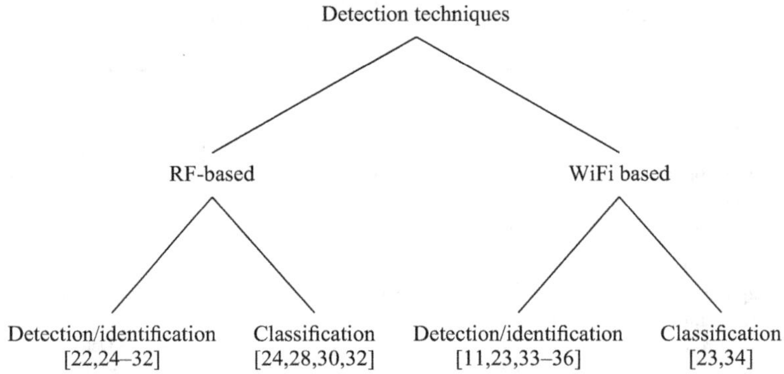

Figure 3.2 Classification of work done on drone detection and classification using RF/WiFi techniques

The research community have put their great efforts in this recently exploited field from different aspects. In the following, relevant literature work is reviewed with respect to methods incorporated and detection system features under the category of drone detection.

3.3.1.1 Detection in RF-based approach

Data presented in Table 3.2 cover the methods and procedures used in the designing of RF technique-based detection system. Capturing drone propeller signals and eavesdropping the communication signals methods employed in [26] to confirm the drone presence in the vicinity. By analyzing the signal behaviors in both time and frequency domains, variations in the signal indicating the drone detection were recorded. It is observed experimentally in the literature that Received Signal Strength (RSS) is higher than the noise level when a drone is present in a specific defined distance range and vise versa. Further, to validate the physical characteristics because of drone's motion, two different indoor experiments were carried out on DJI Phantom and Parrot Bebop drones by Nguyen *et al.* [28]. In the first experiment, different behaviors of drone movement were observed by collecting and processing the data. With this experiment, drone shifting and vibration were confirmed. Frequency till 40 Hz reflected body shifting and the frequency between 50 Hz and 70 Hz reflected body vibration movements. In the second experiment, a transmitting antenna was attached to the drone and a receiving antenna was used to capture the signals from the drone. The antenna sent a tone pulse at the frequency of 2.4 GHz and the signal was captured to analyze the changes in the received RSS of the signal. More specifically, the purpose of this test was to analyze the received signal for the drone body shifting and body vibration. The results confirmed the drone movements at the same frequencies as observed by the first experiment by listening to the signal using Software-Defined Radio (SDR), demodulating it to the baseband, and then applying Fourier analysis. Toward detecting the drones in the Non-Line-of-Sight (NLoS) signals, Raja [29] designed the detection

Table 3.2 State-of-the-art comparison for RF-based drone detection systems

Ref.	Method/ approach	Techniques & algorithm	Environment	Range	Outcome/ performance
[22]	By eavesdropping communication between drone & its controller	USRP RF sensor, spectrum analyzer SVM classifier	Outdoor campus	100m	Detection probability: 73.3% False alarm probability: 2.8%
[24]	By capturing uplink communication between drone and its controller	Wavelet transform detector NCA & Naive Bayes approach kNN, DA, SVM, and NN classifiers	Indoor	Up to 130 m	At 10 dB SNR & 80 m distance: Detection accuracy = 84% Optimum classifier: SVM with accuracy rate ≈96.84%
[25]	By capturing UAV communication signals and frequency estimation approaches	Spectrum accumulation, statistical fingerprint analysis	Outdoor	Up to 2800 m	Single UAV: Detection rate is 90% at 2800 m Two UAVs: Detection error is 0.2%
[26]	By eavesdropping communication between drone & its controller. By capturing the drone propeller signal	Time–Frequency analysis	Urban area & open field	Up to Max: 50 m	Detection confirmed by observing drone signature
[27]	By extracting cyclostationarity signatures from drone communication and AoA estimation by pseudo-Doppler principle By capturing & analyzing the micro-Doppler signatures	Cyclostationairty feature detector Micro-Doppler feature estimation Time–Frequency analysis	Open campus	Not specified	Detection of drone was confirmed by presenting time & frequency plots
[28]	By capturing the signal because of drone body vibration and body shifting	Wavelet transform Detector, Fourier analysis	Urban, suburban and campus	Urban area: 100m sub-urban area: 600m Inside the campus: 200 m	Detection accuracy: 89.4% (urban area) Accuracy: 84.9% (sub-urban area) Able to classify 07 different type of mentioned drones
[29]	Based upon drone motions and vibrations patterns	S-SVM, LSTM	Residential	WiFi range	Detection accuracy: 97.02%

(Continues)

Table 3.2 (*Continued*)

Ref.	Method/ approach	Techniques & algorithm	Environment	Range	Outcome/ performance
[30]	By capturing UAV communication signals for detection and identification	Fourier analysis, DNN model	Indoor	WiFi range	Drone detection only: 99.7% Detection & type identification: 84.5% Detection, type identification & its flight mode: 46.8%
[31]	By capturing the uplink communication signals between drone & its controller	Demodulation techniques, Time–Frequency analysis	Not specified	Not specified	Presence & identification of drone type is claimed based upon time–freq: analysis
[32]	Based upon drone communication signals	PCA, AC-WGAN classifier	Both indoor & outdoor	Up to 200 m	Classification accuracy: 98% at 5 dB SNR (indoor) Classification accuracy: 62.5%–98% at 200 m distance (outdoor)
[40]	Based upon antenna array with beam sweeping scheme	Likelihood ratio test	Air-to-ground channel	100 m	Numerical results for minimum detection error probability were presented
[38]	By sensing UAV transmission signals & converting them into image	Energy detector. VGGNet and ShuffleNetV2	Low SNR environment	1000 m × 1000 m square	At Tx power = 100 mW & no. of sensors = 900: VGGNet classification accuracy: 83% ShuffleNetV2: classification accuracy: 92%
[39]	By incorporating UAV communication signal	Empirical mode decomposition. Back propagation NN model	Both indoor and outdoor	Up to 15 m and 3 km	Indoor detection rate: Up to 90% Outdoor detection rate: 82%

USRP, Universal Software Radio Peripheral.

system based on NLoS signal, where received RF signal is classified as Line of Sight (LoS) and NLoS first by using Supervised-Support Vector Machine (S-SVM) model. From these NLoS signals, drone motion and vibration signatures were extracted for the detection of drones using Long Short-Term Memory (LSTM) Neural Network (NN) model.

To eavesdrop on the drone communication signals, Nguyen *et al.* [26] performed experimental work using Surveillance Drones (SDrs). During the experiment, the range of the receiver was set around 20–100Hz to capture the signals since the drone communicates with its controller at 30 times/s [37], as this frequency is distinguishable with the other related WiFi-based operating devices. Cyclostationarity signature of drones and Pseudo Doppler principle were used by Fu *et al.* [27] to detect and identify the ground-controlled drones. Downlink video communication signals from the drone to its controller, which are based upon Orthogonal Frequency-Division Multiplexing (OFDM), were captured and processed further to complete the task of detection. In addition to detection of drone, Al-Sa'd *et al.* [30] proposed a system to identify the type and operating mode of a drone by creating the database with such information in prior. Three drones were used in the experiment, Parrot Bebop, Parrot AR Drone, and DJI Phantom 3.

In a surveillance area, authorized drones could also be present for different purposes transmitting communication signals. So, if an illegal drone is entered in the area then the detection of that illegal drone could be complicated. Keeping this in view, Chen *et al.* [38] utilized the concept of Wireless Sensor Networks (WSN), multiple sensors were implemented to sense and transmits the spectrum data to the fusion center where energy data from the multiple sensors is converted into the images to feed to the VGGNet and ShuffleNetV2 Deep Learning (DL) models for the actual detection and classification of legal/illegal drones. Towards detection of UAVs at larger distances in both outdoor and indoor environments, Zhang *et al.* [39] designed an Artificial Neural Network (ANN) based detection system with a success rate of more than 82% up to a distance of 3 km. Apart from this, to increase the detection range, Yang *et al.* [25] deployed high gain antennas at the receiver side and covered a range up to 2,800 m in the outdoor environment. With the enhancement in technology and continuously growing demand for drone surveillance systems, the accuracy of the detection system has become a paramount factor. Since missing detection could be more malicious and threatened than false detection. Therefore, to avoid any missing detection, Shi *et al.* [22] has proposed a system named it as ADS-ZJU for performing detection by integrating RF along with both audio and video techniques. Further, Fu *et al.* [27] carried out research work in the case of autonomous drones where micro-Doppler signature-based drone detection was employed when the drone is not sending any signals. In this case, time–frequency analysis was made using spectrogram plotting and the reason for adopting this approach was the non-stationary and complicated nature of micro-Doppler signatures created by the drone propeller rotation.

It is well understood that drone communication takes place both in uplink and downlink channels, therefore drones can not only be identified using the information sent to the controller but also by reversing this communication path [31]. Therefore, in this connection, Haluza and Čechák[31] set up the proposal in a way that identifies the frequency first for few drones which their controller used to send the control signals and then sets receiving station with monitoring antenna to capture the radio signals. In other words, the uplink frequency method of data transmission is used. Once those signals are captured, then the decoding process is applied. However, the decoding process is dependent upon the specific device used (Specific drone and its receiving frequency). Authors in [24] used the uplink channel to capture RF signals for the detection of drones in which the controller is sending information to the drones. Raw signals were split into frames and transformed into the wavelet domain to remove the bias from signals, and hence the size of the data was reduced to be processed. A naive Bayes approach based on Markov models was implemented to check for the presence of a UAV in each frame. Fourteen different UAV controllers, operating at 2.4 GHz, were included in the experimental setup and 100 RF signals were received from each controller to evaluate the designed algorithm.

Towards realization of the detection of multiple drones, Yang *et al.* [25] used Spectrum Accumulation (SA) and Statistical Fingerprint Analysis (SFA) techniques to confirm the simultaneous presence of two UAVs placed at a distance of 2,400 m and 2,500 m. Generally, approaches employed for the detection of drones are based upon UAV communication signals. Researchers even have decoded those signals to identify the mode of operation. However, towards a more practical approach, covert communication is being used between the UAV and its controller. To confirm the detection in this situation, Hu *et al.* [40] proposed a system in which UAV detection is formulated as a binary detection problem that either UAV transmits a covert message or it does not.

Different types of drones available have different signatures due to their features, operating frequency, size, structure, make & model, etc. and these parameters have an impact on the force generated by the propellers and hence the effect on the drone vibration frequencies. So, using physical patterns, seven different types of drones were classified by Nguyen *et al.* [28]. To make this happen, an experimental setup was performed to collect the motion data. Based on the central frequency of vibration determined from the collected motion data, frequency of drone vibration was observed and frequencies were different for all seven types of drones ranging from 20 Hz to 140 Hz.

In the context of drone detection systems, Machine Learning (ML) algorithms are employed with two tasks mainly which are feature extraction and classification of drone patterns [27]. With the help of training and classification, a specific type of drone or a particular feature of any specif drone may be obtained. The majority of the work presented in the literature has combined the drone detection and classification systems, for example, [30,32]. Three different types of drones, Parrot Bebop, Parrot AR, and DJI Phantom 3 were detected first, and then a specific type of drone was classified using Deep Neural Network (DNN) model w.r.t specific features stored in the database in [30]. Further, the problem of drone detection and classification is also solved using Auxiliary classifier Wasserstein Generative Adversarial Networks (Ac-WGANs) in [32], where UAVs' features are detected first and then identified using WGANs approach along with the classification of UAVs when no any prior information, such as protocol, Service Set Identifiers (SSIDs), or Media Access Control (MAC) addresses, of the received wireless signal is available. In continuation of work done in the context of classification, Ezuma *et al.* [24] proposed a system in which both tasks of detection and classification of drones were performed. In the classification phase, features were selected using Neighborhood Component Analysis (NCA) to minimize the computational cost and the selected features were passed to different ML algorithms k-Nearest Neighbors (kNN), Discriminant Analysis (DA), Support Vector Machine (SVM) & NN for classification.

3.3.1.2 Detection in WiFi based approach

Data presented in Table 3.3 cover the methods and procedures used in the designing of WiFi-based detection systems. Using the model of attack structure, which comprises approaching, surveillance, and escaping of drone, Birnbach *et al.* [11] designed a detection system in which statistical features inherent to drones are derived for their movement and proximity. The proposed approach works based on RSS measurement

Table 3.3 State-of-the-art comparison for WiFi based drone detection systems

Ref.	Method/ approach	Techniques & algorithm	Environment	Range	Outcome/ performance
[11]	By incorporating an attacking model for detection. By capturing the packets of UAV with video transmission.	Standard deviation-based algorithm & analysis	Residential	Up to 62 m in LoS	Detection confirmed using graphical analysis when value of standard deviation is crossed a certain threshold.
[23]	By eavesdropping the communication signals. WiFi statistical fingerprint analysis.	WiFi sniffer. Random tree, Random forest, SMO, and Logic Boost classifiers.	Working	WiFi coverage	Optimum classifier: Random tree with 96% precision.
[33]	By eavesdropping the communication signals. WiFi statistical fingerprint-based approach with analysis of the impact of video streaming flow.	WiFi sniffer. NB, RF, and J48 classifiers.	Working	WiFi coverage	Optimum classifier: Random tree with a true-positive rate of 97%.
[34]	By eavesdropping the encrypted drone communication signals. Packet by Packet analysis & utilization of critical features from WiFi traffic.	WiFi packet sniffer. One-versus-all logistic regression with re-weighted ℓ_1-norm technique.	Campus	WiFi coverage	Identification accuracy is 85.7%–95.2% within 0.15–0.35 s.
[35]	By eavesdropping the encrypted communication between drone & its controller.	Wireless probe sniffer. Trees-J48, Random Forest and NNs.	Indoor	WiFi coverage	Drone state identification rate: 93% in 3.71 s.
[36]	Based upon mobile crowd-sensing approach. By capturing WiFi beacon information and the MAC addresses of drones.	WiFi signals sensing using Smart devices.	Campus	Up to 350 m	Detection rate is 86.7%.

received through the WiFi receiver. In continuation of the work done in the context of WiFi approach, WiFi statistical fingerprint method-based detection of unauthorized amateur drone detection is proposed in [23,33]. In [33], detection is confirmed both in the presence and absence of video transmission from the drones. Detection strategy followed in these systems is given as follows: Wireless data (WiFi) are sniffed using Wireshark software, which is dumped into the pcap file. This wireless data is called flow, identified by a pair of MAC addresses (source and destination) and is taken at a specific interval for a short period to realize real-time capturing and detection. These flows are partitioned and processed to extract the special features and these features are referred to as fingerprints. Extracted features are further processed for actual drone detection using ML classifiers. For instance, Random tree, Random forest, Sequential Minimal Optimization (SMO) and Logic Boost are employed in [23]. The mobile crowdsensing approach is used in [36] rather than a traditional detection approach such as images, acoustic, and radars. The detection algorithm is designed with the help of WiFi beacon information and the MAC addresses. The participation of consumers in this system is proposed by designing the auction-based online incentive mechanism in which participants should bid first, set of participants would be selected as auction winners who shall be assigned to complete the sensing task and upload the required data to the platform. A database of MAC addresses and SSIDs of drones manufactured by commercial companies is established and whenever any MAC address/SSID is observed by the participants, it is compared with database for confirmation of drone detection.

It is not necessary that the communication the drones make with their controllers is always un-encrypted, it could be encrypted to secure the drones detection or identification from any surveillance system. To deal with the detection with encrypted WiFi traffic, Alipour-Fanid *et al.* [34] proposed the system to detect and identify the drones even if their transmission is encrypted. It was assumed that communication between the drone and its controller has been encrypted using WPA2 security protocol and the information that can be extracted from such encrypted protocol is MAC addresses of source and destination, packet size, and also packet arrival time. Additional number of packets received and processed caused accurate information to be gained, but at the same moment, identification and detection time are compromised and this makes a trade-off between detection accuracy and delays in the detection. Therefore, a delay-aware UAV identification cost function is formulated, which is optimized for receiving the optimal number of flows for the detection of UAVs. In conjunction with this, Sciancalepore *et al.* [35] designed a system to detect and identify the state of the drone movement over an encrypted WiFi network. Packet size and packet inter-arrival time were analyzed from the flows captured. Six features were generated, namely inter-arrival time, packet size, and mean and standard deviation of these two, which were further sent for classification purpose to the Random tree classifier. Drones having an open-source operating system Arducopter are used in the experimental analysis, as the same open-source firmware is being used by more than 30 drones, and this also makes the system generalize in detecting as much number of drones before knowing their prior frequency patterns.

Classification concept has been implemented in drone detection system by means of two ways, either it is used to distinguish between drone and background flows to detect the drone, named as a binary classifier, or it has been employed in order to distinguish a particular drone among different make and models after its detection. In the former approach, traffic of different sources is gathered, and then the decision of drone or no drone is made, for example, in [23,33] where no drone case is termed as background flows. However, in the latter approach, a database is built up containing a different type of data for multiple drones, for example, their make and model, etc. So, once the drone is detected the classifiers are used to distinguish or identify a particular type of drone. The mentioned latter approach does not need to be valid to drones only, it can also be used to classify the controller of the drones or even any RF/WiFi source.

3.3.2 Techniques and algorithms

Designing the drone detection system involves detection, and classification of drones, in general. To complete these tasks, different experimental and simulation-based setups have been proposed to obtain different results for analysis and validation purposes. Different tools and techniques incorporated in all stages of the drone detection system are highlighted in this section based upon the extracted data presented in Tables 3.2 and 3.3 for RF and WiFi techniques, respectively. Because the selection and implementation of detection techniques is a paramount factor in the drone detection systems. Before focusing on the implemented techniques in the reviewed articles, the background of spectrum sensing techniques is presented.

Spectrum sensing and analysis concepts have crucial importance in the detection system as the first task is to detect any RF signal and then its spectrum analysis to determine not only the frequency but also RSS of the RF signal. Different spectrum-sensing techniques have been presented in the literature in the context of Cognitive Radio (CR), which are Energy Detector (ED) [41], Cyclostationarity Feature Detector (CFD) [42], Wavelet Transform-based Detector (WTD) [43], and Matched Filter-based Detector (MFD) [44]. The most common and simple approach is the ED because of its low computational and implementation complexities. However, it is less promising in differentiating the interferences from RF signal and noise, also having poor performance in low Signal-to-Noise Ratio (SNR) values. Because, in the sensing and detection of RF signals, SNR value is very important as it is desired to sense and capture the RF waves at very low SNR values. So, among these mentioned techniques, CFD is the most promising technique in detecting the primary RF signal from interferences and noises at low SNR values but at the cost of high computational and implementation complexities [45]. Cyclic Correlation Function (CCF), instead of power spectral density, is used for the detection of signals in the CFD technique. Cyclic spectrum intelligence technique was proposed in [45] by exploiting cyclostationarity features to mitigate the RF interferences in wideband radios. Accuracy versus complexity-based comparison for the commonly used sensing and detection techniques (CFD, ED, MFD, waveform-based sensing, and radio frequency identification) is made in [46].

RF category

Cyclostationarity sensing-based drone detection was performed in [27], where cyclostationarity of modulated signals, WiFi like OFDM signals, is exploited by detecting the spectral peaks and signal repeating patterns, and hence determining CAF and Spectral Correlation Density (SCD) in the low SNR values as it is a suitable tool for the implementation of the passive drone detection. WTD and Frequency-based Detector (FD) were used to identify the unique signature patterns of drone body vibration and body shifting in [28]. WTD detected the sudden shifts of drone body while FD captured the maximum frequency peak for drone body vibration in the range of 50–22 Hz. Further, WTD approach and Discrete Haar Wavelet Transform Detector (DHWTD) have been implemented in [24] because of the ability of WTD to detect in low SNR values. NN-based algorithm has been employed in [29] for feature extraction and classification purpose. S-SVM classifier is implemented in order to differentiate between LoS and NLoS signals from the RSS values, and LSTM-NN model to confirm the drone presence. As per Zhao *et al.* [32], to classify the UAVs, SVM, and kNN classifiers requires a large number of samples for feature extraction in order to train the model, and in comparison to these, AC-WGANs provide optimum performance for the same purpose. Different software tools have been used for the implementation of these techniques and algorithms, such as MATLAB® [28,30], Labview & Python [30], and SDR [26,28].

WiFi category

In WiFi-based drone detection, WiFi packet sniffer Wireshark is used in almost all of the reviewed articles. Drone identification was performed by employing Random tree, Random forest, SMO, and logic Boost ML classifiers in [23], while NB, Random forest, and J48 classifiers were employed in [33]. To realize the quick drone detection, ℓ_1-norm technique was used to eliminate the extra features and the one-versus-all logistic regression function was used to further finalize the features to be used for detection in [34]. Further, Trees-J48, Random forest, and NN were used in [35] for the detection of drones. Eavesdropping the drone communication, statistical features extraction, training of models, recognition through classifiers, these are different stages of drone detection system incorporated in the WiFi fingerprint-based approach, and all these stages require certain tools and techniques to be implemented. Apart from WiFi sniffers and ML/DL classifiers, open-source tools have been used to implement these classifiers, for example, Weka in [23].

3.3.3 Performance evaluation

Various solutions have been offered in the literature considering different scenarios and analyzing the aforementioned characteristics as indicated in Tables 3.2 and 3.3. In the following, we separately discuss both of the RF/WiFi techniques.

3.3.3.1 RF-based systems

The more detection rate and accuracy, the more reliable the detection system. In other words, the system is capable enough to not only avoid missing of targets but also detects it correctly. Therefore, detection with accuracy is the paramount consideration in the design of these systems, but it is influenced by multiple factors, such as the type of environment, the distance between transmitter and receiver, maximum range, SNR value, type of technique employed, etc. It is a general observation that detection performance is degraded when the experiments are carried out in the outdoor environment as compared to indoor. As result, the detection range is also compromised. For example, RSS of the received signal was decreasing beyond the distance of 50 m in the outdoor environment as observed by Nguyen *et al.* [26] while Nguyen *et al.* [28] performed experiments in three different environments named urban, suburban, and inside the campus with a detection distance of 100 m, 600 m, and 200 m, respectively. In some cases, the detection range can be increased by using sophisticated hardware but at the same time environment, especially urban area with NLoS conditions, influences the performance. Toward increasing the detection range, distances up to 3 km and 2,800 m achieved in [25,39] in the outdoor environments. It is important to note that both of these higher values were obtained in the LoS conditions, so the performance of the system would be degraded if NLoS scenarios are incorporated. However, almost the same detection rate was observed in these two and a maximum of two drones were tested during the experimental setup. On the contrary, the detection performance in [24] was almost the same, but the distance achieved was only up to 130 m and they had incorporated fourteen different drone controllers. So, it can be said that there is a trade-off among detection rate, number of drones incorporated, and indoor/outdoor environment, and also a trade-off between maximum range and type of environment together with LoS and NLoS conditions. In parallel, the cost of the system hardware cannot be ignored. The impact of NLoS conditions on the detection performance was analyzed experimentally by Raja [29] and succeeded in achieving the detection performance up to 97%, and the designed system was also cost-effective as implemented by using Raspberry pi modules.

It can be noted from Table 3.2 that ML techniques are not employed when drone propeller and drone physical characteristics-based detection methods are implemented, for example, [26,28]. In comparison to this, all of the systems have used ML techniques where the drone communication signals approach is employed for detection purposes. Further, it can also be observed from Table 3.2 that systems without employing ML techniques and algorithms have used graphical tools to confirm the presence of a drone or its controller. The presence and identification of a drone type is claimed based upon time–frequency analysis in [31], and in the same way, Fu *et al.* [27] also used time and frequency plots to confirm the presence of drones. However, quantitative results are also presented in this category, for example, [28], where detection accuracies of 89.4% in the urban area and 84.9% in suburban area were achieved. The majority of the systems in the non-ML-based category have used LoS conditions and their detection process is fully dependent upon the patterns of drones and their controllers. ML classifiers have been used to recognize the drones from background flows as well as for distinguishing the particular type of drone and its controller. The performance of classifiers used in the latter case is discussed here. Classification accuracy of such classifiers lies in the range between 84% and 98% as indicated in Table 3.2. It can be noted that the DNN model is found with an accuracy of 84.5% with three different drones in [30] and 98% accuracy with four different drones in [32]. Higher accuracy was observed in [32], because the author's database was limited to four different UAVs, which must be increased in order to classify the different types of drones.

3.3.3.2 WiFi-based systems

Performance of the WiFi technique-based system design is higher as compared to RF-based systems as it can be noted by comparing Tables 3.3 and 3.2. Accuracy of detection is almost greater than 90% in all the designed systems except for [36], which is based upon the crowdsensing approach and the detection rate observed is 86.7%. In [23], results for experimental setups were presented in terms of True Positive Rate (TPR), the precision of UAV class, and classification as terrestrial drone and background traffic. Among four classifiers, Random tree classifier was found with optimum performance and the average precision of the designed algorithm was achieved greater than 96%. Moreover, in [33], system performance is evaluated in terms of TPR, positive predicted value, and accuracy. Proposed algorithms allowed to have TPR of 96% for all the considered scenarios. The designed system is able to efficiently detect and identify the drone even in the absence of video streaming data. If an unauthorized drone goes beyond the range of guardian WiFi in [23,33], then it is not possible to detect the intruder drone as the designed algorithm can only detect the drones that are WiFi operated.

3.4 Drone localization system

3.4.1 Overview

In the context of a drone surveillance system, detection of the intruders is not enough, it is necessary to localize the drone in the unauthorized vicinity if the target is not only to monitor but to interdict the detected intruders. Therefore, not only finding the location but the accurate one for these intruders is the primary and necessary requirement of any drone surveillance system. As per Guvenc and Chong [47], localization is carried out in two phases, in general. One, through the estimation of the distance between two nodes, is called accurate ranging. Second is the determination of an unknown node from the known nodes by an already measured range between known nodes, called localization. Localization problem is broadly categorized as either range-based measurement or range-free measurement in WSN, the former approach is preferred due to their good position estimation [48,49] and the basic concept lies in the category of ranging based localization is the Trilateration.

Ranging-based localization provides good accuracy, but the main concern is how to estimate the distance. In order to do so, different techniques used in the literature are Time of Arrival (ToA) (e.g., [50]), Time Difference of Arrival (TDoA) (e.g., [51]), RSS (e.g., [48,52,53]), and Direction of Arrival (DoA) (e.g., [54,55]). Normally, ToA, TDoA, and RSS techniques are used for the ranging-based localization, while DoA is employed in ranging-free measurements [49] and two receivers are sufficient for determining the position of the transmitter as achieved in [56]. To decide which measurement method to implement for the localization is based upon the trade-off made between localization accuracy and complexity of each technique. For example, the accuracy of localization is higher in ToA and TDoA but synchronization is mandatory and the process of timing and synchronization is complicated [48]. On the other hand, the RSS technique is a low cost, less processing approach, and doesn't require additional hardware for implementation, but the accuracy of localization is compromised [48]. Like ToA and TDoA techniques, DoA (also called Angle of Arrival, AoA) also requires cooperation and coordination between receivers to perform the triangulation task and hence localization [57]. Generally, RSS-based ranging technique is more attractive and dominant as compared to other methods/techniques as it neither requires coordination between nodes nor time synchronization between transmitter and receiver [49]. Usually, this method requires LoS scenario between transmitter and receiver. Nevertheless, this method has also been utilized in NLoS conditions as given in [49].

Data related to work done on drone localization is presented in Table 3.4. Further, classification of work done based upon the localization approach and nature of the surveillance station is shown in Figure 3.3. It has been observed that decision for

Table 3.4 State-of-the-art comparison for drones localization in anti-drone system

Ref.	Source signal	Method/ approach	Techniques & algorithm	Environment	Area/ altitude	Outcome/ performance
[58]	RF	RSS	RSS-based distance estimation	Outdoor	Coverage radius upto 1000 m	Localization error = 100 m at: Optimum altitude of 800 m Distance between SDr is 300 m.
[22]	RF	RSS & DoA	Least Squares and Maximum Likelihood Criteria	Outdoor campus	Not specified	Not specified
[36]	WiFi	RSS-based crowd-sensing approach (2-D localization)	Log-normal shadowing model & Maximum likelihood estimation	Campus	60 m distance Altitude: 10 m	Shelter free: MSE = 1 m With shelter: MSE = 10 m
[56]	RF	AoA	Triangulation	Outdoor campus	30–150 m distance Altitude: 20 m	Drone: Avg: error = 12.2° & 12.71 m Drone controller: Avg: error = 9.9° & 11.36 m
[59]	RF	Joint TDoA & AoA	Non linear Least square estimator SDP with interior point methods	NLoS urban	100 m × 100 m × 100 m	MSE = 4.2 m (3D) MSE = 3 m (2D)

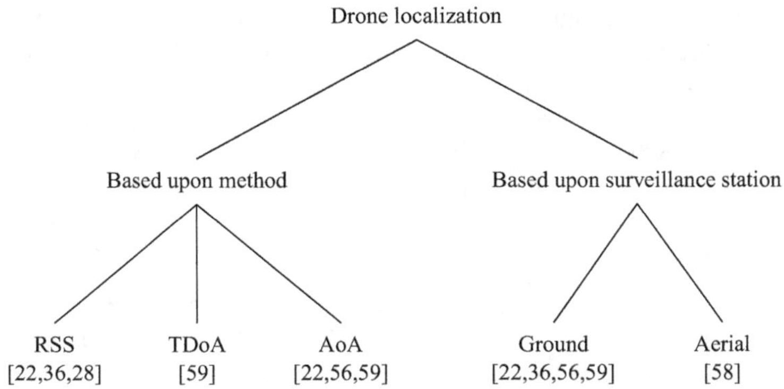

Figure 3.3 Classification of work done on RF/WiFi-based drone localization system with respect to the incorporated methods and nature of the surveillance station

selection of these techniques for drone localization is not only dependent upon the merits and demerits of these but is also significantly influenced by the nature of surveillance station, environment, LoS/NLoS conditions, and position of and distance between receivers. It can also be noted that ToA approach has not been used yet as shown in Figure 3.3 as per reviewed studies; this could be because of requirements of ToA approach as it requires a common clock, precise sampling rate, and strict time synchronization between transmitter and receiver. As in the case of localization of drone by the drone surveillance system, the drone is transmitter but non-cooperative, so the criteria of ToA may not be achieved in this perspective.

3.4.2 Methodology

In the scenario where a drone is remotely operated using its controller, the requirement to find not only the location of a drone but also the location of its controller is paramount. Toward providing solutions in such scenarios, authors in [56] put their efforts in exploring and designing a system that can not only determine the direction and location of the detected drone but also determine the direction and location of the controller with whom the drone is communicating; triangular relation was used for determining their respective locations. The system utilized RF-based signal analysis and AoA techniques together to determine the peak in RF signal related to a drone or its controller. Bebop 2 and DJI Phantom 4 Pro were used for testing and analysis purposes. To tackle the problem of drone localization in urban environments, [22,36,59] added some significant contribution. Specifically, Xu *et al.* [59] addressed the localization problem in NLoS conditions using only three receivers by incorporating joint TDoA and AoA approaches. Apart from this, RSS and DoA measurements from the received RF signal were employed in [22] to locate the intruders. Mobile crowdsensing concept-based drone localization was implemented in [36] in which the authors used RSS information from the WiFi network to locate the detected drones in the urban environment. In conjunction with localization of drones from ground stations, Azari [58] performed the localization of amateur drones in the no-fly zone through RSS-based distance estimation from the aerial station (using SDrs) with optimum altitude compromising on LoS and path loss. Consideration of higher altitude also results in good LoS but at the same time higher path loss, therefore, optimum altitude for the UAV swarm has been highlighted.

3.4.3 Techniques and algorithms

For the position and localization of the detected drones, Yang *et al.* [36] used a log-normal shadowing model and maximum likelihood estimation algorithm while least-square and maximum likelihood criteria were used in [22]. Similarly, the least-square estimator was also used by Xu *et al.* [59] along with SDP with interior point methods. Two nodes were used by Nguyen *et al.* [56] at the receiving side and to identify the direction and location of the drone and its controller, a triangular approach was implemented. Further, in the case of the aerial station where three UAVs were positioned in the equilateral triangle shape, RSS-based distance estimation was employed in the localization of amateur drone. To explore, identify, and select the particular technique are the key parameters for the implementation of the drone localization system. The nature of the problem to be solved has also a great impact on the choice of a particular algorithm and technique. For example, Lower-power Computational-Cost Residual weighting algorithm (LCC-Rwgh) was used in WSN network to mitigate the NLoS problem in localization, at the same time, MultiDimensional Scaling (MDS) and Taylor series linearization-based WLS iterative algorithm can also be used to mitigate the NLoS error, but these two are computationally expensive and not suitable in the case of WSN network [60]. All of the techniques used so far to localize the amateur drones are sampling-based and it can be noted that ML algorithms have not been incorporated specifically for the same.

3.4.4 Performance evaluation

The performance of the designed localization system is presented in Table 3.4. Because the dynamics of a UAV system are very fast, having a static target assumption for its localization cannot be reflected at all. So, in this situation, the precision of localization can be obtained in an acceptable neighborhood range rather than an exact and precise one. If we observe the outcome of the drone localization system, authors have obtained localization error up to 100 m, though the assumption and existence of other parameters are always there. Analyzing the performance of the drone localization system reflects the deficiencies to be improved in a particular design. Most attractive results have been observed in the category of WiFi technique using RSS approach in [36] indicating Mean Square Error (MSE) of 1 m in shelter-free environment, while MSE of 10 m in the opposite scenario. These striking results compromised with the number of receivers, which are eight, and the distance achieved is 60 m. However, the decreasing distance would result in a more attractive outcome, but the coverage area would be decreased. Though optimum results have been observed, it can locate only those drones that are WiFi operated and those whose WiFi beacons can be captured. Further, Nguyen *et al.* [56] observed a localization error of 12.71 m and 11.36 m for drones and their controllers, respectively, by utilizing the AoA approach. Though, only two monitoring nodes, with a distance of 50 m between them, were used for localization, they have experienced a larger error. Moreover, field experiments were conducted with experiencing the distance up to 150 m, but experiments were carried out in LoS scenario only. In [58], NLoS effects and impact of channel characteristics on localization performance were addressed to fly the SDrs above than the amateur drones. Interestingly, NLoS and channel characteristics impact were overcome by employing the aerial stations, but the approach proposed is purely environment dependent, which may not provide the same results if operated in a different environment. Further, to maintain the SDr at higher altitudes at all the time in their operational state is itself challenging.

3.5 Drone-tracking system

Drone tracking and localization are interrelated with each other. In general, information on location coordinates is obtained in localization, and the sequence of such location coordinates is considered as tracking. The localization information is considered as the basis and input for tracking, and it can also be the source for minimizing the localization error up to some extent as it contains localization information in prior [12]. In the literature, both aerial and ground-based stations have been deployed for the tracking objective. As per Azari *et al.* [58], more sensor nodes at the ground station are needed to be deployed, especially in a highly dense area with challenging propagation environments for the detection and tracking of intruders. Therefore, the same task can be better performed by the SDrs being operated with the passive ability, and in result, detection and tracking can be achieved efficiently in terms of greater range, higher accuracy, LoS scenario, higher SNR, and quick deployment. Therefore, attention has been given to developing and designing an intruder-tracking system using single UAV or with the help of UAV swarm in recent years. Classification of work done on drone tracking based upon the approach incorporated and nature of the surveillance station is shown in Figure 3.4. It can be noted that localization-based approach is followed by both aerial and ground-based surveillance stations. Further discussions upon the method and procedure are discussed in respective sections.

3.5.1 Methodology

As per reviewed articles for the tracking mentioned in Table 3.5, it is observed that till now tracking task has been performed using only RF approach. All shortlisted studies have used localization-based mechanism for realization of drone tracking. In the ground surveillance station, DoA- based localization approach was adopted in [56] to identify the direction and location of drone and its respective controllers. Data of AoA measurements were collected and gathered, hence trajectory tracking of drone and its controller was achieved. Alternatively, in the category of the aerial station, the concept of quick and accurate drone tracking of moving RF source, a malicious

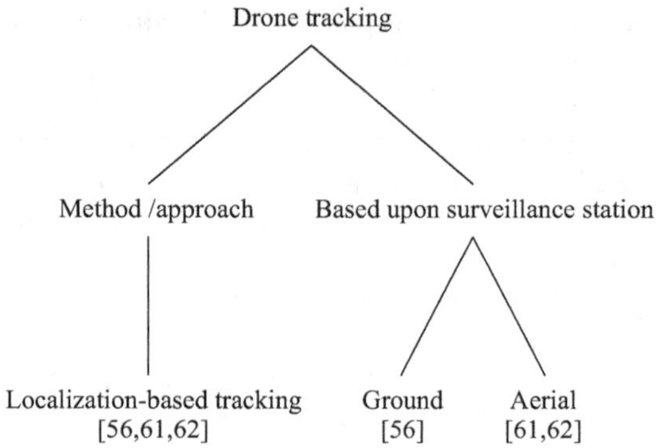

Figure 3.4 Classification of work done on RF/WiFi-based drone tracking system with respect to the approaches incorporated and nature of the surveillance station

Table 3.5 State-of-the-art comparison for drones tracking in anti-drone system

Ref.	Source signal	Method/ procedure	Techniques/ algorithm	Environment	Range	Outcome
[56]	RF	Based upon DoA localization of target	Triangulation	Outdoor campus	30–150 m	Graphical analysis of 100 angle measurements confirming tracking of identified locations
[61]	RF	RSS-based estimation of target position & path planning of UAV swarm	Receding horizon approach & FIM optimization. (D-optimality criterion)	Not specified	141 m	Tracking confirmed via estimation error & incremental graph of D-optimality criteria Error: Up to 10 m
[62]	RF	RSS-based estimation of target position & path planning of UAV swarm	Steepest descent posterior CRLB (D-optimality criterion)	Not specified	Not specified	MSE is below 50 Incremental graph of D-optimality criteria

drone, with the help of other UAVs using receding horizon approach is highlighted in [61]. The same work is also discussed in [63]. The position of targeted UAV was estimated using the Extended Kalman Filter (EKF) approach and the tracking model was formulated using receding horizon maximization technique and Fisher

Information Matrix (FIM) predictive model. In addition, tracking of drone has an intermittent nature: Koohifar *et al.* [62] used UAV swarm to track the intermittent drone where trajectory of UAV swarm is planned with the help of steepest descent posterior Cramer-Rao Lower Bound (CRLB) and bioinspired heuristic path planning approaches.

3.5.2 Techniques and algorithms

Two indispensable techniques for tracking amateur drones have been highlighted by Ding *et al.* [12], which are context-aware tracking and behavior prediction. The former one is based on trajectory filtering for the intruders and the latter one is through mining the huge historical data on intruder drone activities. Among the reviewed articles, Nguyen *et al.* [56] have achieved tracking by gathering the collected data of localization and for that triangulation was performed. Path planning algorithms implemented in the reviewed literature are receding horizon path planning algorithms [61] and both bioinspired and steepest descent posterior CRLB in [62]. Among multiple factors that could be the basis for the selection of a particular algorithm, environmental factors are also considered and incorporated. In real-time scenarios, it is not possible to have noise-free RSS data available from the source at all the time and surely noisy data would have a significant effect on the output performance. For example, a known path loss model was considered in [61] with the assumption that the same RSS measurement is available to each UAV, however, this is not the case in real scenarios. Also, the case of intermittent RF signals from the source was addressed in [62], but the continuous moving target with this nature may have different RSS values with other environmental factors such as the presence of any obstacle. Similar to localization, ML algorithms and techniques are also missing in the drone-tracking literature.

3.5.3 Performance evaluation

Koohifar *et al.* [61,62] used estimation error and D-optimality criteria as evaluation parameters. In addition, Koohifar *et al.* [62] also presented computational time and complexity analysis. In terms of estimation error, Koohifar *et al.* [61] achieved 31% better results using the receding horizon approach as compared to the bioinspired method. However, the authors in [61] have assumed the known transmit power and known path loss exponent, which is not necessarily the case in reality. In practical scenarios, signal power strength is unknown because of noises and other environmental factors. So, in such scenarios, it is necessary to approximate the signal strength first and then the realization of tracking may be implemented. Therefore, the designed system may be investigated with practical considerations of Received Signal Strength Indication (RSSI) measurements. Steepest descent approach and bioinspired heuristic approaches have been used in [62], where the performance of bioinspired heuristic was degraded in terms of D-optimality as UAVs approached the target; in other words, the steepest descent approach has a good value of power, which has resulted in good tracking. However, the addition of processing noise did not have a significant impact on the performances of these two approaches. Although the authors considered intermittent RF transmitter, the transmitter's power, frequency, and antenna gain are already known in prior.

Based upon the discussion on the collected and extracted data, it can be highlighted that designing and deployment of the drone-tracking system is dependant upon the type of surveillance system, and this would depend upon the application where the surveillance system is deployed, for instance, domestic or commercial. Both types of surveillance systems are based upon localization, so the fundamental task is to have a good position estimation of the target. The more an accurate position estimation is obtained, the more improvement in tracking performance can be achieved. It is also observed from the studies that the nature of system design, difficulties, and challenges are broadly dependent upon the type of surveillance station deployed. Although aerial stations have advantages as compared to ground one in terms of coverage, LoS feasibility, long range, and also others, but at the same time extra complexities also raised like path planning, obstacle avoidance, communication, coordination with each other, etc.

3.6 Research trends and challenges

Toward designing of a reliable, robust, and versatile drone surveillance system, a number of research opportunities and open issues need to be addressed. Some of these are highlighted here.

3.6.1 Detection

It has been observed, based upon reviewed articles, that drone detection system is not universal for all types of drones. A system designed for one drone may not be applicable to other types of drones. Therefore, the research community started to design the system with only one type of drone, extended it, and improved its design by incorporating different types of drones. On one hand, this approach seems fruitful in terms of the diversity of the drone surveillance system being improved day by day. On the other hand, the system also suffers in incorporating the big data, especially when the task involves not only to detect the drone but also to identify its type and operating mode of the drone. Keeping this in view, it can be said that the drone surveillance system cannot truly detect the drones until and unless they do not have prior information of drones in their dataset. A surveillance system capable of detecting all types of drones must require huge dataset for all types of drones, especially the detection systems that employ ML algorithms. With the advancement in technology, a drone sent for malicious attacks could be the autonomous drones that are preprogrammed and follow their trajectory with help of GPS/satellites. Further research needs to be done in more detail in terms of the detection of different autonomous drones and their classification and incorporation of the latest tools and techniques.

High computational time may create a delay in detection performance and possibly miss the target as the intruder may fly away quickly by completing the malicious task or attack. Therefore, a system with fast processing along with maintaining maximum detection accuracy must be ensured. A compact and hybrid system must be designed keeping in view not only the relevant influencing parameters but also consider the cost-effective features, and consider whether real and easy deployment by the end-users could be made possible. Further, it is also observed that the accuracy of detection and classification of systems depend upon multiple factors such as SNR, number of output classes, number of data samples collected, length of timing window, and set of data obtained from different UAVs and their controllers. Classifiers implemented yet in the category of RF/WiFi techniques have a different performance with respect to these factors. So, the choice of classifier and its performance also significantly depend upon the nature of the problem as well as the type and quantity of data collected.

3.6.2 Localization

The number of receivers used with an appropriate localization approach significantly affects the localization performance. It can be observed in studies [36,56,59] that as the number of receivers have increased, performance has also improved. Though more receivers improve the localization accuracy, at the same time solutions become costly. Because the approach presented in [36] is purely dependent upon the number of participants and the data shared by them. If the task is to look after the privacy and security concerns of a small area, then some alternate approaches may also be incorporated as it would not be enough to rely on the number of participants who are actually outside that area. Most of the systems designed so far have considered the constant speed of the incorporated drones, and the direction of drones is also determined using the DoA approach based on angle information as reviewed in the literature. Determining the speed of targeted drones along with their position estimation would not only be beneficial for localization but also fruitful in tracking them. Although direction finding with the AoA approach under the RF approach with constant and known speed of the drone is presented, this task needs to be investigated with other localization approaches and unknown and variable speeds, by not only using RF technique but using WiFi as well.

It is also reviewed that RSSI needs minimum hardware requirements as compared to time-based localization techniques but compromises on the localization accuracy. Nevertheless, time-based techniques require additional hardware for their practical realization. Therefore, an optimum solution needs to be designed and implemented for drone localization with sophisticated hardware, which may contribute to improving localization accuracy but at the same time maintaining the cost-effective system design solution. In addition, impact on localization performance because of different environmental and signal propagation factors, localization of multiple drones, and incorporation of ML/DL approaches need to be investigated in detail.

3.6.3 Tracking

It is indicated in the literature that tracking of intruder drones is achieved by performing localization over time. Because an intruder is a moving target, it will change its position in a certain time, therefore localization must be performed by the surveillance station within that time frame so that efficient tracking of the intruders could be realized. As far as implementation of a drone-tracking system using ground station is concerned, tracking tasks can be visualized as monitoring tasks of detecting intruders. Although accurate position estimation would be necessary again in order to start the tracking of the target. The number of ground stations deployed for the target positioning and localization purely depends upon the deployment area requirement. One or more than one station could be mounted and tracking tasks can be carried out accordingly. One important thing is the selection of suitable hardware that may sense the signals effectively in the vicinity and defined boundary, for instance, the direction or omnidirectional antennas and sensors with a movable or fixed platform that can point out the movement of the target. In parallel, consideration of cost-effective design cannot be ignored. Further, the selection of suitable techniques and algorithms remains a significant challenge on its own. In ground station scenario, if multiple base stations are formed, then problems of different stations must be explored. For instance, coordination and communication between base stations, underlying communication protocols, centralized system or distributed architecture, accurate tracking with minimum error, realization of tracking with respect to the speed of the target, and all these challenges need to be addressed in-depth for the practical realization of drone-tracking system. In addition, tracking of multiple drones and smart implementation of these systems with the help of ML/DL algorithm is also appealing.

3.7 Summary

Because of the potential importance of drone surveillance systems, this survey chapter comprehensively explores the work done in the category of RF/WiFi techniques because of their significant importance over the rest of the available techniques. The incidents and threats because of the enormous increase of UAV applications in different areas are highlighted. Three main subsystems (drone detection, drone localization, and drone tracking system) of drone surveillance systems are covered in detail. Methodologies and approaches employed, tools and techniques incorporated for their realization, and evaluation of obtained performance is discussed in each subsystem. Also, future research trends in these three systems are outlined, which are worth exploring.

References

[1] Valavanis KP and Vachtsevanos GJ. *Handbook of Unmanned Aerial Vehicles*. 1st ed. Springer Publishing Company: Berlin; 2015.

[2] Zeng Y, Zhang R and Lim TJ. Wireless communications with unmanned aerial vehicles: opportunities and challenges. *IEEE Communications Magazine*. 2016; 54(5):36–42.

[3] Kaleem Z and Rehmani MH. Amateur drone monitoring: State-of-the-art architectures, key enabling technologies, and future research directions. *IEEE Wireless Communications*. 2018; 25(2):150–159.

[4] Hanscom A and Bedford M. Unmanned aircraft system (UAS) service demand 2015–2035. Literature review and projections of future usage, Res Innov Technol Admin, US Dept Transp, Washington, DC, USA, Tech Rep DOT-VNTSC-DoD-13-01. 2013.

[5] Shakhatreh H, Sawalmeh AH, Al-Fuqaha A, *et al.* Unmanned aerial vehicles (UAVs): A survey on civil applications and key research challenges. *IEEE Access*. 2019;7:48572–48634.

[6] Mozaffari M, Saad W, Bennis M, *et al.* A tutorial on UAVs for wireless networks: Applications, challenges, and open problems. *IEEE Communications Surveys Tutorials*. 2019;21(3):2334–2360.

[7] McFarland M. Airports scramble to handle drone incidents [database on the Internet]. *CNN Business*. 2019 [cited 2020 Jan 08]. Available from: https://edition.cnn.com/2019/03/05/tech/airports-drones/index.html.

[8] Sheldon L. Woman terrified by drone outside her window [database on the Internet]. *KIRO 7*. 2014 [cited 2020 Jan 01]. Available from: https://www.kiro7.com/news/woman-terrified-drone-outside-her-window/81721261/.

[9] Solodov A, Williams A, Al Hanaei S, *et al.* Analyzing the threat of unmanned aerial vehicles (UAV) to nuclear facilities. *Security Journal*. 2018; 31(1):305–324.

[10] Reuters. Greenpeace slams Superman-shaped drone into nuclear plant [database on the Internet]. *NewYork Post*. 2018 [cited 2020 Jan 05]. Available from: https://nypost.com/2018/07/03/greenpeace-slams-superman-shaped-drone-into-nuclear-plant/.

[11] Birnbach S, Baker R and Martinovic I. Wi-fly?: Detecting privacy invasion attacks by consumer drones. *NDSS Symposium*. 2017;p. 1–15.

[12] Ding G, Wu Q, Zhang L, *et al.* An amateur drone surveillance system based on the cognitive Internet of Things. *IEEE Communications Magazine*. 2018;56(1):29–35.

[13] Unlu E, Zenou E, Riviere N, *et al.* Deep learning-based strategies for the detection and tracking of drones using several cameras. *IPSJ Transactions on Computer Vision and Applications*. 2019;11(1):7.

[14] James J, Ford JJ and Molloy TL. Learning to detect aircraft for long-range vision-based sense-and-avoid systems. *IEEE Robotics and Automation Letters*. 2018;3(4):4383–4390.

[15] Nam SY and Joshi GP. Unmanned aerial vehicle localization using distributed sensors. *International Journal of Distributed Sensor Networks*. 2017;13(9):1550147717732920.

[16] Yue X, Liu Y, Wang J, *et al.* Software defined radio and wireless acoustic networking for amateur drone surveillance. *IEEE Communications Magazine*. 2018;56(4):90–97.

[17] Anwar MZ, Kaleem Z and Jamalipour A. Machine learning inspired sound-based amateur drone detection for public safety applications. *IEEE Transactions on Vehicular Technology*. 2019;68(3):2526–2534.

[18] Chang X, Yang C, Wu J, *et al.* A surveillance system for drone localization and tracking using acoustic arrays. In: *2018 IEEE 10th Sensor Array and Multichannel Signal Processing Workshop. SAM*; 2018. p. 573–577.

[19] Zhao Y and Su Y. Cyclostationary phase analysis on micro-Doppler parameters for radar-based small UAVs detection. *IEEE Transactions on Instrumentation and Measurement*. 2018;67(9):2048–2057.

[20] Lü M, Yi J, Wan X, *et al.* Experimental research of drone monitoring using multi-static passive radar. *The Journal of Engineering*. 2019; 2019(20):6795–6798.

[21] Taha B and Shoufan A. Machine learning-based drone detection and classification: State-of-the-art in research *IEEE Access*. 2019;7:138669–138682.

[22] Shi X, Yang C, Xie W, *et al.* Anti-drone system with multiple surveillance technologies: Architecture, implementation, and challenges. *IEEE Communications Magazine*. 2018;56(4):68–74.

[23] Bisio I, Garibotto C, Lavagetto F, *et al.* Unauthorized amateur UAV detection based on WiFi statistical fingerprint analysis. *IEEE Communications Magazine*. 2018;56(4):106–111.

[24] Ezuma M, Erden F, Anjinappa CK, *et al.* Micro-UAV detection and classification from RF fingerprints using machine learning techniques. In: 2019 *IEEE Aerospace Conference*; 2019. p. 1–13.

[25] Yang S, Qin H, Liang X, *et al.* An improved unauthorized unmanned aerial vehicle detection algorithm using radiofrequency-based statistical fingerprint analysis. *Sensors*. 2019;19(2):274.

[26] Nguyen P, Ravindranatha M, Nguyen A, *et al.* Investigating Cost-effective RF-Based Detection of Drones. In: *Proceedings of the 2nd Workshop on Micro Aerial Vehicle Networks, Systems, and Applications for Civilian Use*. ACM; 2016. p. 17–22.

[27] Fu H, Abeywickrama S, Zhang L, *et al.* Low-complexity portable passive drone surveillance via SDR-based signal processing. *IEEE Communications Magazine*. 2018;56(4):112–118.

[28] Nguyen P, Truong H, Ravindranathan M, *et al.* Matthan Drone presence detection by identifying physical signatures in the drone's RF communication. In: *Proceedings of the 15th Annual International Conference on Mobile Systems, Applications, and Services*. ACM; 2017. p. 211–224.

[29] Raja AV. Uncover the power of multipath: Detecting NLOS Drones using low-cost WiFi devices [PhD Dissertations and Master's Theses]. Embry–Riddle Aeronautical University; 2019.

[30] Al-Sa'd MF, Al-Ali A, Mohamed A, *et al.* RF-based drone detection and identification using deep learning approaches: An initiative towards a large open source drone database. *Future Generation Computer Systems*. 2019;100:86–97.

[31] Haluza M and Čechák J. Analysis and decoding of radio signals for remote control of drones. In: *2016 New Trends in Signal Processing (NTSP)*. IEEE; 2016. p. 1–5.

[32] Zhao C, Chen C, Cai Z, *et al.* Classification of small UAVs based on auxiliary classifier Wasserstein GANs. In: *2018 IEEE Global Communications Conference (GLOBECOM)*. IEEE; 2018. p. 206–212.

[33] Bisio I, Garibotto C, Lavagetto F, *et al.* Blind detection: Advanced techniques for WiFi-based drone surveillance. *IEEE Transactions on Vehicular Technology*. 2018;68(1):938–946.

[34] Alipour-Fanid A, Dabaghchian M, Wang N, *et al.* Machine learning-based delay-aware UAV detection over encrypted Wi-Fi traffic. In: *2019 IEEE Conference on Communications and Network Security (CNS)*. IEEE; 2019. p. 1–7.

[35] Sciancalepore S, Ibrahim OA, Oligeri G, *et al.* Detecting drones status via encrypted traffic analysis. In: *Proceedings of the ACM Workshop on Wireless Security and Machine Learning*; 2019. p. 67–72.

[36] Yang G, Shi X, Feng L, *et al.* CEDAR: A cost-effective crowdsensing system for detecting and localizing drones. *IEEE Transactions on Mobile Computing*. 2019.

[37] Piskorski S, Brulez N, Eline P, *et al.* AR. drone developer guide parrot SDK 2.0. Tech Rep. 2012.

[38] Chen H, Wang Z and Zhang L. Collaborative spectrum sensing for illegal drone detection: A deep learning-based image classification perspective. *China Communications*. 2020;17(2):81–92.

[39] Zhang H, Cao C, Xu L, *et al.* A UAV detection algorithm based on an artificial neural network. *IEEE Access*. 2018;6:24720–24728.

[40] Hu J, Wu Y, Chen R, *et al.* Optimal detection of UAV's transmission with beam sweeping in covert wireless networks. *IEEE Transactions on Vehicular Technology*. 2019;69(1):1080–1085.

[41] Chin WL, Li JM and Chen HH. Low-complexity energy detection for spectrum sensing with random arrivals of primary users. *IEEE Transactions on Vehicular Technology*. 2015;65(2):947–952.

[42] Kozłowski S. Implementation and verification of cyclostationary feature detector for DVB-T signals. *IET Signal Processing*. 2016;10(2): 162–167.

[43] Zhao Y, Wu Y, Wang J, *et al.* Wavelet Transform for spectrum sensing in Cognitive Radio networks. In: *2014 International Conference on Audio, Language and Image Processing. IEEE*; 2014. p. 565–569.

[44] Zhang X, Gao F, Chai R, *et al.* Matched filter based spectrum sensing when primary user has multiple power levels. *China Communications*. 2015;12(2):21–31.

[45] Toma A, Nawaz T, Gao Y, *et al.* Interference mitigation in wideband radios using spectrum correlation and neural network. *IET Communications*. 2019;13(10):1336–1347.

[46] Yucek T and Arslan H. A survey of spectrum sensing algorithms for cognitive radio applications. *IEEE Communications Surveys and Tutorials*. 2009;11(1): 116–130.

[47] Guvenc I and Chong C. A survey on TOA based wireless localization and NLOS mitigation techniques. *IEEE Communications Surveys and Tutorials*. 2009;11(3):107–124.

[48] Tomic S, Beko M and Dinis R. RSS-based localization in wireless sensor networks using convex relaxation: Noncooperative and cooperative schemes. *IEEE Transactions on Vehicular Technology*. 2015;64(5):2037–2050.

[49] Gong Z, Li C, Jiang F, *et al.* Design, analysis, and field testing of an innovative drone-assisted zero-configuration localization framework for wireless sensor networks. *IEEE Transactions on Vehicular Technology*. 2017;66(11):10322–10335.

[50] Lanzisera S, Zats D and Pister KS. Radio frequency time-of-flight distance measurement for low-cost wireless sensor localization. *IEEE Sensors Journal*. 2011;11(3):837–845.

[51] Xiong J, Sundaresan K and Jamieson K. Tonetrack: Leveraging frequency-agile radios for time-based indoor wireless localization. In: *Proceedings of the 21st Annual International Conference on Mobile Computing and Networking*. ACM; 2015. p. 537–549.

[52] Ouyang RW, Wong AK and Lea C. Received signal strength-based wireless localization via semidefinite programming: Noncooperative and cooperative schemes. *IEEE Transactions on Vehicular Technology*. 2010;59(3): 1307–1318.

[53] Wang G and Yang K. A new approach to sensor node localization using RSS measurements in wireless sensor networks. *IEEE Transactions on Wireless Communications*. 2011;10(5):1389–1395.

[54] Wang A, Ji X, Wu D, *et al.* GuideLoc: UAV-assisted multitarget localization system for disaster rescue. *Mobile Information Systems*. 2017;2017.

[55] Pack DJ, DeLima P, Toussaint GJ, *et al.* Cooperative control of UAVs for localization of intermittently emitting mobile targets. *IEEE Transactions on Systems, Man, and Cybernetics, Part B (Cybernetics)*. 2009;39(4): 959–970.

[56] Nguyen P, Kim T, Miao J, *et al.* Towards RF-based localization of a drone and its controller. In: *Proceedings of the 5th Workshop on Micro Aerial Vehicle Networks, Systems, and Applications*; 2019. p. 21–26.

[57] Soltanaghaei E, Kalyanaraman A and Whitehouse K. Multipath triangulation: Decimeter-level wifi localization and orientation with a single unaided receiver. In: *Proceedings of the 16th Annual International Conference on Mobile Systems, Applications, and Services*. ACM; 2018. p. 376–388.

[58] Azari MM, Sallouha H, Chiumento A, *et al.* Key technologies and system trade-offs for detection and localization of amateur. *IEEE Communications Magazine*. 2018;56(1):51–57.

[59] Xu C, Wang Z, Wang Y, *et al.* Three passive TDOA-AOA receivers based flying-UAV positioning in extreme environments. *IEEE Sensors Journal*. 2020.

[60] Pagano S, Peirani S and Valle M. Indoor ranging and localisation algorithm based on received signal strength indicator using statistic parameters for wireless sensor networks. *IET Wireless Sensor Systems*. 2015;5(5): 243–249.

[61] Koohifar F, Kumbhar A and Guvenc I. Receding horizon multi-UAV cooperative tracking of moving RF source. *IEEE Communications Letters*. 2017;21(6):1433–1436.

[62] Koohifar F, Guvenc I and Sichitiu ML. Autonomous tracking of intermittent RF source using a UAV swarm. *IEEE Access*. 2018;6:15884–15897.

[63] Guvenc I, Koohifar F, Singh S, *et al.* Detection, tracking, and interdiction for amateur drones. *IEEE Communications Magazine*. 2018;56(4):75–81.

Chapter 4

Cybersecurity for the L-band Digital Aeronautical Communications System (LDACS)

Nils Mäurer[1], Thomas Gräupl[1], and Corinna Schmitt[2]

Today's analog voice-based air–ground communication system for tactical aircraft guidance is suffering from the VHF band's increasing saturation in high-density areas. The air–ground communication infrastructure is therefore undergoing digitization to ensure the sustainable growth of the air transportation system in the coming decades. As safety and security are strongly interrelated in aviation, strong cybersecurity is the foundation and enabler for digitalization in aviation. One of the new air-ground data links that shall enable this transformation is the L-band Digital Aeronautical Communication System (LDACS). It will be the primary long-range terrestrial data link of the future IP-based aeronautical telecommunications network. In this chapter, we describe the design process, draft, and the state-of-the-art cybersecurity architecture for LDACS.

4.1 Introduction

Air Traffic Communications (ATC) is the backbone for safe and secure civil air traffic enabling aerial transport of 4.5 billion passengers and 61.3 million tonnes uplift in 2019 [1]. Up to 2020, civil air traffic has been growing constantly at a compound rate of 5.8% per year [2] and despite the severe impact of the COVID-19 pandemic, air traffic growth is expected to resume very quickly in post-pandemic times [1,3]. With the growth of civil air traffic together with the increasing demand for data-requiring digital services for aircraft guidance and the business operation of airlines, communication increases as well. To cope with this growth, Air Traffic Management (ATM) systems must make more efficient use of its dedicated, limited spectrum making digitization of ATM services unavoidable [4].

The entire industry is currently undergoing such a digital transformation and one area that is mostly affected by this is Communication, Navigation and

[1]Communications Department, Institute of Communications and Navigation (KN), German Aerospace Center (DLR), Wessling, Germany

[2]Research Institute Cyber Defence (CODE), University of the Bundeswehr Munich, Neubiberg, Germany

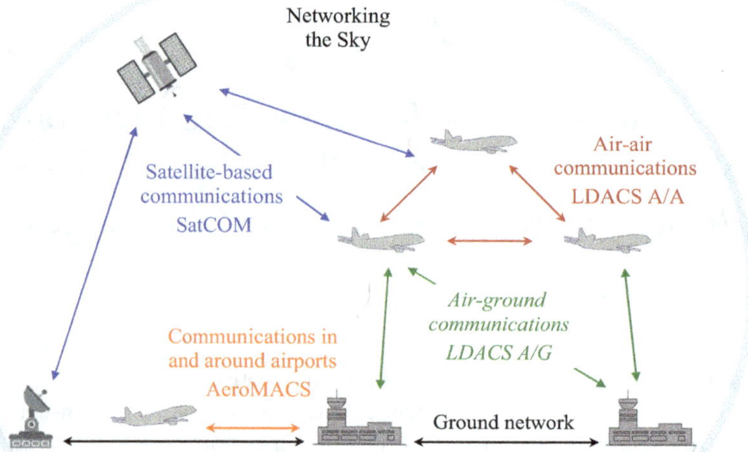

Figure 4.1 The Future Communication Infrastructure (FCI) with new data links such as LDACS and AeroMACS [7,8]

Surveillance (CNS). The Single European Sky ATM Research (SESAR)* program in the European Union (EU) and NextGEN† in the US have been tasked with the development of new technologies to create an aeronautical, digital Future Communications Infrastructure (FCI). Candidates in the FCI are Aeronautical Mobile Airport Communications System (AeroMACS) for airport communications (Airport (APT) and Terminal Maneuvering Area (TMA) domain), SatCOM for Oceanic Remote Polar (ORP) domains, and L-band Digital Aeronautical Communications System (LDACS) for long-range terrestrial aeronautical communications [5,6]. LDACS will be the focus of this work. Figure 4.1 shows the general structure for the FCI.

As safety and security are strongly interrelated in aviation, strong cybersecurity is the foundation and enabler for digitalization in aviation. The World Economic Forum, the International Civil Aviation Organization (ICAO), as well as the cybersecurity research community agree with that of [9–12]. Unfortunately, cybersecurity for CNS is still not realized in most deployed systems [13–16]. In part, this is because of the requirements for aeronautical data links: (1) low latency and (2) low additional security data overhead [17].

Another part is the difficulty and cost of implementing cybersecurity in a system, once it has been designed, prototyped, released and possibly deployed. The further down in that chain, the less likely and the more expensive it is to integrate security measures. One excellent example is VHF Data Link mode 2 (VDLm2) [18].

*https://www.sesarju.eu/, Oct. 10, 2021
†https://www.faa.gov/nextgen/, Oct. 10, 2021

VDLm2 is a digital data link based on Very High Frequency (VHF). It was invented in the 1990s and provides a data rate of 31.5 kbps by using Differential 8 Phase Shift Keying (D8PSK), Carrier Sense Multiple Access (CSMA) in the 118–137 MHz band [19]. Despite being around for decades, the requirement documents [20,21] or specifications [19,22] of the data links VHF or VDLm2 do not fulfill any of the information security definitions of confidentiality, integrity, availability, authenticity, accountability, non-repudiation or reliability defined by RFC 4949 [23]. Even newer aeronautical information systems, such as Automatic Dependent Surveillance Broadcast (ADS-B), a Global Navigation Satellite System (GNSS)-dependent surveillance technology used by aircraft to automatically broadcast their GNSS-based position, which is mandatory since 2020, are mainly known for its insecurity [13,24–27] by information security standards. One of the few data links in the aeronautical ecosystem, which has a dedicated cybersecurity architecture, is the FCI candidate for airport communications, AeroMACS. The system is majorly based on the IEEE 802.16 WiMAX standard [7], which has a security sub-layer integrated in its protocol stack. These selected examples show a clear picture: Information security or cybersecurity is still scarcely integrated into aeronautical communications.

In this chapter, we describe the design process, draft, and the state-of-the-art of the cybersecurity architecture for the ground-based digital communications system LDACS. LDACS has been designed with security in mind and shall introduce state-of-the-art cybersecurity to aeronautics.

In Section 4.2, we introduce LDACS, relevant technical details, and previous work on the cybersecurity of LDACS. In Section 4.3, we list requirements and objectives for the cybersecurity architecture for LDACS, which we present in Section 4.4. For evaluation purposes, we evaluate the cybersecurity additions in a mathematical model, a software simulation of LDACS, and real flight trials in Section 4.5 before concluding in Section 4.6.

4.2 Background on LDACS

LDACS is a ground-based digital communications system for flight guidance and communications related to safety and regularity of flight [5,6]. It has been mainly developed in Europe and is currently under standardization by ICAO [28,29].

4.2.1 System characteristics

LDACS has its origin in merging parts of the B-VHF [30], B-AMC [31,32], TIA-902 (P34) [33], and WiMAX IEEE 802.16e technologies [34]. In 2007, the spectrum for LDACS was allocated at the World Radio Conference (WRC), which is shown in Figure 4.2.

It was decided to allocate the spectrum next to Distance Measuring Equipment (DME), resulting in an inlay approach between the DME channels for LDACS as illustrated in Figure 4.3.

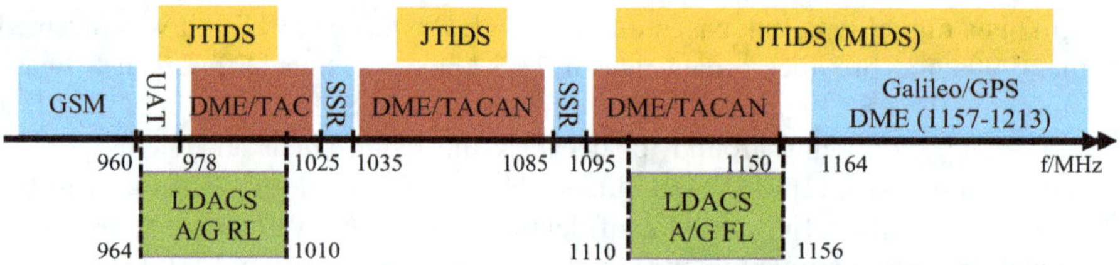

Figure 4.2 Frequency assignment for LDACS at WRC 2007, next to DME

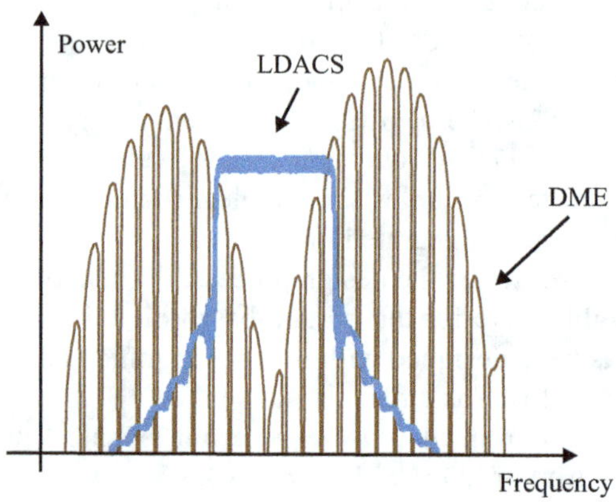

Figure 4.3 Inlay approach for LDACS in between the DME bursts

Furthermore, LDACS uses 4G technology to remain highly flexible and scalable and efficient in coding, supporting adaptive coding and modulation. Additionally, it applies Frequency Division Duplexing (FDD) because of the limited bandwidth available with the inlay approach. Besides all these, LDACS supports seamless handovers, data and voice transmissions, Quality of Service (QoS), has a navigation and surveillance extension, and an Air-to-Air (A2A) link is currently being developed.

4.2.2 *Communication functionality*

LDACS was especially designed for ATC and ATM applications, such as Controller–Pilot Data Link Communications (CPDLC), Automatic Dependent Surveillance-Contract (ADS-C), full 4D trajectories exchange, and real-time weather information, such as the Graphical Weather Service (WXGRAPH). It is envisioned that Ground Based Augmentation System (GBAS) functionality will also be provided using LDACS in the future as well [35,36]. The underlying enabler for all those applications are the main LDACS parameters listed in Table 4.1. Thus, LDACS covers current Air Traffic Services Air Traffic Services (ATS), Aeronautical Operational

Table 4.1 Main parameters for LDACS

Number of sub-carriers	**64 (50 used)**
Bandwidth	625/488 kHz
Subcarrier spacing	9.765625 kHz
OFDM symbol duration	102.4 μ
Guard interval	$(4.8 + 12.8)$ μ
Net data rate	470 kbits 2.82 Mbit/s

Control (AOC) data and also future applications, enables new concepts (i.e., sectorless ATM) and have at least an order of magnitude more net capacity than the currently used terrestrial links such as the VDLm2 system [29].

The latest extension of LDACS communications capabilities – A/A communication – is researched in the German national project IntAirNet [6]. The goal is to establish direct A/A communications between aircraft in communication range allowing infrastructure-less aeronautical network for ad hoc networks between aircraft. However, this research is only at an early stage, thus we will focus solely on the LDACS A/G link.

4.2.3 LDACS network entities

Different entities are included in LDACS Network in order to allow communication in the system. Figure 4.4 depicts them and illustrates the established communication links.

Up to 512 Aircraft Stations (AS) communicate to an LDACS Ground Station (GS) in the Reverse Link (RL). GS communicates to AS in the Foward Link (FL). GSs are controlled by a Ground Station Controller (GSC). The GSC connects the LDACS sub-network to the global Aeronautical Telecommunications Network Aeronautical Telecommunications Network (ATN) to which the corresponding ATS and Aeronautical Operational Control (AOC) end systems are attached.

4.2.4 LDACS protocol stack

For AS and GS, we can identify different layers and entities in the LDACS protocol stack, namely Physical Layer (PHY), Medium Access Layer (MAC), Data Link Service (DLS), LDACS Management Entity (LME), Voice Interface (VI) and Sub-Network Protocol (SNP) and, for the GSC, we identify the LME, as illustrated in Figure 4.5.

The physical layer provides the means to transfer data over the radio channel (R1 interface). The LDACS ground station supports bidirectional links to multiple aircraft under its control. The forward-link direction (ground-to-air) and the reverse-link direction (air-to-ground) are separated by FDD. Forward link and reverse link use a 500-kHz channel each. The ground station transmits a continuous stream of

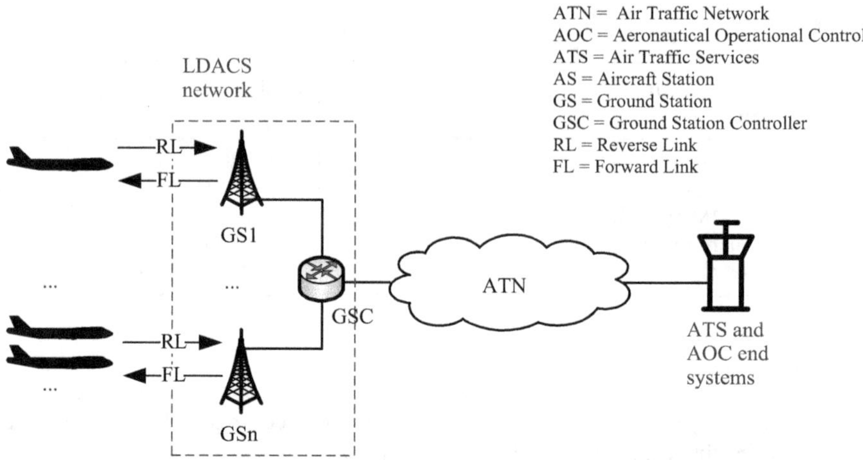

Figure 4.4 Network architecture of LDACS [37]

Orthogonal Frequency-Division Multiplexing (OFDM) symbols on the forward link. In the reverse link, different aircraft are separated in time and frequency using a combination of Orthogonal Frequency-Division Multiple Access (OFDMA) and Time Division Multiple Access (TDMA). Aircraft thus transmit discontinuously on the reverse link with radio bursts sent in precisely defined transmission opportunities allocated by the ground station [17].

The data-link layer provides the necessary protocols to facilitate concurrent and reliable data transfer for multiple users. The LDACS data link layer is organized in two sub-layers: the MAC sub-layer and the Logical Link Control (LLC) sub-layer. The MAC sub-layer manages the organization of transmission opportunities in slots of time and frequency. The logical link control sub-layer provides reliable and acknowledged point-to-point logical channels between the aircraft and the ground station using an automatic repeat request protocol.

Within the LDACS data link layer, two entities are of special interest to us: the *LDACS Management Entity* (*LME*) and the *Sub-Network Protocol* (*SNP*).

The main task of the LME is to perform configuration, resource management, and mobility management of LDACS. The mobility management service in the LME provides support for registration and de-registration (cell entry and cell exit of aircraft), scanning channels of neighbouring cells and handover between cells. It also manages the addressing of aircraft within cells. The resource management service is responsible for link maintenance (power, frequency and time adjustments). In Figure 4.5, we see that these functionalities are provided from within the LME using the “Radio Link Function,” while user data and communications to the SNP are managed using the “Sub-Net Functions.” The SNP glues the LDACS network together and works as a connector to the network layer. It provides end-to-end user plane and control connectivity between the aircraft, ground station, and ground-station controller within the LDACS sub-network.

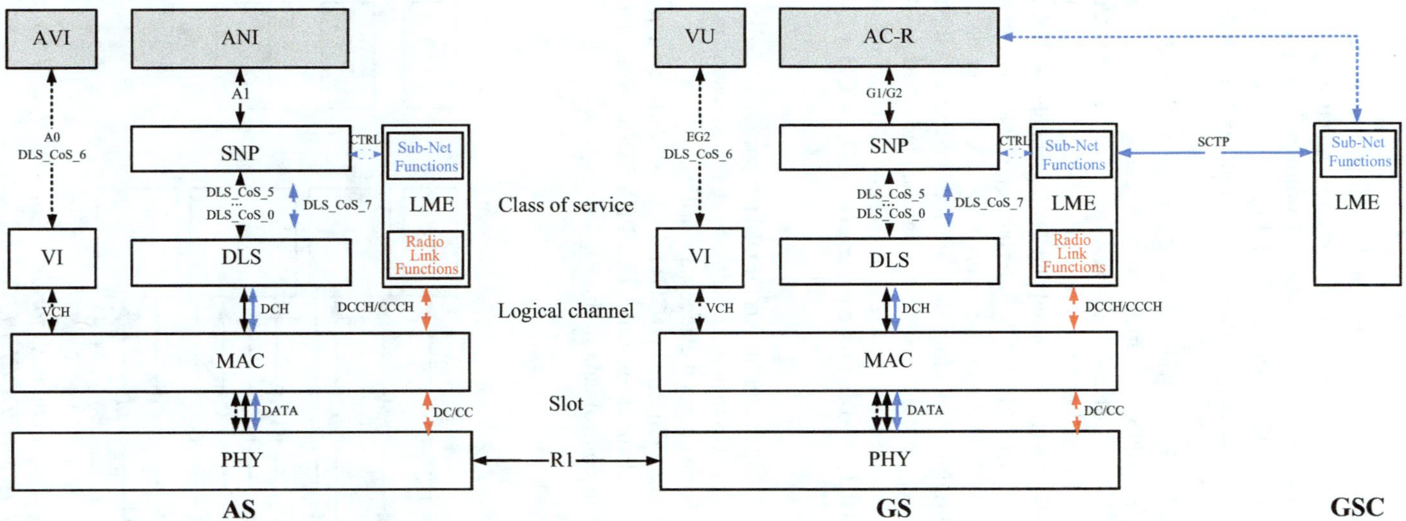

Figure 4.5 *The LDACS sublayer is embedded in the FCI (IPv6, voice and control traffic) and consists of Physical layer (PHY), Medium Access Layer (MAC), Data Link Service layer (DLS), and Voice Interface (VI), all located in the logical link control sublayer and finally the Sub-Network Protocol layer (SNP). The LDACS Management Entity (LME) serves as a cross-layer entity between MAC, DLS, and SNP layers and is the single LDACS relevant entity in the GSC. Voice data are transmitted to higher layers using the Airborne Voice Interface (AVI) on AS side and Voice Unit (VU) on GS side. Data are passed to higher layers using the Airborne Network Interface (ANI) in the AS and using the Access-Router (AC-R) in the GS.*

4.2.5 Interfaces, data flow, and logical channels

LDACS internal control data are exchanged between AS and GS over the radio link (R1) and up within the protocol stacks using Common Control (CC) and Dedicated Control (DC) slots between PHY and MAC, Common Control Channel (CCCH) and Dedicated Control Channel (DCCH) logical control channels between MAC and LME. Thus, all critical LDACS control functions are handled by the LME.

User data are depicted in black in Figure 4.5 and travel from ground-based Aeronautical Network Service Provider (ANSP) or airline servers using the AC-R over the G1/G2 link into the SNP. The DLS offers different Classes of Service (CoS), depending on the priority of the user packet and using these (DLS_CoS_0...5) into the DLS. Between DLS and MAC, the Data Channel (DCH) logical channels transport user data to the lower layer, and using the DATA interface on slot level, user data are handed down from the MAC to PHY layer. Between AS and GS, user data are transmitted then using the radio link (R1), before they are received and handed up between the different protocol layers in the AS, before being handed up to the ANI using the A1 interface. In Figure 4.6, we see the corresponding control channels, with a deeper description of the user (DCH) and control channels (RACH, BCCH, CCCH, DCCH) as follows:

As the **FL Data Channel (DCH)** is held in continuous OFDM transmission, dedicated for the deliverance of user data, the GS locally allocates FL channel resources (i.e. FL PHY-SDUs) within slots and manages the access priorities.

In contrast to the FL DCH, the **RL Data Channel (DCH)** uses a bandwidth on demand scheme. Each AS has to request channel resources (RL PHY-SDUs) from the GS before they can send any user or control data in the RL data channel.

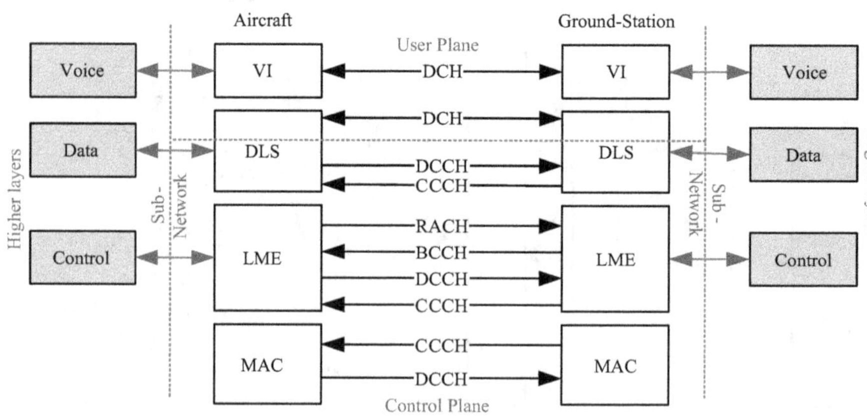

Figure 4.6 Overview of LDACS logical channels for user data (DCH) and control data (BCCH, RACH, CCCH, DCCH) [29]

As the **Data Control Channel (DCCH)** is used in RL only, by any AS to convey MAC/Logical Link Layer (LLC) control messages to the GS, while each AS has its own DCCH so that none other than this specific aircraft can send on that channel.

The **Common Control Channel (CCCH)** is used in FL only and only by one GS in order to announce, e.g. the MAC slot layout and to perform resource allocation in the FL/RL to the AS. Also, the GS may send control messages on this channel.

The **Random Access Channel (RACH)'s** purpose is predominantly for AS to make cell entry requests. Only AS may use it.

The **Broadcast Control Channel (BCCH)** is used to send cell configuration information and mobility management commands to the AS using broadcast messages. Only the GS can use it and it reaches all the aircraft listening to the same broadcast Subscriber Access Code (SAC).

4.2.6 LDACS frame structure

In the FL direction, each Super Frame (SF) starts with a Broadcast Channel (BC) slot, where the GS announces its existence to the AS and sends physical parameters for link establishment. The rest of the FL SF is split into four Multi Frame (MF), each containing nine OFDM frames and each frame comprises three FL Physical Layer-Service Data Unit (PHY-SDU). Every FL PHY-SDU can be used to transmit FL user data or CC data, in which GS can allocate resources to an AS. Details about the LDACS frame structure are depicted in Figure 4.7.

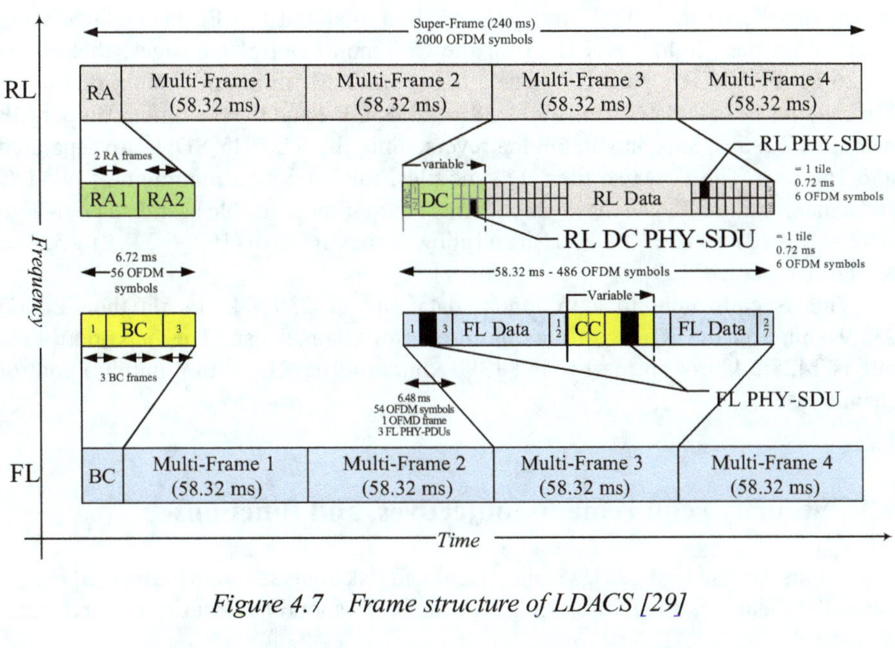

Figure 4.7 Frame structure of LDACS [29]

In the RL, a SF starts with a Random Access (RA) slot, where AS can request access to an LDACS cell and continues with four MFs. Each RL MF is constructed from 162 RL PHY-SDU equivalent to OFDMA tiles. They are used for two purposes, namely (1) to transmit DC data, which are used by an AS to request the allocation for resources allowing them to send on the RL, and (2) to transmit RL user data.

4.2.7 LDACS cell entry procedure

Once a GS is securely connected to the aeronautical ground network using the GSC, it starts sending a broadcast message, the System Identification Broadcast (SIB) message, containing relevant information such as network identification, physical parameters such as channel frequencies and more. When an AS enters the cell served by a GS, it receives the SIB and sends a CELL_RQST message in reply. The CELL_RQST message contains a "unique address identifying the LDACS radio" [29]. When the GS receives the CELL_RQST message, a CELL_RESP message is sent back to the AS, informing the AS about its System Area Code (SAC). The SAC is a local and temporary address for the AS in the cell. After this exchange of control channel messages, both communication parties are informed about LDACS specific addresses, timing, frequency, and power values and can start the user data communication.

4.2.8 LDACS data rate

Data are transported in the DCH using different FL PHY-SDUs and RL PHY-SDUs of different sizes. Depending on coding and modulation, thus the channel quality, the FL PHY-SDUs sizes range from 728 to 3296 Bit, and the RL PHY-SDUs range from 112 to 528 Bit. With 27 frames per MF in total and 1–8 FL PHY-SDUs being reserved on the Flight Level (FL) for the Common Control messages, this leaves 19–26 FL PHY-SDUs per MF for data transport. The minimum amount of Bit per MF can thus be calculated with $19 * 728 = 13,832$ Bit and the maximum Bit per MF with $26 * 3296 = 85,696$ Bit. On the reverse link, the RL PHY-SDUs are separated into 162 tiles. The first two tiles are sync tiles, followed by a minimum of two DC and a maximum of 32 DC tiles, which limits the minimum usable user data per MF to $(162 - 2 - 32) * 112 = 14,336$ Bit and allows a maximum of $(162 - 2 - 2) * 528 = 83,424$ Bit per MF [29].

This is equivalent to a minimum data rate of 230.5 kbit/s on the FL and 238.9 kbit/s on the RL, with maximum control channel use. The maximum data rate is 1428.3 kbit/s on the FL and 1390 kbit/s on the RL, with minimum control channel use.

4.3 Security requirements, objectives, and functions

In previous works [17,37,38], several threat and risk analyses were performed resulting in the identification of assets, threats, and corresponding security requirements, objectives, and functions. LDACS must follow the same protection goals as any other aeronautical data link within the FCI. These are [11,39]:

Safety means the system must sufficiently mitigate attacks, which contribute to safety hazards.

Flight regularity means the system must sufficiently mitigate attacks, which contribute to delays, diversions or cancellations of flights.

Protection of business interests means the system must sufficiently mitigate attacks which result in financial loss, reputation damage, disclosure of sensitive proprietary information, or disclosure of personal information.

The next step was to identify assets and objectives: Anything that someone places value upon is regarded as asset. For LDACS five assets were identified: (1) hardware, (2) software, (3) link, (4) data, and (5) services [11].

LDACS Hardware is responsible for the execution of LDACS software, providing relevant functionality (i.e. supporting any of LDACS possible CNS services). Furthermore LDACS relevant information is stored on it. LDACS hardware refers to AS, GS, and GSC but also to an Authentication, Authorization and Accounting (AAA) server, e.g. integrated within the GSC, Access Routers, the links between entities and the respective LDACS specific internal and shared network and routers.

LDACS Software enables LDACS CNS capabilities. Thus, we need to make sure that a software component of the devices or sub-system is not corrupt, has no errors or other defects. Also, we have to prevent wrong installation or configuration of the software components.

LDACS Radio Link required radio links, accurate time synchronization along with LDACS control data and radio communications connections enabling LDACS to transmit send and receive data using that link are assets. Most important here is preventing unauthorized access, altered hardware, eavesdropping, jamming, and spoofing. However, we will not introduce hardware protection mechanisms, such as regular quality checks, access limitations to special hardware, and control of personal working on that hardware, or physical layer robustness mechanisms, such as frequency hopping, pilot symbol scrambling, but rather focus on providing protocol-based security from the MAC layer up.

LDACS Data, relevant for an error-free execution of LDACS services, are an asset. These include (but are not limited to) (1) identities of communication entities, (2) LDACS control data, (3) user data, (4) confidential data, only accessible for legitimate users and entities only, (5) any cryptographic material, (6) configuration data or (7) data relevant for navigation services.

LDACS Services such as system management, announcement and routing, mobility and authentication service, are required for an operational baseline for LDACS. As use case, at least 21 high critical ATS user data services and 14 high critical AOC data services [37,38] were identified to be provided by LDACS. As new functions in ATS and AOC services can be introduced on a frequent basis, this work can only contribute to highlighting already existing safety relevant services with regard to LDACS. Examples of these are the ATC Clearance (ACL), Data Link Logon (DLL), Flight Plan Consistency (FLIPCY), Flight Plan Data (FLT-PLAN), Network Connection NETCONN or Network Keep Alive NETKEEP service.

This analysis leads to the identification of five security objectives for LDACS in [17], which were extended to nine objectives in the official LDACS Standards and Recommended Practices (SARPS) [40]. These are to (1) protect availability and continuity of service, (2) protect integrity, (3) authenticity for user and control plane messages in transit, (4) provide non-repudiation of origin, (5) confidentiality for user plane messages in transit, (6) mutual entity authentication, (7) authorize explicitly permitted actions of users or entities, (8) prevent the propagation of intrusions within LDACS domains and towards external domains, and (9) protect against service attacks to a level consistent with the application service requirements.

With these guidelines and objectives in mind, security functions were defined in [11,41], following the definitions from RFC 4949 [23]:

Entity Authentication: We need to integrate functions for mutual *entity identification*, *authentication authorization* and *accounting* for every participant within the LDACS sub-net, thus preventing any unauthorized access or use of LDACS.

Key Management: LDACS shall include functions for secure *key generation*, *key agreement*, *key derivation*, *key access* and *key destruction*.

(User) Data Confidentiality: We suggest using strong *symmetric encryption* for user data encryption, because of low computational overhead and fast operation times. After a master key has been negotiated between each communicating party and an encryption key derived from it, incoming user messages from the air traffic network can be encrypted.

Data Integrity: As several threat-and-risk analyses for LDACS were revealed, data-in-transit integrity is one of the most important security property for wireless communications systems [17,37,38], as because of the wireless nature of the communication medium, it is inherently easy to eavesdrop on messages, modify or delete them.

Data Origin Authentication: Some data on the link, such as entity authentication related data, shall include a *data origin proof*.

System Integrity: *Self-checks* and checks at startup of systems/devices of LDACS shall detect manipulation or errors in the behaviour of the security mechanisms.

Robustness: LDACS shall support functions ensuring *reliability* and *robustness* to mitigate jamming, spoofing, interference or DoS attacks.

Secure Logging: Mechanisms for security and non-security relevant logging, together with regular checks and possibly including the digital signature of the logging device ensure secure logging of actions within the LDACS radio devices.

Physical Access: LDACS shall provide physical security commensurate to the data it contains. This includes zeroing out keys and other secret data in emergency cases.

4.4 A comprehensive cybersecurity architecture for LDACS

In this chapter, we combine the inner workings of LDACS from Section 4.2 and cybersecurity requirements and functions from Section 4.3 together to present a comprehensive cybersecurity architecture for LDACS.

4.4.1 Placements of Security Functionality in Protocol Stack

In [17,37,41], suitable placement of security functionality within the LDACS protocol stack was discussed.

We argue that placing protection mechanisms in the LME and SNP entities within the protocol stack will be most efficient in securing LDACS. MAC and DLS will also receive new tasks (e.g. measures for control channel protection). Security endpoints for secure user data communication and primary entity authentication are the AS and GSC, while the control data plane will be protected between GS and AS. Lastly, GSC and GS will establish a secure connection, prior to any aircraft being able to successfully connecting to the LDACS network. With these measures, we can achieve user plane end-to-end security from GSC to AS, provide entity authentication among all parties, and introduce key negotiation and derivation functions between relevant parties.

4.4.2 Trust

The LDACS security concept requires all entities in an LDACS network to authenticate to each other to ascertain that only trusted participants can use the system. To establish trust within the network, there are multiple ways to achieve this, where we mention only two of them here:

(a) Public Key Infrastructure (PKI):

The general idea of a PKI is the attempt to solve the problem of having to trust a communication's partner identity claim.

A PKI can solve this problem via involving a trusted third party who verifies the identities of the parties who wish to engage in communication via issuing a digital certificate.

The most commonly used digital certificates are X.509 [42] certificates, which contain (1) the issuing Certificate Authority (CA), the (2) CA digital signature, (3) version number, (4) serial number, (5) owner, (6) owner's public key, (7) validity period, (8) certificate usage, and (9) signature algorithm. As aviation operates worldwide, a hierarchical PKI will have to be deployed with several sub-CAs being distributed over the world.

Basically, there are two proposals on how to achieve worldwide trust coverage [43]:

One root CA is installed per geographic region and then it performs cross-certification with distributed root-CAs of all other geographic regions around the world. Subdomains can exist within ATM organizations. Here trust emerges from the assured trustworthiness of each regional root CA cross-certifying other and being cross-certified by other regional CAs. This approach is depicted in Figure 4.8.

The other idea is to have one worldwide (probably offline) root CA, hosted by a trusted worldwide entity, such as ICAO, with several regions sub-CAs distributed around the world. That way, the ICAO hosted root CA serves as trust bridge, as seen in Figure 4.9. However, a PKI comes with some drawbacks for digital aeronautical communications:

1. Massive rollout, management, and revocation of certificates are required.
2. A root of trust has to be declared and accepted by state actors worldwide potentially requiring secure cross-certification among all countries worldwide respecting political situations and regulations in aviation. Thus, the infrastructure must map to the political reality of aviation which is, that a small number of

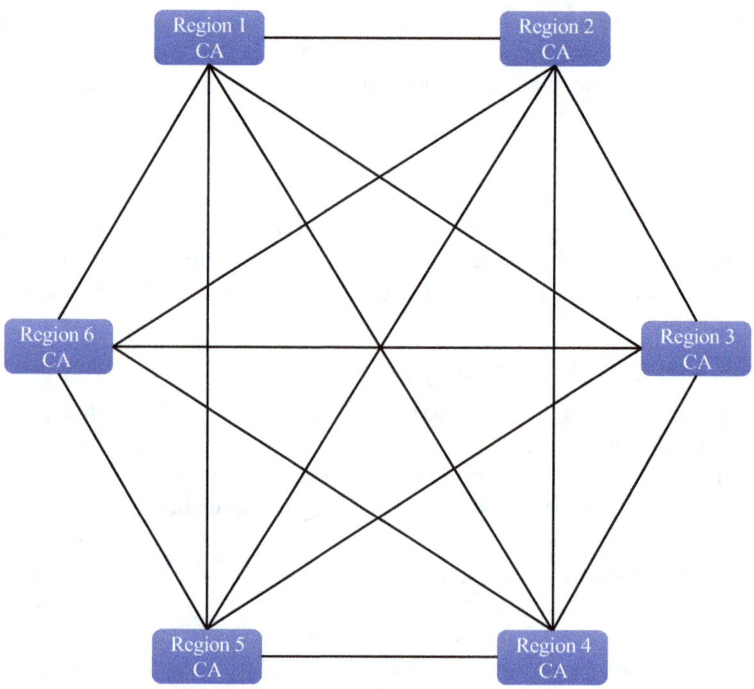

Figure 4.8 Worldwide aeronautical PKI – cross-certification [7,8]

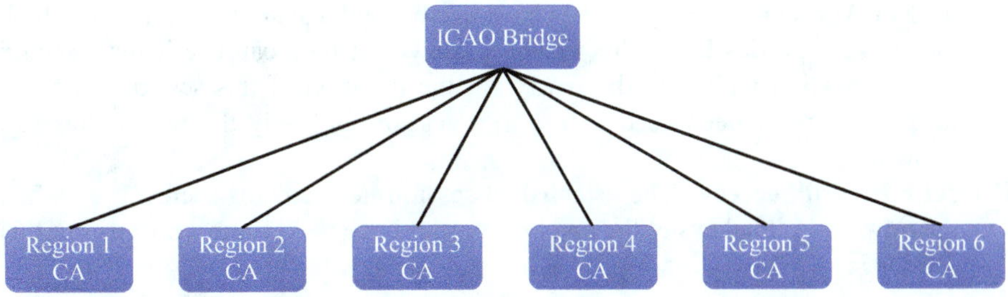

Figure 4.9 Worldwide aeronautical PKI – ICAO trust bridge

state actors capable of securing critical infrastructure with limited trust toward the outside. All this makes a PKI possibly not the best solution for a aeronautical trust framework.

With the two drawbacks, mentioned PKI may be a challenging solution for an aeronautical trust framework. But keeping in mind that digital data links for civil aeronautical traffic (i.e. AeroMACS) use a PKI as their trust solution [44], it looks promising to use PKI also for LDACS.

The advantage of this way forward is the possible alignment of the LDACS PKI concept with the AeroMACS PKI [44], which is already realized and operational. Furthermore, all entities within the FCI should remain interoperable, providing a seamless multi-link concept for aeronautical data, which makes the PKI-based trust solutions the most likely one.

(b) Physical Unclonable Function (PUF) Challenge-Response Pair (CRP): The concept of Physical Unclonable Function (PUF) based trust lies within the property of unclonability of PUFs and thus the uniqueness of PUFs. Hence, a PUF can be interpreted as a unique device's fingerprint, an enabler to create a unique set of CR pairs and a strong random number generator. With a secure database on ground where Challenge Response Pairs (CRP) are stored (and possibly unique per authentication round), a MAKE scheme can be established and trust incorporated into the system based on the trust placed unto the first transmission of CRP and secrecy of CRP. However, the PUFs would have to be installed within the radio hardware during a secure manufacturing process.

4.4.3 Mutual authentication and key exchange (MAKE)

Depending on the method how trust is incorporated into the system, there are different approaches for Mutual Authentication and Key Exchange (MAKE) procedures. Overall all procedures need to fulfill the following three objectives:

Mutual Authentication: Both parties can be sure of the identity of the other and that both actually participated in this interaction.

Secure Key Agreement: Both parties have established a shared session key, which means both parties know this key and know that they can use it for a secure communication with the other party for the duration of this session. The key must have never been used before in a session and only the two parties can know it.

Perfect Forward Secrecy: The established session key remains secret, even when the private signing keys of the involved parties have been compromised after this session.

4.4.3.1 Station-to-Station (STS)-MAKE

The origin of LDACS mutual authentication and key exchange protocol, first mentioned in [41], is a variation of the STS protocol [45]. Since the publication of [41], we investigated different STS variants and protocol 5.25 "Modified STS protocol" in [45] proves to be more secure and concise than that mentioned in [41]. It prevents the possibility of a Man-in-the-Middle attack during the exchange of the key material by signing the respective material with the help of exchanged or pre-stored public key certificates of the respective communication partner. However, in order to ensure trust in public keys from the respective communication partner, a PKI is required [41]. With a PKI and certificates in place, the modified STS protocol becomes a good candidate for mutual authentication and key agreement for LDACS.

The LDACS STS-MAKE protocol is illustrated in Figure 4.10 and consists of the following five steps:

Step 1 – Start of STS: After cell entry is done, and the DCH of LDACS is open for authentication purposes, the GSC chooses a secret x and calculates its Diffie–Hellman public key $t_{GSC} = g^x$.

Step 2 – Server Hello Key Exchange: The GSC sends the *ServerHelloKeyExchange* message to the AS. The AS chooses a secret y and calculates its Diffie–Hellman public key t_{AS}. It then calculates the static Diffie–Hellman shared key $S_{AS,GSC} = (g^x)^y \bmod p$ and the shared session key $K_{AS,GSC} = KDF(S_{AS,GSC})$ via a predefined Key Derivation Function (KDF) and creates its own signature $Sig_{AS}(t_{AS}, t_{GSC}, ID_{AS}, ID_{GSC})$.

Step 3 – Client Hello Key Exchange: The AS sends now its Diffie–Hellman public key t_{AS} and its signature to the GSC in the *ClientHelloKeyExchange* message. The GSC verifies the AS signature $Sig_{AS}(t_{AS}, t_{GSC}, ID_{AS}, ID_{GSC})$. If that verification passes, at this point the AS is authenticated to the GSC. The GSC proceeds to generate $S_{AS,GSC} = (g^y)^y \bmod p$ and $K_{AS,GSC} = KDF(S_{AS,GSC})$ via a predefined KDF. Now both parties have a shared session key $K_{AS,GSC}$. It then builds another signature tag $Sig_{GSC}(N_{GSC}, t_{GSC}, t_{AS}, ID_{GSC}, ID_{AS})$.

Step 4 – Server Key Exchange Finished: The GSC sends its nonce N_{GSC} and signature $Sig_{GSC}(N_{GSC}, t_{GSC}, t_{AS}, ID_{GSC}, ID_{AS})$ in the *ServerKeyExchangeFinished* message to the AS. There the AS verifies the GSC signature $Sig_{GSC}(N_{GSC}, t_{GSC}, t_{AS}, ID_{GSC}, ID_{AS})$. If that verification passes, at this point, the GSC is authenticated to the AS.

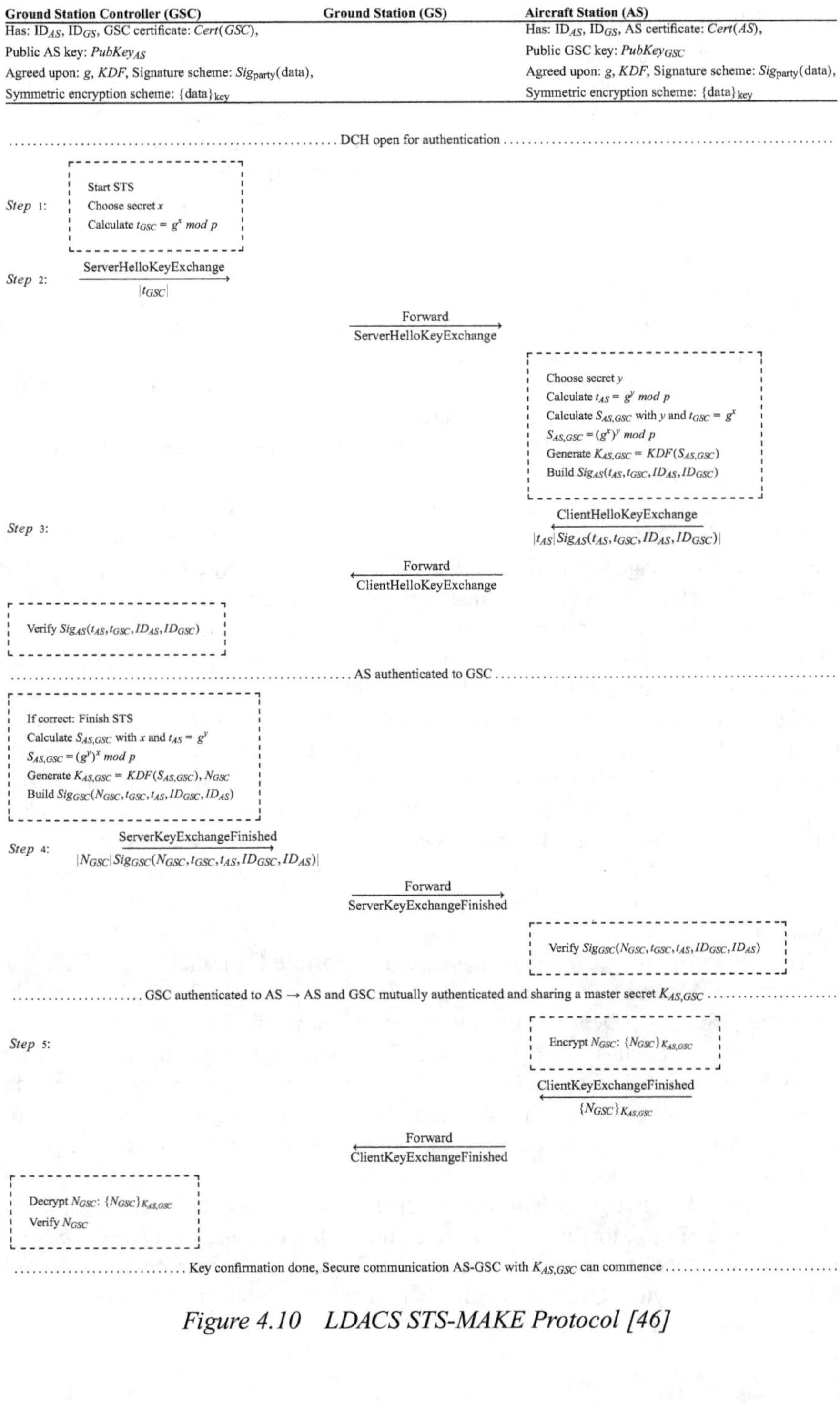

Figure 4.10 LDACS STS-MAKE Protocol [46]

Step 5 – Client Key Exchange Finished: To attain key confirmation, the AS encrypts the nonce N_{GSC}, $\{N_{GSC}\}_{K_{AS,GSC}}$ and sends that in the *ClientKeyExchangeFinished* message to the GSC. At the GSC, the N_{GSC} nonce is decrypted and verified. If that verification step is successful, key confirmation of the key $K_{AS,GSC}$ is achieved.

Currently, three different DHKE approaches are considered for the STS-MAKE procedure: Classic, ephemeral DHKE, Elliptic Curve Diffie–Hellmann (ECDH), and Supersingular Isogeny Diffie–Hellman (SIDH). The original DHKE was first published in 1976 and is based on the discrete logarithm or Diffie–Hellman problem [47]. Given a cyclic group G of prime order n, a generator g of G, and elements $g^x, g^y \in G$, find g^{xy}. Because of the possibility of Man-in-the-Middle attacks [48], authenticated DHKE schemes (e.g., STS, Internet Key Exchange (IKEv2), and IKE version 2 (IKEv2) protocols [45]) were invented.

Elliptic curve cryptography [49] enabled smaller key sizes, resulting in the ECDH protocol [50]. Based on the conjectured difficulty of finding isogenies between supersingular elliptic curves [51], SIDH finally represents a post-quantum robust version of the DHKE [52,53].

We define data sizes for the authentication and key agreement messages here for the STS for LDACS protocol. For signature lengths, we assume a total length of 64 Bytes for a message signature, produced by current signature procedures such as EdDSA-Ed25519 [54] or even post-quantum procedures such as rainbow [55].

All messages have a header consisting of *TYPE*, *ID*, *UA*, and *PRIO* fields. *TYPE* is a 4-Bit long field and clarifies the message type, *ID* is 12-Bit long and denotes the ID of that message, *UA* is a 28-Bit long Unique Address field, containing the LDACS specific addresses of AS and GS, and finally a 4-Bit long *PRIO* field signifies the priority this particular message has. We collect all these fields into the *header* resulting in a 48-Bit length. A nonce N is of length 96 Bytes. t_{GSC} and t_{AS} are the Diffie–Hellman public keys of the respective entities and have different sizes, depending on the choice of the Diffie–Hellman procedure. The sizes for the Diffie–Hellman public key of the GSC t_{GSC} are $\{DHKE = 3072|ECDH = 256|SIDH = 2624\}$. The sizes for the Diffie–Hellman public key of the AS t_{AS} are $\{DHKE = 3072|ECDH = 256|SIDH = 2640\}$.

The *ServerHelloKeyExchange* message, responsible for initiating the STS protocol between AS and GSC, consists of the *header* and the Diffie–Hellman public key of the GSC t_{GSC}. Depending on the size of the public keys, the sizes for the *ServerHelloKeyExchange* are $\{3120, 304, 2672\}$ Bytes. The key exchange message AS to GSC is denoted as *ClientHelloKeyExchange* and consists of the *header*, the Diffie–Hellman public key of the AS t_{AS}, and an AS signature Sig_{AS}. Depending on the size of the Diffie–Hellman public keys, the sizes for *ClientHelloKeyExchange* are $\{3632, 816, 3200\}$ Bytes.

The *ServerKey ExchangeFinished* consists of the *header*, a nonce N_{GSC}, and a GSC signature Sig_{GSC}, totalling 652 Bytes. Finally, the *ClientKeyExchangeFinished* finishes the protocol and simply contains a header and the encrypted nonce N_GSC, resulting in 144 Bytes. Overall, this leads to a total amount of authentication bits summarized in Table 4.2.

Table 4.2 Message sizes for STS-MAKE in bit

Message	STS-DHKE	STS-ECDH	STS-SIDH
Step 2	3120	304	2672
Step 3	3632	816	3200
Step 4	656	656	656
Step 5	144	144	144
Total	7552	1920	6672

Note that the STS-MAKE for LDACS was proven to fulfill the three objectives for LDACS MAKE protocols, mentioned at the beginning of the chapter, using the model checker Tamarin [56].

4.4.3.2 Physical unclonable function-based mutual authentication and key exchange Physical Unclonable Function based Mutual Authentication and Key Exchange (PMAKE)

As mentioned in Section 4.4.2, because of several reasons, PKI-based MAKE protocols might hold several disadvantages because of the political reality in aviation, that is, a small number of dominant state actors are capable of securing critical infrastructure and have limited trust towards others.

To address this problem, we proposed the use of PUFs within LDACS radios to generate device unique CRPs, used in PMAKE for mutual authentication. PUFs use device with unique random patterns, which are introduced in the manufacturing process to differentiate chips and make them uniquely identifiable. In other words, a PUF can be interpreted as a unique device's fingerprint, an enabler to create a unique set of cr pairs and a strong random number generator.

Instead of the establishment of a PKI, in PMAKE it is assumed that a (1) very mobile node (aircraft) has to be equipped with a PUF during the construction process of the specific LDACS radio device and (2) an initial CRP has to be exchanged between aircraft and ground-based secure verification database in a secure environment. At the end of this initial exchange, the secure database within the GSC securely stores the CRP $< C_{AS_0}, R_{AS_0} >$ and the AS stores $< C_{AS_0} >$. The main part of the protocol is depicted in Figure 4.11 and consists of the following steps:

Step 1: The AS, upon receiving such a beacon, generates a random number r_{AS} and depending on the respectively chosen DHKE procedure, it calculates t_{AS} and $\alpha = HMAC_{R_{AS_0}}(\mathrm{ID}_{AS}, \mathrm{ID}_{GS}, t_{AS})$. It then responds with $|t_{AS} \oplus C_{AS_0}|\alpha|\mathrm{ID}_{AS}|$.

Step 2: Once the GS receives the response to the beacon message, it appends its ID to the message and forwards $|t_{AS} \oplus C_{AS_0}|\alpha|\mathrm{ID}_{AS}|\mathrm{ID}_{GS}|$ to the GSC.

Step 3: With the help of the previously stored tuple $< C_{AS_0}, R_{AS_0} >$, the GSC can compute the Diffie–Hellman public key of the AS $t_{AS} = t_{AS} \oplus C_{AS_0} \oplus C_{AS_0}$ and

Table 4.3 Notations used in the PMAKE scheme

Notation	Definition
msg1 $\oplus$ msg2	XOR operation on msg1 with msg2
msg1 | msg2	Concatenation operation on msg1 with msg2
PUF_A	Physical Unclonable Function of entity A
$HMAC_K(msg)$	Hash-based Message Authentication Code with key K and input data *msg*
HKDF(K)	HMAC Key Derivation Function (HKDF) with input K
C_{A_i}	i-th Challenge for PUF from entity A
R_{A_i}	i-th Response from PUF from entity A
ID_A	Identifier of entity A
r_A	Random integers of entity A "Ephemeral private key"
t_A	Ephemeral public key of entity A
g	Public Diffie–Hellman parameters
$S_{AS,GSC}$	Static Diffie–Hellman key shared between AS and GSC
$K_{AS,GSC}$	Session key for AS-GSC communications
$\{msg\}_K$	Symmetric encryption of data *msg* with key K
N_A	Nonce of entitiy A

$\alpha' = HMAC_{R_{AS_0}}(ID_{AS}, ID_{GS}, t_{AS})$. It then checks whether $\alpha' == \alpha$ match. If that is the case, the AS has authenticated to the GSC. Then the GSC generates a random number r_{GSC} of its own and again in dependence on the previously agreed DHKE procedure, calculates t_{GSC}. Now the shared AS-GSC key $S_{AS,GSC}$ can be calculated using the secret of the GSC r_{GSC} and the Diffie–Hellman public key of the AS t_{AS}. With that, the GSC calculates the session key $K_{AS,GSC}$ using the HKDF and $S_{AS,GSC}$. Finally, a new challenge C_{AS_1} is chosen by the GSC and two new MAC tags are calculated. β is used to conceal C_{AS_1}, whereas γ serves as an authentication proof about the GSC for the AS. It finally sends $|\beta \oplus C_{AS_1}|t_{GSC} \oplus t_{AS}|\gamma|ID_{GSC}|$ to the GS.

Step 4: The GS forwards that message to the AS.

Step 5: First, the AS calculates the Diffie–Hellman public key of the GSC through $t_{GSC} = t_{GSC} \oplus t_{AS} \oplus t_{AS}$. To be able to decipher C_{AS_1}, β' is calculated using the AS by reconstructing R_{AS_0} and using previously established values t_{GSC}, t_{AS}, ID_{GSC}, ID_{GS}, and ID_{AS}. As $C_{AS_1} = \beta \oplus C_{AS_1} \oplus \beta'$, the AS successfully received the new challenge C_{AS_1}. It then calculates its own value for $\gamma' = HMAC_{R_{AS_0}}(C_{AS_1})$ and compares $\gamma' = \gamma$. If they match, the GSC has authenticated to the AS. Furthermore, the verifiable integrity and return of t_{AS} proves to the AS that the GSC actually participated in the protocol. Now the AS calculates the shared key $S_{AS,GSC}$ with r_{AS} and t_{GSC} and derives the session key $K_{AS,GSC} = HKDF(S_{AS,GSC})$. Using the AS PUF, a new response R_{AS_1} is generated to the new challenge C_{AS_1} using $C_{AS_1} \rightarrow PUF_{AS} \rightarrow R_{AS_1}$. It then

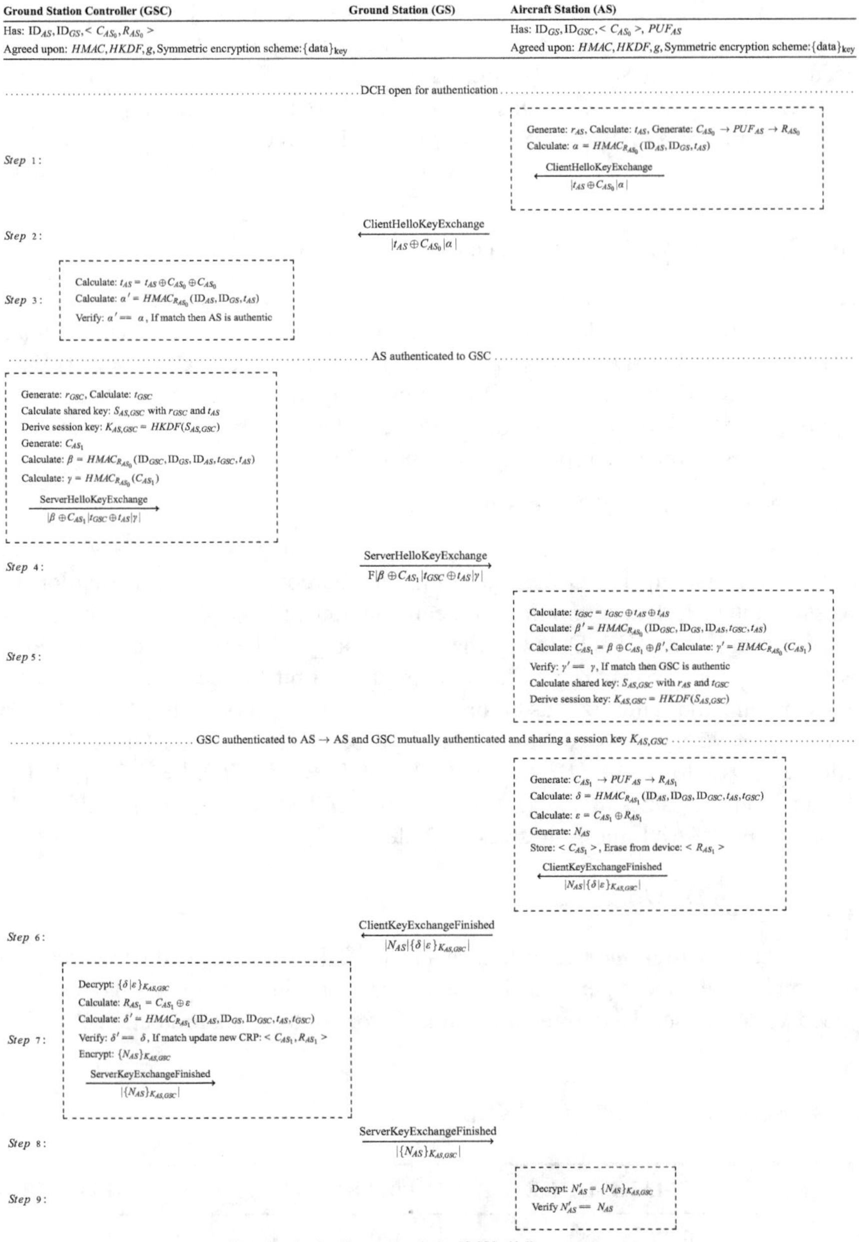

Figure 4.11 PMAKE protocol [57]

calculates $\delta = HMAC_{R_{AS_1}}(\mathrm{ID}_{AS}, \mathrm{ID}_{GS}, \mathrm{ID}_{GSC}, t_{AS}, t_{GSC})$ that will be used by the GSC as proof for the authenticity and correctness of the new response R_{AS_1}. $\varepsilon = C_{AS_1} \oplus R_{AS_1}$ is used to conceal the response R_{AS_1} during transport. For key confirmation purposes, AS generates a nonce N_{AS}. At this point, the AS securely stores C_{AS_1} and erases R_{AS_1} from memory. As AS and GSC have previously agreed upon suitable encryption algorithms, the AS sends N_{AS} in the clear and δ and ε encrypted with $K_{AS,GSC}$ back to the GSC.

Step 6: The GS forwards that message to the GSC.

Step 7: The GSC decrypts the message with the agreed upon encryption algorithm and key $K_{AS,GSC}$. It then computes $R_{AS_1} = C_{AS_1} \oplus \varepsilon$. It then calculates $\delta' = HMAC_{R_{AS_1}}(\mathrm{ID}_{AS}, \mathrm{ID}_{GS}, \mathrm{ID}_{GSC}, t_{AS}, t_{GSC})$ and checks whether $\delta' == \delta$. If that is the case, the GSC can be sure of the authenticity of the response R_{AS_1} and the participation of AS in the protocol. It updates the current tuple for that AS to $< C_{AS_1}, R_{AS_1} >$. It then encrypts N_{AS} and sends it back to the AS.

Step 8: The GS forwards that message to the AS.

Step 9: Finally, AS decrypts $N'_{AS} = \{N_{AS}\}_{K_{AS,GSC}}$ and compares $N'_{AS} == N_{AS}$. If they match, AS is assured that GSC also holds the shared key, key confirmation was successful and user data communication can commence.

Concerning PMAKE's Data Overhead in analogy to the previous description of message sizes for STS-MAKE, we assign amounts of bits to the different messages.

We assume the same sizes as before, so *header* = 48 Bytes, all IDs = 28 Bytes, nonces N_{AS} = 96 Bytes, MAC tag (cf., $\alpha, \beta, \gamma, \delta, \varepsilon$) = 128 Bytes, Diffie–Hellman public key sizes for t_{GSC}: $\{DHKE = 3072|ECDH = 256|SIDH = 2624\}$, Diffie–Hellman public key sizes for t_{AS}: $\{DHKE = 3072|ECDH = 256|SIDH = 2640\}$. We summarize all PMAKE message sizes in Table 4.4.

4.4.4 Key derivation

Once all parties within the network have successfully authenticated each other, key derivation is necessary to generate different keys for different purposes. For example, we need keys for user data protection and keys for control data protection.

Table 4.4 Message sizes for PMAKE in bit

Message	PMAKE-DHKE	PMAKE-ECDH	PMAKE-SIDH
Step 1	3276	460	2844
Step 3	3404	588	2956
Step 5	400	400	400
Step 7	144	144	144
Total	7224	1592	6344

HKDF

As shown in Figures 4.10 and 4.11, we use the HKDF, a KDF built from Hash-based Message Authentication Codes (HMAC). It uses the "extract-then-expand" paradigm, meaning that it consists of two main phases.

First, the input keying material (here: master key/static Diffie–Hellman shared key) is taken and a fixed-length pseudo-random key is extracted. The extract phase is especially important, if the master key is not sufficiently uniform (e.g. the key is uniform only in a subset of the original key space). In that case, we extract a pseudo-random key from the master key by adding a salt value, which can be any fixed non-secret string chosen at random. In the process, the pseudo-random key becomes indistinguishable from a uniform distribution of bits. In general, HKDF can be used with or without salt value, both variations work; however, the use of salt significantly increases the strength of HKDF. Salt ensures independence between different uses of the hash function, supports "source-independent" extraction, and strengthens the analytical results that back the HKDF design [58].

User Data Protection Keys

Also depending on the algorithm choice for securing user data on the data link, we need either one symmetric key (e.g. for AES-GCM [59], which allows integrity and confidentiality protection) or two symmetric keys for integrity and confidentiality protection.

Control Data Protection Keys

As discussed before in Section 4.2.6, securing the control plane of LDACS proves more difficult than its user plane. The underlying problems are very small chunks of data (e.g. RL DCCH: 83 Bit) and the need that every aicraft within one LDACS cell can read the entire control data plane of LDACS. Thus using individual AS-specific keys for securing the CCCH, DCCH channels is not possible. We do not put additional cryptographic protection on the Random Access Control Channel (RACH) or bcçh, as if any spoofer or attacker sending information within those data links is detected and filtered out during the MAKE procedure of LDACS. As the data on the DCCH is mainly responsible to enable an AS to request data allocations, such that it can send user data and the CCCH being the logical channel, via which those resources are granted, integrity protection of these is most important to prevent any attacker redistributing LDACS resources.

The only possible way to meet all these requirements is via introducing group-key mechanisms for LDACS. We are currently investigating and comparing the suitability of the Group-IKEv2 [60], Chinese Remainder Group Key (CRGK) [61], Central Authorized Key Extension (CAKE) [62] and Logical Key Hierarchy (LKH) protocol [63] for LDACS. This process is ongoing work and has not concluded yet.

4.4.5 *User data security*

After the MAKE procedure and with the key derivation of user keys, the user data plane of LDACS can be secured.

We propose to secure LDACS SN-PDUs, thus Packet Data Uni (PDU)s on the SNP, as their size can vary from 128 to 1536 Byte [29], which makes them possibly the largest PDUs within LDACS. This helps minimizing security data overhead, in case a Message Authentication Code (MAC) tag is attached to the SN-PDU.

Confidentiality Protection
We suggest using symmetric approaches for data encryption, due to low computational overhead and fast operation times. We propose to establish end-to-end encryption, for example, AOC data between GSC and AS. After extensive discussion with representatives of ATC instances, it became apparent that ATS data will probably not be encrypted, as every air traffic controller operating in the area should be able to read that data for safety reasons. Thus, between layer 3 and the LDACS sublayer, some notification of the content of the packet will be necessary. As encryption algorithm, we recommend AES-256-GCM [59] with Galois Counter Mode (GCM) being a mode of operation on symmetric key block. It provides authenticated encryption and decryption operations and it proves robust against currently known quantum-computer-based algorithms [64]. The last property is important, as due to the long service life of aeronautical communications, this rising threat might become dangerous during its life cycle.

Integrity Protection
All user data, sent using LDACS, require some kind of integrity protection mechanism. We propose two mutually exclusive strategies:

1. **AES-GCM** (Galois Counter Mode) is a mode of operation for symmetric key cryptographic block ciphers (i.e. here AES). It supports two operations: authenticated encryption and decryption. If only message authentication is required, the variation Galois Message Authentication Code (GMAC) can be used. After the application of this operation mode, we are left with a ciphertext C of exactly the same length as the plaintext P and an authentication tag T of 128 Bit length. However [65] defines tag sizes of 128, 120, 112, 104, or 96, 64 and 32 Bit length, with a clear recommendation to use 96 Bit or more.
2. **HMAC** Here, the idea is to combine the Keyed-Hash Message Authentication Code (HMAC) message authentication mechanism [66] and combine it with hash-functions from the SHA-3 hash-family [67]. We propose HMAC-SHA3-128, that way, we are left with a plaintext P and a message tag T of 128 Bit length.

4.4.6 Control data security

As described in-depth in Section 4.4.4, securing the control data plane of LDACS proves to be far more complicated than its user plane. Reasons for this are far smaller message sizes (cf., DC: 83 Bit, CC: 728 Bit), the need that every participant within an LDACS cell needs to be able to (1) read its contents and (2) be assured of their authenticity.

That way, it is clear that we need control data integrity and not necessarily confidentiality. This leaves us with the problem that a simple message tag simply does not fit in the small control messages. Even if we shrink the tag size to, e.g. 32 Bit, radically reducing the security level, the capacity of LDACS would be also reduced drastically, as fewer resources could be requested and allocated, because of the reduced configuration data space left within the messages.

Securing FL control plane data is easier than its RL counterpart, as the GS can send a continuous stream of OFDM symbols on the FL and is thus the only entity putting data on the CCCH. Assuming a group key has been negotiated within the group of one GS and all AS within that particular LDACS cell, that key can be used by the GS to add a small message tag to a CC-PDU. Assuming a 64-Bit tag to be sufficiently secure, this still results in a $64/728 = 8.8\%$ overhead.

On the other direction, every AS requests resources on the DCCH, thus every aircraft contributes DC-PDUs and every one is 83-Bit long. With that size, and given that only 8 Bit are currently reserved for padding [29], applying a message tag per DC-PDU is impossible.

This topic, altogether with the investigation of group-key procedure for LDACS, is currently ongoing and will have to be solved before the LDACS security architecture can be finalized.

4.4.7 Changes within the LDACS protocol stack

At the beginning of the chapter, we discussed placement of security functionality within the LDACS protocol stack. After having described the entire process of how trust is handled, options for entity authentication and key exchange, key derivation, user data and control data protection, we want to assign all these tasks to certain entities within the LDACS protocol stack.

In Figure 4.12, we have added security relevant states and functionality, marked in red, within the LDACS protocol stack. As said at the beginning of this chapter, LME and SNP are the two most important, security-related entities. The LME handles the entire connection establishment and now MAKE procedure, together with key negotiation, derivation, handover, and secure logging. It is also responsible for protecting control channel data. The SNP, after the MAKE procedure is over and it has received the user data session key from the LME, handles user data protection.

Also, one important factor is the CoS of LDACS: After initial connection establishment and during the MAKE procedure, the DLS and SNP only allow packets of the highest priority (i.e. CoS=7) to pass through to the LME for authentication purposes. No other data packet is able to pass through, before the LME has not successfully completed the entire MAKE procedure, with session keys being deployed at LME for control data protection and SNP for user data protection.

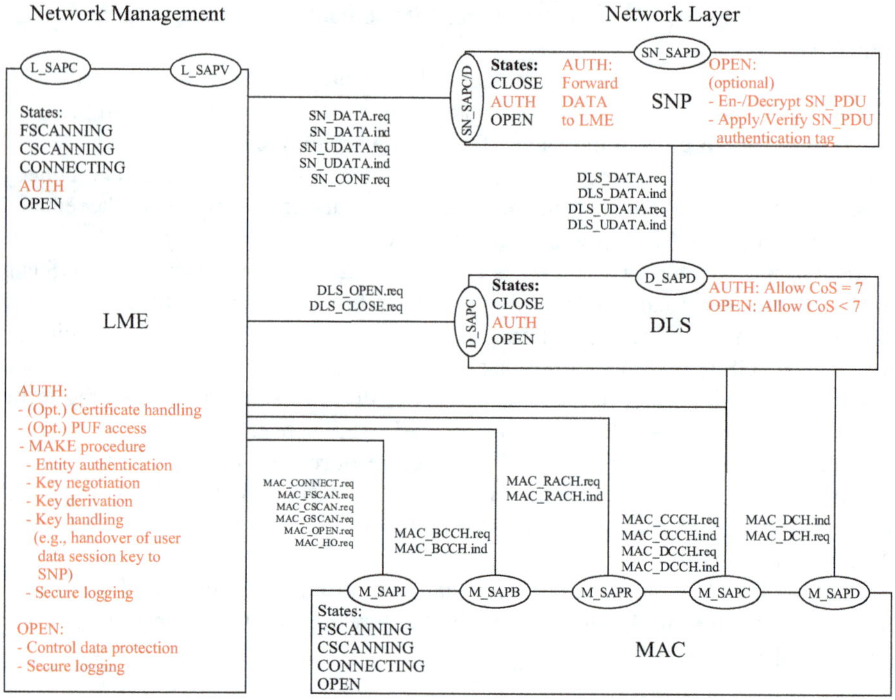

Figure 4.12 Security-related changes within the LDACS protocol stack

4.5 Evaluation of the LDACS cybersecurity architecture

The cybersecurity measures of LDACS were evaluated in [11,46,57,68]. Here, we want to present evaluations (1) in a theoretical model for LDACS security data and latency overhead, (2) modelled in an event-based simulation framework Framework for Aeronautical Communications and Traffic Simulations 2 (FACTS), and (3) actual flight trials.

4.5.1 Theoretical model for LDACS security latency overhead

In 2015, Gräupl *et al.* [69] presented a full methodology on how to emulate latencies for user data in the forward and reverse link of LDACS depending on the Bit Error Rate (BER) and message size.

Taking re-transmissions into account, the FL latency can be calculated with

$$L_{FL}(t) = m_{FL}(t) + (1 + \delta_{RX}(1 + n)) \times d_{MF} \tag{4.1}$$

and the RL latency with

$$L_{RL}(t) = m_{RL}(t) + (2 + \delta_{RX}(N + 3)) \times d_{MF}. \tag{4.2}$$

Table 4.5 Parameter values for latency timing for the LDACS MAC protocol [46]

Forward Link Model (Eq. 4.1)		**Reverse Link Model (Eq. 4.2)**	
Parameters	**Values**	**Parameters**	**Values**
d_{MF}	60 ms	d_{MF}	60 ms
$m_{FL}(t)$	Time until start of next FL MF: Every 1–60 ms modelled by $U(1, 60)$	$m_{RL}(t)$	Average time until start of next MAC cycle: $\#AS/32 \times d_{MF} + wait$ $wait$ modelled by $U(1, 60)$
n	Average amount of MF after transmission until next DC slot is scheduled for AS in MAC-cycle: $n = \#AS/32$	N	Average amount of MFs after transmission until next DC slot is scheduled for AS: $N = (\#AS/32 - 3) \mod \#AS/32$
BER	$10^{-6}, 10^{-5}$		
P	$P(\{\text{no error in packet}\}) = (1 - BER)^l$ $P(\{\text{error in packet}\}) = 1 - ((1 - BER)^l)$		

In equation (4.1), we use $m_{FL}(t)$ to classify the time until the start of the next CC frame, $\delta_{RX} \in \{0, 1\}$ to indicate a re-transmission, d_{MF} denotes the length of a MF and n is derived from the length of the reverse link medium access cycle from forward link perspective.

In equation (4.2), we use $m_{RL}(t)$ to denote the time until the start of next DC slot, $\delta_{RX} \in \{0, 1\}$ to indicate a re-transmission, d_{MF} denotes the length of a MF and N is derived from the length of the reverse-link medium access cycle from reverse-link perspective.

We model $\delta_{RX} \in \{0, 1\}$ as stochastic process based on the packet error rate. Given a BER, we can calculate the packet error rate based on the length of a packet l: $P(\{\text{no error in packet}\}) = (1 - BER)^l$. Thus, the opposite event that a packet indeed contains an error is: $P(\{\text{error in packet}\}) = 1 - ((1 - BER)^l)$. These two probabilities decide the value of δ_{RX}, whether a re-transmission is necessary and, thus, an error appeared in the packet or not.

For more details on the model, refer to [46,69].

With these equations, we calculate LDACS authentication latencies based on the amount of AS within an LDACS cell. These results are shown in Figure 4.13.

Note that because of the similar message sizes, PMAKE and STS-MAKE have almost equal authentication latencies. Both authentication procedures finish quicker with smaller message sizes at higher BER, thus ECDH finishes faster than SIDH, which in turn finishes faster than DHKE. The requirements document DO-350A imposes a $RCTP_{CSP} = 10s$ threshold for RCP 130/A1 message types [70], meaning that all authentication and connection establishments must be completed below the

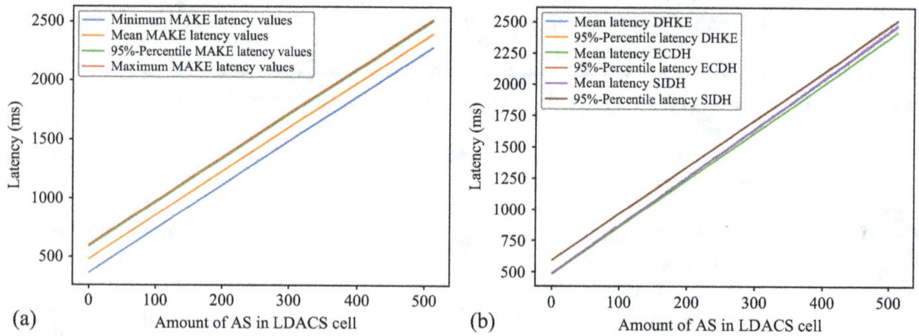

Figure 4.13 Authentication baseline latency at BER = 0 vs. authentication latency at worst-case scenario of BER = 10^{-5}

10^{-s} threshold [29]. As we clearly see in Figure 4.13, we always remain below this threshold, even when an LDACS cell is full.

4.5.2 Model within FACTS2 simulation framework

Another important way of evaluating changes within the LDACS protocol is rapid prototyping with software-based simulations. FACTS [71] is a simulation framework based on modern, service-oriented software architecture: Simulation services organized in a parallelized tool-chain of loosely coupled software services split by the separation of concerns. It allows for the simulation of infrastructure, generation of aeronautical data traffic, simulation of flight patterns or parsing of real-world flight patterns, simulation of arbitrary aeronautical data links and protocols and more. Every tool is tasked with one task (e.g. simulating the LDACS data link) and multiple tools can be piped together, using Unix pipes on the command line. In Figure 4.14, we see first worldwide air traffic movement, then FACTS simulates realistic air traffic data, simulates LDACS as underlying data link and outputs a report on the performance of the data link.

In [68], we used FACTS to demonstrate a proof of concept of the user data security architectural concept of LDACS. Results proved that the implementation concept with the "auth" state within protocol entities and different classes of service, with the highest priority class reserved for authentication purposes, works as intended. Thus, authentication, key exchange, derivation, and handover were performed and SN-PDUs could be secured using the AES-256-GCM algorithm and the negotiated, shared session keys as described in Section 4.4.7.

4.5.3 Real-world demonstration of LDACS security features

Lastly, during the German national project Migration towards COm/NAV capabilities of LDACS (MICONAV) in March/April 2019, we could demonstrate some security features of LDACS in the worldwide first LDACS flight trials. Figure 4.15 shows the

(a) Graphical output of worldwide air traffic

Application Layer - Services

Offered Load, Throughput, Loss

Note that the table below shows the offered load without taking CANCELED packets into account.

	Avg. Offered Load (kbps)		Avg. Throughput (kbps)		Loss (% of offered load)		Loss (Byte)	
	FL	RL	FL	RL	FL	RL	FL	RL
all	174.89	26.65	174.88	26.65	0.01	0.00	1358	0
(secure) ads-c	0.00	1.72	0.00	1.72	0.00	0.00	0	0
(secure) cpdlc	0.01	0.02	0.01	0.02	0.00	0.00	0	0
key_exchange	3.65	3.70	3.65	3.70	0.00	0.00	0	0
secure GBAS	171.24	0.00	171.22	0.00	0.01	0.00	1358	0
secure audio	0.00	21.22	0.00	21.22	0.00	0.00	0	0

Latency

Latency	95%-required		min. (ms)		avg. (ms)		95%-percentile (ms)		99%-percentile (ms)		max. (ms)		std. deviation	
	FL	RL	FL	RL	FL	RL	FL	RL	FL	RL	FL	RL	FL	RL
all	-	-	25.00	82.00	102.24	166.92	119.00	226.00	640.04	250.00	703.00	399.00	86.30	38.27
(secure) ads-c	-	-	0.00	92.00	0.00	124.13	0.00	166.00	0.00	190.00	0.00	194.00	0.00	21.54
(secure) cpdlc	-	-	59.00	82.00	70.83	120.67	84.25	151.40	85.65	165.48	86.00	169.00	10.54	21.12
key_exchange	-	-	25.00	82.00	431.57	207.19	677.00	250.00	690.80	256.36	703.00	258.00	252.54	41.69
secure GBAS	-	-	45.00	113.00	96.31	113.00	118.00	113.00	616.91	113.00	703.00	113.00	66.61	0.00
secure audio	-	-	0.00	146.00	0.00	181.23	0.00	207.00	0.00	210.00	0.00	399.00	0.00	20.28

(b) Output of LDACS measurement report

Figure 4.14 Two important capabilities of FACTS (a) visual output of air traffic movement and (b) measurement reports on the performance of tested data links [71]

research aircraft used in the flight trials, a Falcon 20-E5, together with the position of the L-band antenna.

In Figure 4.16(a) and (b), we show the LDACS GS demonstration setup and, in Figure 4.16(c), the AS LDACS demonstration equivalent.

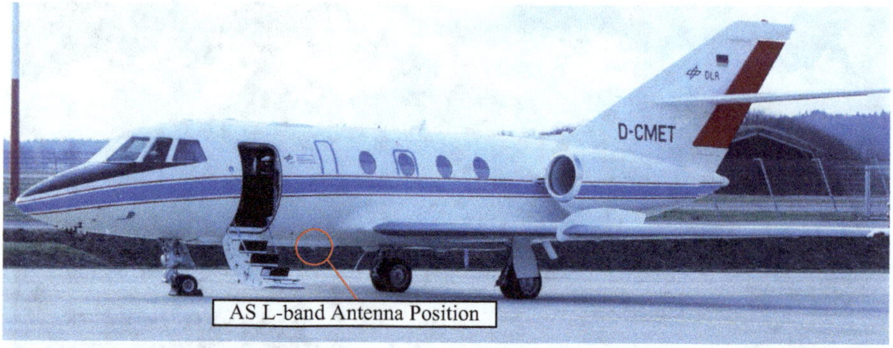

Figure 4.15 DLR's research aircraft Falcon 20-E5 (D-CMET) [36]

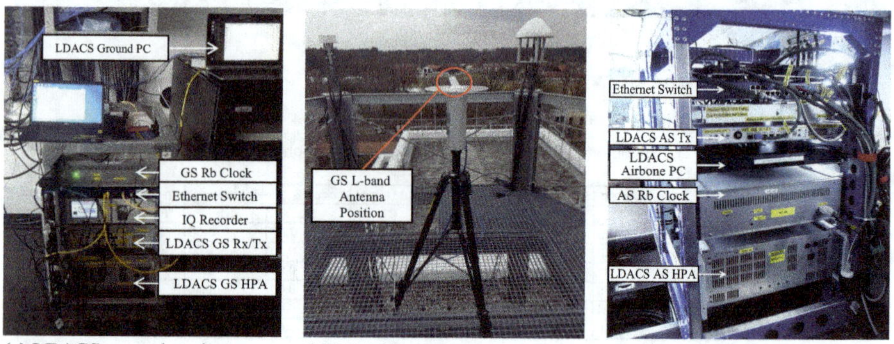

(a) LDACS ground station (b) LDACS antenna (c) Airborne equipment

Figure 4.16 Overview of airborne and ground LDACS equipment [36]

With pre-installed certificates at the AS and GS, we demonstrated a key encapsulation procedure based on the post-quantum robust public-key cryptosystem [72]. Using this negotiated session key as a result, we encrypted any actual data, such as text messages, CPDLC messages, ADS-C messages for AS positioning with the aforementioned AES-256-GCM symmetric cryptosystem. With data gained during the actual flight, we can confirm that the chosen MAKE procedure itself remained below the required $RCTP_{CSP} = 10s$ threshold for RCP 130/A1 message types [70]. Overall, we performed 19 key exchanges and observed a mean key exchange time of 508.15 ms and 95 percentile of 513.13 ms. This matches the values presented in Figure 4.13 with just one AS in an LDACS cell very well and is a strong indicator for the accuracy of the theoretical model and the FACTS simulation.

In Figure 4.17, we see the console output of a successful key exchange and with secured communications channel opened, aboard the Falcon 20-E5 during MICONAV.

Figure 4.17 Console output during after successful key exchange. All further communication afterwards were secured using AES-256-GCM

4.6 Conclusions

This work presents a summary of the design of the LDACS cybersecurity concept. We introduce the FCI and the current state of cybersecurity in digital aeronautical communications, which is that especially on legacy data links (i.e. ADS-B, VDLm2), there is no information security based on the definitions from the Internet security glossary RFC 4949. We continue with a presentation of relevant technical details about LDACS and then with threat and risk analyses, defining security requirements, objectives and functions. We start the section on the LDACS cybersecurity architecture with different trust approaches, PKI- and PUF CRP-based approaches, then continue with two MAKE procedures based on these trust solutions. We propose the use of AES-GCM for user data protection and state out reasons why integrity is more important than confidentiality. On the security for control plane data, we are investigating the use of group-key procedure, however this is still a work in progress. Finally, we finish with a complete integration of the security additions within the LDACS protocol stack. In the evaluation, we show that both MAKE procedures finish far below the required latency threshold and only put up to 1 kB of additional security data on the link. Also, using analytical evaluation, a software event-based simulation, and flight trials, we present a first proof-of-concept of LDACS cybersecurity.

Future work aims to solve the open problem of control channel security of LDACS by investigating different group-key procedures and different approaches on guaranteeing integrity of control data. Also, proving the semantic correctness and soundness of LDACS protocol security with model checkers such as Tamarin, Scyther, or ProVerif is an important step in the validation of LDACS cybersecurity features. Finally, further flight trials are planned for 2022, where the entire cybersecurity concept can be validated in flight trials on more advanced LDACS prototype hardware.

All candidates within the FCI require strong cybersecurity features. We agree with this requirement as we strongly believe that cybersecurity is the enabler in the future automation of civil air traffic. As LDACS is the worldwide first truly integrated CNS system, and thus an important part of the digitization process, this work on cybersecurity for LDACS is an essential pillar for the final standardization and deployment of LDACS.

Abbreviations

A2A	Air-to-Air
AC-R	Access-Router
ADS-B	Automatic Dependent Surveillance Broadcast
ADS-C	Automatic Dependent Surveillance-Contract
AeroMACS	Aeronautical Mobile Airport Communications System
ANI	Airborne Network Interface
ANSP	Aeronautical Network Service Provider
AOC	Aeronautical Operational Control
APT	Airport
AS	Aircraft Station
ATC	Air Traffic Communications
ATM	Air Traffic Management
ATN	Aeronautical Telecommunications Network
ATS	Air Traffic Services
AVI	Airborne Voice Interface
BC	Broadcast Channel
BER	Bit Error Rate
CA	Certificate Authority
CC	Common Control
CCCH	Common Control Channel
CNS	Communication, Navigation and Surveillance
CoS	Classes of Service
CPDLC	Controller–Pilot Data Link Communications
CRP	Challenge-Response Pair
CSMA	Carrier Sense Multiple Access
D8PSK	Differential 8 Phase Shift Keying
DC	Dedicated Control
DCCH	Dedicated Control Channel
DCH	Data Channel
DHKE	Diffie-Hellman Key Exchange
DLS	Data Link Service
DME	Distance Measuring Equipment
ECDH	Elliptic Curve Diffie-Hellmann

FACTS	Framework for Aeronautical Communications and Traffic Simulations 2
FCI	Future Communications Infrastructure
FDD	Frequency Division Duplexing
FL	Flight Level FL Foward Link
GBAS	Ground Based Augmentation System
GNSS	Global Navigation Satellite System
GS	Ground Station
GSC	Ground Station Controller
HKDF	HMAC Key Derivation Function
HMAC	Keyed-Hash Message Authentication Code
ICAO	International Civil Aviation Organization
IKEv2	Internet Key Exchange
KDF	Key Derivation Function
LDACS	L-band Digital Aeronautical Communications System
LLC	Logical Link Control
LME	LDACS Management Entity
MAC	Medium Access Layer
MAKE	Mutual Authentication and Key Exchange
MF	Multi Frame
MICONAV	Migration towards COm/NAV capabilities of LDACS
OFDM	Orthogonal Frequency-Division Multiplexing
OFDMA	Orthogonal Frequency-Division Multiple Access
ORP	Oceanic Remote Polar
PDU	Packet Data Unit
PHY	Physical Layer
PHY-SDU	Physical Layer-Service Data Unit
PKI	Public Key Infrastructure
PMAKE	Physical Unclonable Function based Mutual Authentication and Key Exchange
PUF	Physical Unclonable Function
QoS	Quality of Service
RA	Random Access
RACH	Random Access Control Channel
RL	Reverse Link
SAC	System Area Code
SARPS	Standards and Recommended Practices
SESAR	Single European Sky ATM Research
SF	Super Frame
SIB	System Identification Broadcast

SIDH	Supersingular Isogeny Diffie–Hellman
SNP	Sub-Network Protocol
STS	Station-to-Station
TDMA	Time Division Multiple Access
TMA	Terminal Maneuvering Area
VDLm2	VHF Data Link mode 2
VHF	Very High Frequency
VI	Voice Interface
VU	Voice Unit

References

[1] IATA. Economic Performance of the Airline Industry. International Air Transport Association (IATA); 2020. [Online]. Available: https://www.iata.org/en/iata-repository/publications/economic-reports/airline-industry-economic-performance—november-2020—report/.

[2] ACI. Economic Performance of the Airline Industry. Airports Council International (ACI); 2019. [Online]. Available: https://store.aci.aero/product/annual-world-airport-traffic-report-2019/.

[3] Iacus SM, Natale F, Santamaria C, *et al.* Estimating and projecting air passenger traffic during the COVID-19 coronavirus outbreak and its socio-economic impact. Safety Science. 2020;129:104791. Available from: http://www.sciencedirect.com/science/article/pii/S0925753520301880.

[4] Slim M, Mahmoud B, Pirovano A, *et al.* Aeronautical communication transition from analog to digital data: A network security survey. Computer Science Review. 2014;11-12:1–29.

[5] Schnell M, Epple U, Shutin D, *et al.* LDACS: Future aeronautical communications for air-traffic management. IEEE Communication Magazine. 2014;52(5):104–110.

[6] Schnell M. Update on LDACS – The FCI Terrestrial Data Link. In: 19th Integrated Communications, Navigation and Surveillance Conference (ICNS). New York, NY, USA: IEEE; 2019. p. 1–10.

[7] Kamali B. AeroMACS: An IEEE 802.16 Standard-based Technology for the Next Generation of Air Transportation Systems. John Wiley & Sons; 2018.

[8] Gräupl T, Rihacek C, Haindl B. LDACS A/G Specification. Oberpfaffenhofen, Germany: German Aerospace Center (DLR); 2019.

[9] Strohmeier M, Schäfer M, Pinheiro R, *et al.* On perception and reality in wireless air traffic communication security. IEEE Transactions on Intelligent Transportation Systems. 2016;18(6):1338–1357.

[10] Bernsmed K, Fr C, Meland PH, *et al.* Security requirements for SATCOM datalink systems for future air traffic management. In: 2017 IEEE/AIAA 36th Digital Avionics Systems Conference (DASC). IEEE; 2017. p. 1–10.

[11] Mäurer N, Schmitt C. Towards Successful Realization of the LDACS Cybersecurity Architecture: An Updated Datalink Security Threat- and Risk Analysis. In: 19th Integrated Communications, Navigation and Surveillance Conference (ICNS). New York, NY, USA: IEEE; 2019. p. 1A2/1–1A2–13.

[12] Hall A, Wingfield J, De Moura G, *et al.* Advancing Cyber Resilience in Aviation: An Industry Analysis. World Economic Forum; 2020.

[13] Costin A, Francillon A; EURECOM. Ghost in the Air(Traffic): On Insecurity of ADS-B protocol and Practical Attacks onADS-B Devices. Black Hat USA. 2012 August;p. 1–10.

[14] Strohmeier M, Lenders V, Martinovic I. On the security of the automatic dependent surveillance-broadcast protocol. IEEE Communications Surveys & Tutorials. 2014;17(2):1066–1087.

[15] Stelkens-Kobsch TH, Hasselberg A, Mühlhausen T, *et al.* Towards a more secure ATC voice communications system. In: 2015 IEEE/AIAA 34th Digital Avionics Systems Conference (DASC). IEEE; 2015. p. 4C1–1.

[16] Harison E, Zaidenberg N. Survey of Cyber Threats in Air Traffic Control and Aircraft Communications Systems. In: Cyber Security: Power and Technology. Springer; 2018. p. 199–217.

[17] Bilzhause A, Belgacem B, Mostafa M, *et al.* Datalink Security in the L-band Digital Aeronautical Communications System (LDACS) for Air Traffic Management. Aerospace and Electronic Systems Magazine. 2017;32(11): 22–33.

[18] ICAO. LDACS White Paper–A Roll-out Scenario. International Civil Aviation Organization (ICAO); 2019.

[19] ICAO. Doc 9776 – Manual on VHF Digital Link (VDL) Mode 2. International Civil Aviation Organization (ICAO); 2015. [Online]. Available: http://www.icscc.org.cn/upload/file/20190102/Doc.9776-EN Manual on VHF Digital Link (VDL) Mode 2.pdf.

[20] RTCA. DO-225, VHF Air-Ground Communications System Improvements Alternatives Study and Selection of Proposals for Future Action. Radio Technical Commission for Aeronautics (RTCA); 1994.

[21] RTCA. DO-281C – Minimum Operational Performance Standards (MOPS) for Aircraft VDL Mode 2 Physical Link and Network Layer. Radio Technical Commission for Aeronautics (RTCA); 2018.

[22] ICAO. Annex 10 – Aeronautical Telecommunications – Volume III – Communication Systems. International Civil Aviation Organization (ICAO); 2007. [Online]. Available: https://store.icao.int/en/annex-10-aeronautical-telecommunications-volume-iii-communication-systems.

[23] Shirey R. Internet Security Glossary, Version 2. RFC Editor; 2007. 4949. Available from: https://www.rfc-editor.org/rfc/rfc4949.txt.

[24] Kunkel R. Village DA, editor. Air Traffic Control: Insecurity and ADS-B [YouTube, Date video uploaded: 16.01.2011, [Video file]]. DEFCON 28; 2011. [Online]. Available: https://www.youtube.com/watch?v= aU8Npy Yf9wY [Accessed: Oct. 10, 2021].

[25] McCallie D, Butts J, Mills R. Security analysis of the ADS-B implementation in the next generation air transportation system. International Journal of Critical Infrastructure Protection. 2011;4(2):78–87.

[26] Strohmeier M, Schäfer M, Lenders V, *et al.* Realities and challenges of nextgen air traffic management: the case of ADS-B. IEEE Communications Magazine. 2014;52(5):111–118.

[27] Wu Z, Shang T, Guo A. Security issues in automatic dependent surveillance-broadcast (ADS-B): A survey. IEEE Access. 2020;8:122147–122167.

[28] (ICAO) ICAO. Finalization of LDACS Draft SARPs – Working Paper WP05 including Appendix. Montreal, Canada: ICAO; 2018.

[29] Gräupl T, Mäurer N, Schmitt C. FACTS2: Framework for Aeronautical Communications and Traffic Simulations 2. In: Proceedings of the 16th ACM International Symposium on Performance Evaluation of Wireless Ad Hoc, Sensor & Ubiquitous Networks; 2019. p. 63–66.

[30] Brandes S, Schnell M, Rokitansky CH, *et al.* B-VHF – Selected Simulation Results and Final Assessment. In: 25th Digital Avionics Systems Conference (DASC). New York, NY, USA: IEEE; 2006. p. 3A4/1–3A4/12.

[31] Rokitansky CH, Ehammer M, Gräupl T, *et al.* B-AMC A System for Future Broadband Aeronautical Multi-Carrier Communications in the L-Band. In: 36th Digital Avionics Systems Conference (DASC). New York, NY, USA: IEEE; 2007. p. 4D2/1–4D2/13.

[32] Schnell M, Brandes S, Gligorevic S, *et al.* B-AMC – Broadband Aeronautical Multi-carrier Communications. In: 8th Integrated Communications, Navigation and Surveillance Conference (ICNS). New York, NY, USA: IEEE; 2008. p. 4D2/1–4D2/13.

[33] Haindl B, Rihacek C, Sajatovic M, *et al.* Improvement of L-DACS1 Design by Combining B-AMC with P34 and WiMAX Technologies. In: 9th Integrated Communications, Navigation and Surveillance Conference (ICNS). New York, NY, USA: IEEE; 2009. p. 1–8.

[34] Ehammer M, Gräupl T. AeroMACS – An Airport Communications System. In: 30th Digital Avionics Systems Conference (DASC). New York, NY, USA: IEEE; 2011. p. 4C1/1–4C1/16.

[35] Felux M, Gräupl T, Mäurer N, *et al.* Transmitting GBAS Messages via LDACS. In: 2018 IEEE/AIAA 37th Digital Avionics Systems Conference (DASC). IEEE; 2018. p. 1–7.

[36] Mäurer N, Gräupl T, Bellido-Manganell MA, *et al.* Flight Trial Demonstration of Secure GBAS via the L-band Digital Aeronautical Communication System (LDACS). IEEE Aerospace and Electronic Systems Magazine. 2021; 1–19.

[37] Mäurer N, Bilzhause A. Paving the Way for an IT Security Architecture for LDACS: A Datalink Security Threat and Risk Analysis. In: 18th Integrated Communications, Navigation and Surveillance Conference (ICNS). New York, NY, USA: IEEE; 2018. p. 1A2/1–1A2–11.

[38] Zelkin N, Henriksen S. L-Band Digital Aeronautical Communications System Engineering–Initial Safety and Security Risk Assessment and Mitigation. Cleveland, OH, USA: National Aeronautics and Space Administration, TT

Corporation Advanced Engineering & Sciences Division; 2011 (accessed July 5, 2019).

[39] Mäurer N, Gräupl T, Schmitt C. L-band Digital Aeronautical Communications System (LDACS). Internet Engineering Task Force; 2020. draft-ietf-raw-ldacs-05. Work in Progress. Available from: https://datatracker.ietf.org/doc/html/draft-ietf-raw-ldacs-05.

[40] International Civil Aviation Organization (ICAO). Finalization of LDACS Draft SARPs – Working Paper WP05 including Appendix. Montreal, Canada: International Civil Aviation Organization (ICAO); 2018.

[41] Mäurer, N and Bilzhause, A . A Cybersecurity Architecture for the L-band Digital Aeronautical Communications System (LDACS). In: 37th Digital Avionics Systems Conference (DASC). New York, NY, USA: IEEE; 2018. p. 1–10.

[42] Cooper D, Santesson S, Farrell S, *et al.* Internet X. 509 Public Key Infrastructure Certificate and Certificate Revocation List (CRL) Profile. RFC Editor; 2008. 5280. Available from: https://www.rfc-editor.org/rfc/rfc5280.txt.

[43] R S. Cybersecurity for Global Aviation – A Trust Framework Enabling Global Secure Aviation Interoperability. In: ICNS Conference 2018 Plenary Panel I, Integrated Communications Navigation and Surveillance Conference (ICNS). New York, NY, USA: IEEE; 2016. p. 1–19.

[44] Crowe B. Proposed AeroMACS PKI Specification is a Model for Global and National Aeronautical PKI Deployments. In: WiMAX Forum at 16th Integrated Communications, Navigation and Surveillance Conference (ICNS). New York, NY, USA: IEEE; 2016. p. 1–19.

[45] Boyd C, Mathuria A, Stebila D. Protocols for Authentication and Key Establishment. Heidelberg, Germany: Springer; 2020.

[46] Mäurer, N, Gräupl, T and Schmitt, C . Comparing Different Diffie-Hellman Key Exchange Flavors for LDACS. In: 39th Digital Avionics Systems Conference (DASC). New York, NY, USA: IEEE; 2020. p. 1–10.

[47] Diffie W, Hellman M. New directions in cryptography. IEEE Transactions on Information Theory. 1976;22(6):644–654.

[48] Blake-Wilson S, Menezes A. Authenticated Diffie-Hellman Key Agreement Protocols. In: International Workshop on Selected Areas in Cryptography. Heidelberg, Germany: Springer; 1998. p. 339–361.

[49] Koblitz N. Elliptic curve cryptosystems. Mathematics of Computation. 1987;48(177):203–209.

[50] Baumslag G, Fine B, Kreuzer M, *et al.* A Course in Mathematical Cryptography. Berlin, Germany: Walter de Gruyter GmbH & Co KG; 2015.

[51] Rostovtsev A, Stolbunov A. Public-Key Cryptosystem Based on Isogenies. IACR Cryptology ePrint Archive. 2006;p. 1–19.

[52] Jao D, De Feo L. Towards Quantum-Resistant Cryptosystems from Supersingular Elliptic Curve Isogenies. In: 4th International Workshop on Post-Quantum Cryptography. Heidelberg, Germany: Springer; 2011. p. 19–34.

[53] Jao D, editor. Supersingular Isogeny Key Encapsulation. SIKE – Supersingular Isogeny Key Encapsulation; 2020 (accessed Oct. 10, 2021). https://sike.org/files/SIDH-spec.pdf.

[54] Josefsson S, Liusvaara I. Edwards-Curve Digital Signature Algorithm (EdDSA). RFC Editor; 2017. 8032. Available from: https://www.rfc-editor.org/rfc/rfc8032.txt.

[55] Ding J, Schmidt D. Rainbow, a New Multivariable Polynomial Signature Scheme. In: 3rd International Conference on Applied Cryptography and Network Security. Heidelberg, Germany: Springer; 2005. p. 164–175.

[56] Schmidt B, Meier S, Cremers CJF, *et al.* Automated Analysis of Diffie–Hellman Protocols and Advanced Security Properties. In: Chong S, editor. 25th IEEE Computer Security Foundations Symposium, CSF 2012, Cambridge, MA, USA, June 25-27, 2012. IEEE Computer Society; 2012. p. 78–94. Available from: https://doi.org/10.1109/CSF.2012.25.

[57] Mäurer N, Gräupl T, Schmitt C, *et al.* PMAKE: Physical Unclonable Function-based Mutual Authentication Key Exchange Scheme forDigital Aeronautical Communications. In: 2021 IFIP/IEEE Symposium on Integrated Network and Service Management (IM). IEEE; 2021. p. TODO.

[58] Krawczyk H, Eronen P. HMAC-based Extract-and-Expand Key Derivation Function (HKDF). RFC Editor; 2010. 5869. Available from: https://www.rfc-editor.org/rfc/rfc5869.txt.

[59] Salowey J, Choudhury A, McGrew D. AES Galois Counter Mode (GCM) cipher suites for TLS. RFC Editor; 2008. 5288. Available from: https://www.rfc-editor.org/rfc/rfc5288.txt.

[60] Weis B, Smyslov V. Group Key Management using IKEv2 – draft-yeung-g-ikev2-16. RFC Editor; 2020. Available from: https://tools.ietf.org/html/draft-yeung-g-ikev2-16.

[61] Zheng X, Huang CT, Matthews M. Chinese Remainder Theorem Based Group Key Management. In: Proceedings of the 45th Annual Southeast Regional Conference; 2007. p. 266–271.

[62] Guggemos T, Streit K, Knüpfer M, *et al.* No Cookies, just CAKE: CRT based Key Hierarchy for Efficient Key Management in Dynamic Groups. In: International Conference for Internet Technology and Secured Transactions; 2018. p. 25–32.

[63] Sakamoto N. An Efficient Structure for LKH Key Tree on Secure Multicast Communications. In: 15th IEEE/ACIS International Conference on Software Engineering, Artificial Intelligence, Networking and Parallel/Distributed Computing (SNPD). IEEE; 2014. p. 1–7.

[64] Bernstein DJ, Lange T. Post-quantum cryptography. Nature. 2017;549 (7671):188–194.

[65] Dworkin MJ. SP 800-38D–Recommendation for Block Cipher Modes of Operation: Galois/Counter Mode (GCM) and GMAC. National Institute of Standards & Technology; 2007.

[66] Krawczyk H, Bellare M, Canetti R. HMAC: Keyed-Hashing for Message Authentication. RFC Editor; 1997. 2104. Available from: https://www.rfc-editor.org/rfc/rfc2104.txt.

[67] Pritzker P, Gallagher PD. SHA-3 Standard: Permutation-based Hash and Extendable-Output Functions. Information Tech Laboratory National Institute of Standards and Technology; 2014. p. 1–35

[68] Mäurer N, Gräupl T, Schmitt C. Evaluation of the LDACS Cybersecurity Implementation. In: 38th Digital Avionics Systems Conference (DASC). New York, NY, USA: IEEE; 2019. p. 1–10.

[69] Gräupl T, Mayr M. Method to Emulate the L-band Digital Aeronautical Communication System for SESAR Evaluation and Verification. In: 34th Digital Avionics Systems Conference (DASC). New York, NY, USA: IEEE; 2015. p. 1–18.

[70] RTCA/EUROCAE. Safety and Performance Requirements Standard for Baseline 2 ATS Data Communications (Baseline 2 SPR Standard). Washington, DC/Malakoff, France: RTCA/EUROCAE; 2016.

[71] Gräupl T, Mäurer N, Schmitt C. FACTS2: Framework for Aeronautical Communications and Traffic Simulations 2. In: Proceedings of the 16th ACM International Symposium on Performance Evaluation of Wireless Ad Hoc, Sensor & Ubiquitous Networks; 2019. p. 63–66.

[72] McEliece RJ. A public-key cryptosystem based on algebraic. Coding Thv. 1978;4244:114–116.

Chapter 5

Using terrestrial ranging to monitor GNSS for targeted RF interference

Okuary Osechas[1], Omar García Crespillo[1], and Michael Meurer[1]

Global navigation satellite systems (GNSS) have become the primary means of navigation for civil aviation. The development of required navigation performance (RNP) has set new expectations for the availability of GNSS-based integrity services. However, there is a fundamental frontier in the detectability of large-scale faults in methods based on receiver autonomous integrity monitoring (RAIM) or advanced RAIM (ARAIM). RAIM and ARAIM derive integrity guarantees from the consistency of GNSS-based measurements, so that faults with consistent effects across various satellites, such as might be caused by a multi-constellation fault or intentional spoofing, are at risk of being undetectable with RAIM-based monitors. In this chapter, we propose a way of reducing the risk by complementing GNSS with terrestrial measurements.

With alternative position, navigation and timing (APNT) services being developed towards providing integrity [1,2], the prospect of combining APNT measurements with GNSS-based measurements adds a new capability to integrity monitoring. The sensitivity of the detector is driven by the performance of the less accurate of the two systems. Typically, APNT systems are not able to deliver the tight error bounds that aviation users have become accustomed to with GNSS-based services. The upshot being that they are more robust to interference and are, by nature, guaranteed not to introduce hazardously misleading information (HMI) by way of ephemeris faults or Earth-orientation parameters.

Adequate monitoring of the consistency between APNT and GNSS measurements requires a balance between the virtues and drawbacks of each. In this chapter, we make a specific proposal of how that balance can be found, motivated by the observation that we can provide benefit to an existing service (GNSS-based RNP) by leveraging a service that will likely exist in the future (APNT-based RNP).

[1]Department of Navigation, Institute of Communication and Navigation, German Aerospace Center, Oberpfaffenhofen, Germany

5.1 Background

Current efforts of developing an APNT capability throughout the European airspace aim at providing RNP 1 [2]. Assuming that this capability will be provided, its use as a monitor for severe GNSS faults comes as a side benefit at no additional cost. Although providing RFI robustness to en-route services (as RNP can be) may seem counter-intuitive to some readers, this focus is partly motivated by the findings of the Eurocontrol Voluntary ATM Incident Reporting (EVAIR) study. According to the study, 89% of reported "GPS-out" events occur en-route [3], which is exactly where APNT-based detection is most appropriate because of its scope as en-route navigation.

With APNT-based RNP, geographic availability has often been a limiting factor. Following the recent proposal for modular APNT, introduced in [1,4], the possibility of widespread APNT-based RNP appears a realistic option for the future. The modular approach derives its benefits from the flexibility of using diverse measurement systems to increase redundancy. The authors in [1,4] detail how the method can achieve RNP 0.3 coverage for 98% of the German airspace, at 10,000 ft altitude and above.

As such, the APNT position fix provides an opportunity to monitor a GNSS position fix against a more reliable, if less accurate, source of information, which does not share fault modes with GNSS. Figures 5.1 and 5.2 gives an overview of how the method is designed to work. We use the consistency between APNT and GNSS measurements, at the level of both the position solution (in Figure 5.1) and on individual GNSS-based pseudorange measurements (in Figure 5.2), as the base for a

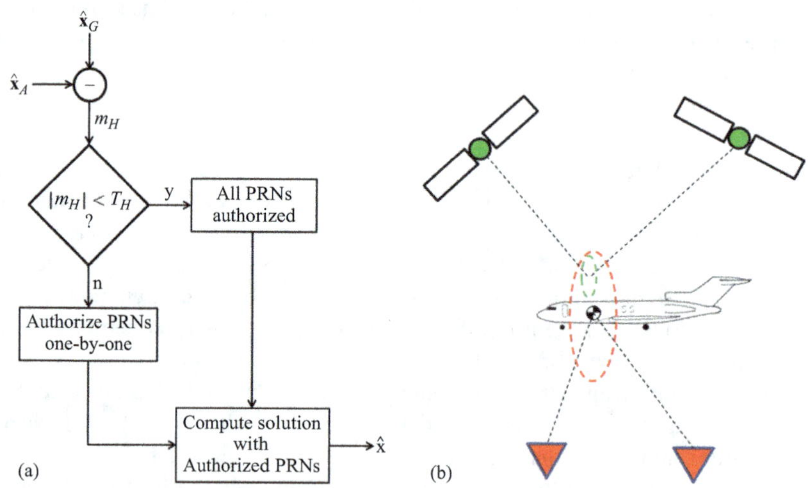

Figure 5.1 (a) Processing of GNSS measurements to assess the consistency of two position solutions. (b) The GNSS-based position solution will typically have a smaller uncertainty (green ellipse), whereas the APNT position will have a greater uncertainty (red ellipse). The challenge of this method lies in reliably detecting anomalous inconsistencies between the two

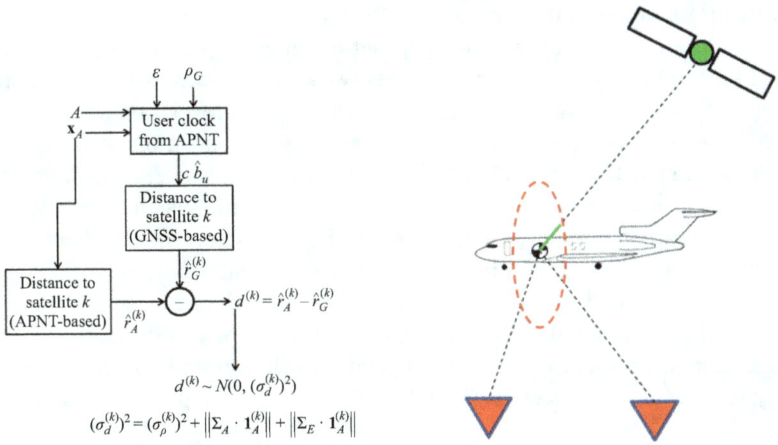

Figure 5.2 Processing of GNSS measurements to assess integrity with respect to severe faults that enables consistency of GNSS measurements with APNT-based position estimates

quantifiable integrity check against faults covering severe ranging errors and one or more GNSS constellation-wide faults.

Note that the screening method makes no distinction between the different fault models that can affect pseudorange measurements. Although the monitor has been designed with intentional RFI/spoofing in mind, it will in fact respond to many other faults. However, in those cases, more sensitive monitors may already exist, which are specifically designed to detect known GNSS fault modes, examples include signal deformation [5] or satellite clock faults [6]. Another point to note is that the higher accuracy of GNSS positioning, compared to APNT systems, is not because of any fundamental limitation; APNT systems based on future radionavigation signals are showing capabilities to reach even higher accuracy levels [7]. The notion of an APNT system that supports precision approach and landing would therefore significantly impact the resilience of PBN to intentional satellite faults.

5.1.1 Prior art

Leveraging consistency checks between GNSS-based position estimates and other sensing principles has been proposed before; for example, by using inertial reference systems [8], in aviation applications, or by comparing GNSS with terrestrial ranging systems, for maritime applications [9].

Basic checks for GNSS spoofing exist, for example, in the form of consistency checks against inertial sensors. Similar checks can be performed against barometric sensors, DME-DME-Inertial systems, or by assessing the consistency of the observed Doppler measurements from all satellites with almanac data. The method described in this chapter is based on a modular approach to APNT, as described in [1], which provides better availability and robustness, compared to the other methods.

In its original conception, modular APNT integrity does not explicitly accommodate satellite-based measurements [1], but it can easily be adapted to do so. The proposal for modular integrity contemplates two different ways of achieving modularity: at a system level and at measurement level. System-level modularity combines position estimates from different systems into a single position solution, weighing each solution with its model covariance. Measurement-level modularity, on the other hand, considers all available measurements from all systems simultaneously, adequately weighted or coupled with a model of the noise on each measurement.

As neither system-level nor measurement-level modularity reflects the situation, where one system is robust to faults but less accurate than another, we propose a hybrid-modular approach that evaluates the consistency of GNSS measurements (in the range domain) with an APNT-based position solution and covariance estimate. An early version of the concept was published in conference proceedings [10].

5.1.2 Assumptions on the APNT system

The performance analysis presented here is based on two fundamental assumptions. We justify the assumptions in this section, reasoning based on feasibility and observations. For the purposes of this chapter, we make the simplifying assumptions that the APNT system:

1. Is significantly less likely to succumb to a spoofing attack than GNSS.
2. Provides the user with clock-bias estimates with an accuracy that compares to the spatial accuracy.

5.1.2.1 Robustness

In protecting users of GNSS-based services from spoofing-induced errors, we observe three reasons that support the assumption that the APNT-based position solution is far less likely to be intentionally modified by spoofing than any GNSS-based solution.

For one, the signal power, both at transmission and reception, of terrestrial sources is higher than that of space-based sources. This fact makes the unauthorized transmissions easier to detect and the hardware complexity of the spoofing attack greater. For another, the hybrid nature of the modular architecture envisioned for future APNT systems [1] further increases the complexity of a spoofing attack on APNT, as it would need to create synchronized spoofing for different radionavigation signals in a coordinated way.

Finally, we observe that no mass-market incentive exists, to the best of our knowledge, to jam or spoof APNT systems, which makes the illicit development of such technologies less likely. This stands in contrast with the reported development of GNSS-deceiving technology used for privacy applications [11].

5.1.2.2 Timing

The method presented in this chapter is based on comparing range estimates computed from different sources. One of the range estimates stems from correcting the pseudorange measurement of a satellite with the estimate of the user clock bias, as derived from APNT. Because the integrity argument of this chapter is based on the independence between variables, the lack of an independent timing source invalidates the argument. If the user clock bias is not available from APNT, it can be calculated from a GNSS-based estimate, but the resulting estimate will become dependent on all satellite measurements considered in the computation of the clock bias.

As presented in [1,4], APNT services will deliver on their promise of providing accurate, reliable timing. We use this assumption to simplify the consistency check between APNT and GNSS measurements. Estimation of the GNSS user clock bias would otherwise reduce the performance of the detection method, but would not render it unfeasible. Investigation of this effect remains a topic for future work.

5.2 Fault mechanism

The method we present was designed to detect multi-satellite faults, such as those realizable from intentional interference, but it is also sensitive to constellation-wide faults. Given the reliance on APNT technologies as a matter of fault detection, that very capability also provides availability of service during GNSS outages, for example, because of jamming if a GNSS fault can be detected but not excluded.

For this reason, we focus on detecting targeted interference below the noise floor, regardless of whether the interfering signal is intended to corrupt GNSS service (as in a spoofing scenario), or whether the intent is different (as might be with GNSS repeaters). This type of error manifests itself as a ranging error on an arbitrary number of navigation satellites, from one to all in view. In principle, the monitor is also sensitive to non-intentional multi-satellite, constellation-wide and multi-constellation faults, as typically considered for ARAIM.

Having established that the APNT position solution is significantly more robust to spoofing and that no equivalent to a constellation fault occurs with terrestrial systems, we propose a method that uses the APNT position fix ($\mathbf{x}_A$) as a reference, with the corresponding model for the covariance matrix Σ_A. This reference can be used to verify whether GNSS-based pseudorange measurements $\rho_G^{(k)}$ are feasible. In contrast with previously established modularity methods, such as measurement-based or system-based modularity [1], we take a hybrid approach. Each GNSS pseudorange measurement is evaluated against the APNT position solution, making GNSS modular at measurement level and APNT modular at the system level.

Throughout the chapter we use a notation similar to that in [12], where the true user position $\mathbf{x}_u$ estimated from GNSS is $\hat{\mathbf{x}}_G$, the satellite geometry matrix is $\mathbf{G}$ and the vector of GNSS observables $\boldsymbol{\rho}_G$. The APNT-based position fix is $\hat{\mathbf{x}}_A$ with covariance matrix (Σ_A) and we assume that the APNT system provides a conservative error model, such that the estimate is unbiased and has a Gaussian distribution:

$$\hat{\mathbf{x}}_A \sim \mathcal{N}(\mathbf{x}_u, \Sigma_A). \tag{5.1}$$

We denote the pseudorange error because of interference on the kth satellite with $E^{(k)}$. Keeping the notation consistent with [12], we obtain the error model for pseudorange measurements:

$$\rho^{(k)} = \left(\mathbf{x}^{(k)} - \mathbf{x}_u\right) \cdot \mathbf{1}^{(k)} + b^{(k)} - b_u + T^{(k)} + I^{(k)} + E^{(k)} + \varepsilon^{(k)}. \tag{5.2}$$

Grouping all $E^{(k)}$ into a vector $\mathbf{E}$ the position error because of interference $\Delta\mathbf{x}_E$ can then be modeled as:

$$\Delta\mathbf{x}_E = \left(\mathbf{G}^T\mathbf{W}\mathbf{G}\right)^{-1} \mathbf{G}^T\mathbf{W}\,\mathbf{E}, \tag{5.3}$$

where $\mathbf{W}$ is the inverse of the measurement covariance matrix.

The method described in this chapter aims to protect users of RNP services when the magnitude of $\Delta\mathbf{x}_E$ becomes hazardous to their operation. One detail that is not yet clear is what the most likely set of $E^{(k)}$ is. In a later part of this chapter, we discuss the implications of two scenarios: the injection of errors into all visible GNSS signals and the injection of errors into the signals of only one constellation. As stated in [3], the feasibility of an intentional RFI incident does not necessarily reflect in a likelihood of it happening. Identifying further scenarios and rating their likelihood of occurrence is an interesting topic for future work, but at this stage we consider it irrelevant whether the injected range errors target specific satellites or constellations, as the method is able to detect individual faults as well as arbitrary combinations thereof.

5.3 Fault monitoring

This section describes two complementary monitors that are designed to detect inconsistencies between the GNSS and APNT systems. First, we discuss a position-domain monitor that relies on comparing the position solutions of both systems. The second is a range-domain monitor that compares two different estimates for the distance between the user and each satellite.

Operationally, the position-domain monitor provides a simple one-shot test of the consistency between the two systems. In terms of performance assessment, the position-domain monitor also has a benefit, as it provides a single metric by which the system performance can be evaluated. The range-domain monitor, on the other hand, provides a more fine-grained tool in detecting and excluding faulted measurements. The operational benefit from this exclusion capability lies in the ability to detect exactly which satellites are affected by a fault, even in the case of multi-satellite, constellation-wide or even multi-constellation faults.

The position-domain consistency check, proposed in [1], requires checking the difference between two position solutions to be below a certain threshold, in order for the two positions to be declared as consistent. Applying this reasoning, we define the monitor $\mathbf{m}$ as the difference between the position as estimated by GNSS $\hat{\mathbf{x}}_G$ and the position as estimated by APNT $\hat{\mathbf{x}}_A$:

$$\mathbf{m} = \hat{\mathbf{x}}_G - \hat{\mathbf{x}}_A, \tag{5.4}$$

where $\mathbf{m}$ is a vector quantity. The horizontal component m_H can be studied independently of the vertical component [13]. Thus, under the fault-free hypothesis and assuming that $\mathbf{m}$ has a Gaussian distribution, we have that its horizontal component m_H is laso Gaussian:

$$m_H \sim \mathcal{N}\left(0, \sigma_{G,H}^2 + \sigma_{A,H}^2\right), \tag{5.5}$$

where $\sigma_{A,H}^2$ is the horizontal part of the APNT position covariance along the principal axis $\sigma_{A,H}^2 = d_{major}\left(\Sigma_A(E,N)\right)$ and $\sigma_{G,H}^2$ is the GNSS counterpart.

Having established the random distribution underlying m_H, we use established methods to compute a threshold value for a given false alarm rate (p_{FA}) and probability of mis-detection p_{MD}. The detection threshold is given implicitly by the cumulative probability ($\mathbf{\Phi}$) of the fault-free distribution at the threshold:

$$p_{FA} = \Phi\left(-T_H; 0, \sigma_{G,H}^2 + \sigma_{A,H}^2\right) + \left[1 - \Phi\left(T_H; 0, \sigma_{G,H}^2 + \sigma_{A,H}^2\right)\right]. \tag{5.6}$$

The expression for the detection threshold becomes:

$$T_H = \mathbf{\Phi}^{-1}\left(1 - \frac{p_{FA}}{2}; 0, \sigma_{G,H}^2 + \sigma_{A,H}^2\right) \tag{5.7}$$

Now as long as $|m_H| < T_H$, the GNSS position solution can be assumed to be free of interference errors. The probability of having an undetected position error because of interference in the validated solution can now be computed as:

$$P_{MD} = \left[\Phi\left(T_H; \|E_H\|, \sigma_{G,H}^2 + \sigma_{A,H}^2\right) - \Phi\left(-T_H; -\|E_H\|, \sigma_{G,H}^2 + \sigma_{A,H}^2\right)\right] \tag{5.8}$$

With this expression, the minimum detectable horizontal error because of interference, for a probability of mis-detection P_{MD}, is:

$$\mathrm{MDE}_H = T_H + k_{MD}\sigma_{G,H}, \tag{5.9}$$

where k_{MD} is the sigma inflation factor associated with P_{MD}. If, on the contrary, the position-domain monitor triggers a fault, it becomes necessary to isolate and exclude the faulted measurements. For that purpose, we inspect each GNSS measurement $\rho^{(k)}$ by checking the consistency of the measurement with the information available from APNT.

5.3.1 Range-domain fault monitor

Having an estimate of the GNSS user clock bias b_G, based on APNT measurements, aids in the computation of the distances between the user and each satellite, $\hat{r}_G^{(k)}$. We define the GNSS-based estimate of the distance from user to satellite $\hat{r}_G$ as the difference between pseudorange and the user clock bias:

$$\hat{r}_G^{(k)} = \rho_G^{(k)} - \hat{b}_u. \tag{5.10}$$

Using the APNT position fix as a reference we can also compute a different observation of that distance, by taking the norm of the difference of positions:

$$\hat{r}_A^{(k)} = \mathbf{1}^{(k)} \cdot (\mathbf{x}^{(k)} - \hat{\mathbf{x}}_A). \tag{5.11}$$

Having two different observations ($\hat{r}_A^{(k)}$,$\hat{r}_G^{(k)}$) of the actual distance ($r^{(k)}$) between user and satellite k, we can assess the difference of the two measurements to decide whether they are consistent or not. We define the monitor $d^{(k)}$ as:

$$d^{(k)} = \hat{r}_G^{(k)} - \hat{r}_A^{(k)}. \tag{5.12}$$

The monitor $d^{(k)}$ is designed to respond to inconsistencies between satellite-based pseudorange measurements and the APNT-based position fix. Conversely, it is also expected to not issue a warning under fault-free conditions. Given that neither $\hat{r}_G^{(k)}$ nor $\hat{r}_A^{(k)}$ are noise-free, a finite detection threshold needs to be defined. The design of the threshold, however, requires modeling the distribution of $d^{(k)}$.

Modeling the probability distribution of $d^{(k)}$ requires models of the individual components. From (5.11), we see that the APNT-based prediction of the distance from the user to satellite k is the projection of the position uncertainties into the line-of-sight direction. As such, the resulting distribution for $\hat{r}_A^{(k)}$ can be modeled as:

$$\hat{r}_A^{(k)} \sim \mathcal{N}(r^{(k)}, (\sigma_e^{(k)})^2), \tag{5.13}$$

where $\sigma_e^{(k)}$ accounts for the errors on satellite ephemeris (Σ_E) and the APNT-based position estimate (Σ_A), both projected into the line of sight. Now σ_e becomes:

$$\sigma_e^2 = \|\Sigma_A \cdot \mathbf{1}^{(k)}\| + \|\Sigma_E \cdot \mathbf{1}^{(k)}\|. \tag{5.14}$$

The distribution of $r_G^{(k)}$, on the other hand, can be modeled with a conventional GNSS overbounding model, as is standard in many integrity services [13]:

$$\hat{r}_G^{(k)} \sim \mathcal{N}(r^{(k)}, (\sigma_\rho^{(k)})^2 + \|\Sigma_E \cdot \mathbf{1}^{(k)}\|). \tag{5.15}$$

With a probability distribution for both components of $d^{(k)}$ we can describe the distribution of $d^{(k)}$ itself, assuming that the errors on terrestrial ranging are independent from the errors in satellite-based ranging. As long as the assumption holds, we have that $d^{(k)}$ has a Gaussian distribution, where the mean is zero and the variance is the sum of the variances of the components:

$$d^{(k)} \sim \mathcal{N}(0, (\sigma_\rho^{(k)})^2 + \|\Sigma_A \cdot \mathbf{1}^{(k)}\| + \|\Sigma_E \cdot \mathbf{1}^{(k)}\|); \tag{5.16}$$

for convenience of notation we also define:

$$\sigma_d^{(k)} = \sqrt{(\sigma_\rho^{(k)})^2 + \|\Sigma_A \cdot \mathbf{1}^{(k)}\| + \|\Sigma_E \cdot \mathbf{1}^{(k)}\|}. \tag{5.17}$$

Knowing the distribution of $d^{(k)}$ enables us to use the reasoning of (5.6) to define a per-satellite detection threshold $T_d^{(k)}$ that satisfies the p_{FA} requirement:

$$p_{FA} = \Phi\left(-T_d^{(k)}; 0, \sigma_d^2\right) + \left[1 - \Phi\left(T_d^{(k)}; 0, \sigma_d^2\right)\right]. \tag{5.18}$$

As with (5.8), we are able to compute a p_{MD} for a given range-domain fault $E^{(k)}$:

$$p_{MD}(E) = 1 - \left[\Phi\left(-T_d^{(k)}; \|E^{(k)}\|, \sigma_d^2\right) - \Phi\left(T_d^{(k)}; -\|E^{(k)}\|, \sigma_d^2\right)\right]. \tag{5.19}$$

5.4 Fault-free operation

In this section, we show an example of the performance achievable by two potential implementations of fault monitoring. A key element in the performance of the consistency monitors is the accuracy and integrity of the APNT system. Improving APNT accuracy will improve the performance of this method. As such, the discussion of achievable sensitivity centers around the evolution of APNT, a topic that has seen some interest in the recent past, as authors have proposed many variations on how to best provide coverage and how to use it to complement GNSS [1,4,9,14–16].

One implementation is based on currently available legacy technology, namely Distance Measuring Equipment (DME), in combination with Inertial Reference Units (IRU) [17]. The second implementation is based on a technology currently being developed for use in next-generation APNT systems: the L-Band Digital Aeronautical Communications System (LDACS) [18].

The currently achievable performance is interesting as a near-term solution to the vulnerability of civil aviation to GNSS tampering. The next-generation solution is interesting as it provides a significantly higher performance than is currently achievable. Although the legacy version will support meaningful detection performance at services such as RNAV 1 or RNP 1, we show how the next-generation solution can protect services similar to RNP 0.3.

5.4.1 Simulation configuration

In predicting the performance of the two proposed implementations we compute the expected position errors from both GNSS and APNT systems over all of Germany. To this end, we assume two operational GNSS constellations: GPS and Galileo. As far as the APNT system we assume that all 159 currently operational DME stations in Germany will be available (Figure 5.3(a)), plus we assume a network of 80 LDACS stations distributed all around Germany (Figure 5.3(b)).

As for the performance of the individual systems that compose our multi-system integrity monitor, we assume the noise characteristics described in Table 5.1. The DME error model is simplified to the worst-case error from the DME standard [19], whereas the error corresponding to LDACS measurements is derived from [20]. For the purposes of computing integrity performance we assume $p_{FA} = 10^{-5}$ and $p_{MD} = 10^{-7}$.

The remainder of this section shows how two different implementations of the proposed method will perform. To this end, we show a detail of the performance at a single reference location, 10, 000 ft above Munich airport

$$(\mathbf{x}_{\text{ref}} = [48.355° \text{ N}, 11.78° \text{ E}, 3048 \text{ m}]^T), \tag{5.20}$$

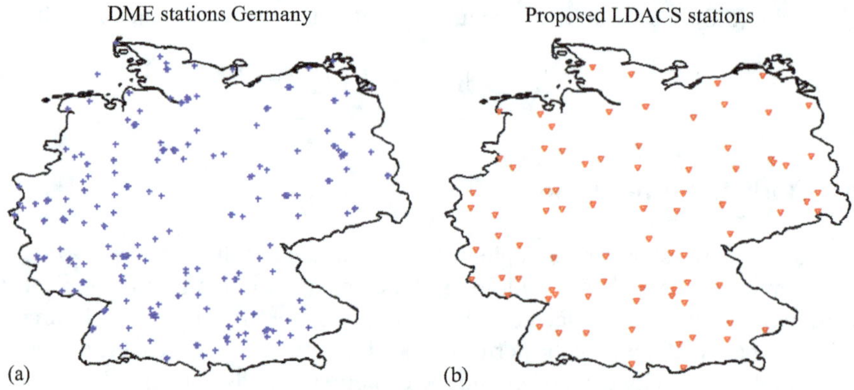

Figure 5.3 (a) Location of currently operational DME stations in Germany. (b) Locations of potential LDACS ground stations in Germany

Table 5.1 Error models for the radionavigation services used in the computations for Figures 5.4 and 5.5, using $p_{FA} = 10^{-5}$ and $p_{MD} = 10^{-7}$

	# of sources	**1σ error**
DME	159	$\sigma_r = 185$ m
LDACS	80	$\sigma_\rho = 10$ m
GPS	24	URA = 2.40 m
Galileo	24	URA = 3.12 m

as well as the performance of the methods at 10, 000 ft, computed over all of Germany and shown in Figures 5.4 and 5.5.

The GNSS constellations were evaluated at the reference position for the epoch of GPS week 1726 at zero seconds of the week. At that time and location the corresponding covariance matrix for the nominal GNSS-based position solution is:

$$\Sigma_G = \begin{bmatrix} 0.76 & -0.96 & 0.08 \\ -0.96 & 0.74 & -0.04 \\ 0.08 & -0.04 & 0.73 \end{bmatrix}^2 m^2, \tag{5.21}$$

which yields $\sigma_{G,H} = 1.05$ m. In turn, the $p_{MD} = 10^{-7}$ implies a $k_{MD} = 5.33$ and therefore $k_{MD} \cdot \sigma_{G,H} = 5.59$ m.

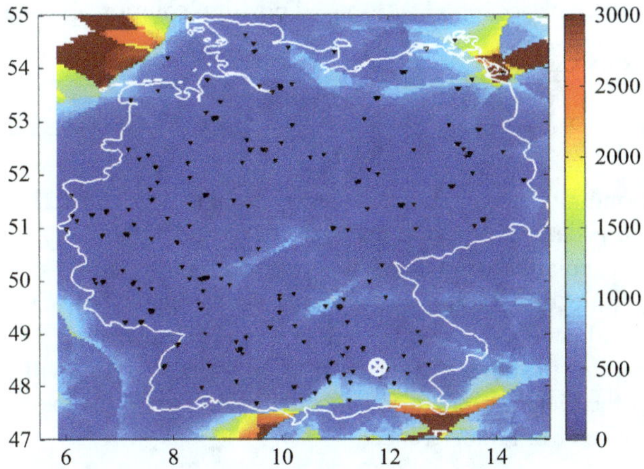

Figure 5.4 Minimum detectable error, according to (5.9) for DME-based APNT. We highlight the position $\mathbf{x}_{ref}$ with a white circle

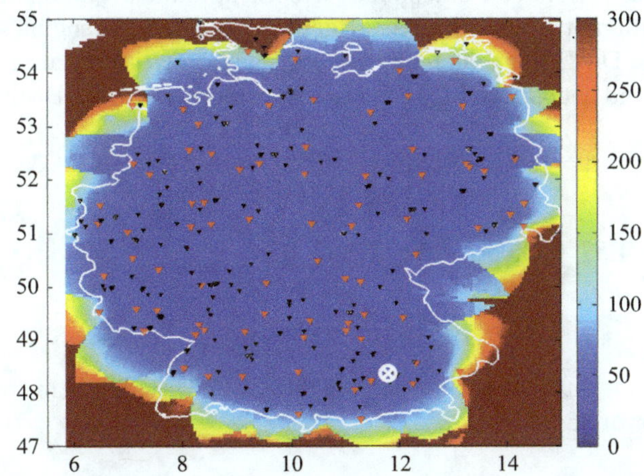

Figure 5.5 MDE, according to (5.9), for DME-LDACS hybrid APNT. Here, too, we highlight the position $\mathbf{x}_{ref}$ with a white circle

5.4.2 Currently achievable performance: DME

Given recent statements by the FAA on the status of its APNT program [17] and the results from other research efforts [2], our first performance benchmark is based on an APNT system is made up of the currently operational DME network. For the purposes of quantifying the benefits of the method we also investigate the performance, assuming DME and IRU capabilities on board.

The covariance matrix for a DME-based position solution, evaluated at $\mathbf{x}_{\text{ref}}$ is:

$$\Sigma_A = \begin{bmatrix} 80.84 & 2.24 & 3.57 \\ 2.24 & 89.88 & -3.37 \\ 3.57 & -3.37 & 151.69 \end{bmatrix}^2 m^2, \tag{5.22}$$

and the corresponding horizontal detection threshold is $T_H = 516.23$ m. Considering that the minimum detectable error for $p_{MD} = 10^{-7}$ is MDE $= T_H + k_{MD}\sigma_{G,H}$, the MDE for this particular situation is MDE $= 521.82$ m.

The geographic distribution of how this reasoning extends to the whole of Germany, also evaluated at 10, 000 ft, is shown in Figure 5.4.

Considering that the horizontal error of an undetectable $\Delta\mathbf{x}_E$ will not exceed 300 m over most of the German territory, it seems feasible that this method will protect RNP-1-like services from interference.

5.4.3 Next-generation APNT: DME-LDACS hybrid

Considering that the detection performance of the proposed method is largely driven by the accuracy of the APNT system, evidently improving the APNT reference will improve error detection. One potential development in the evolution of APNT is the future introduction of LDACS-based navigation. The LDACS ranging signal is more precise than the DME signal and the resulting position estimates are more accurate, even if the ground network of LDACS has fewer stations than the DME network (see Table 5.1).

At $\mathbf{x}_{\text{ref}}$ the position covariance matrix of the hybrid DME-LDACS system, in the ENU frame, is:

$$\Sigma_A = \begin{bmatrix} 4.55 & -0.05 & -0.31 \\ -0.05 & 4.70 & -1.17 \\ -0.31 & -1.17 & 109.72 \end{bmatrix}^2 m^2, \tag{5.23}$$

which leads to a position-domain detection threshold $T_H = 29.73$ m and a minimum detectable error of 35.32 m.

5.5 Faulted operation

In the absence of empirical data from actually induced GNSS faults, quantifying the behaviour of the proposed monitors under faulted conditions requires a hypothesis on how these faults occur. From considerations of the detectability with existing monitors, we derive two different faults that will serve as benchmarks. This reasoning is not meant to be a statement of the likelihood of such faults occurring, instead it serves as a proposal to address the vulnerability of existing systems.

One of the biggest threats to ARAIM-based services stems from coordinated errors injected in multiple PRNs at a time, because ARAIM bases fault detection on consistency checking. Thus, the greatest risk to navigation integrity is an error that affects all visible satellites at the same time. In such a situation ARAIM will not be able to detect the presence of a fault, while our method provides a detection capability.

A further scenario that may be interesting, also from the perspective of fault observability under ARAIM, is the situation where errors are injected into the signals of one constellation, but not other constellations. We illustrate how our detector performs when one of two available constellations is affected. In such situations, ARAIM may be able to detect the inconsistency between the two sets of satellites, but will not be able to resolve the conflict. Instead, the range-domain screening of our method will pick out the satellites that most affect the position solution, providing in turn a sanitized position solution.

5.5.1 Multi-constellation fault

For the example of (5.21), as analyzed at $\mathbf{x}_{\text{ref}} = [48.355°\text{N}, 11.78°\text{E}, 3048\ \text{m}]^T$, we inject a fault $\Delta\mathbf{x}_E$ into the position estimate by computing the vector of individual pseudorange faults $\mathbf{E}$ and applying them to each measurement.

The resulting GNSS-based position estimate $\hat{\mathbf{x}}_G$ now has a systematic error equal to $\Delta\mathbf{x}_E$ that will lead to a mismatch with the APNT-based position estimate $\hat{\mathbf{x}}_A$ and, in turn, to a non-zero component on $\mathbf{m}$ as defined in (5.4).

For the constellation in (5.21), the position-domain error

$$\Delta\mathbf{x} = [100, 0, 0]^T\ \text{m} \tag{5.24}$$

will result from the range-domain errors

$$\begin{aligned}\mathbf{E} = [&4.95, -16.25, -24.43, 61.45, 96.06, 86.87, -22.86, 29.17,\\ &84.79, 30.57, -47.83, -36.65, -46.30, 0.85, 47.66]^T\ \text{m}.\end{aligned} \tag{5.25}$$

The combination of GNSS constellation with the visible APNT ground stations, summarized in the covariance matrix of (5.23), will detect errors of 26.33 m horizontal position error, as discussed above. A position error of 100 m will therefore be approximately $70\sigma_{G,H}$ above the detection threshold and trigger a fault with a very high probability.

The resulting range-domain inspection leads to the range-domain detection thresholds $T^{(k)}$, written in vector form:

$$\begin{aligned}\mathbf{T} = [&29.86, 27.40, 19.73, 26.57, 29.41, 28.66, 21.07, 16.17,\\ &29.22, 20.86, 27.06, 33.08, 29.15, 14.35, 28.80]^T\ \text{m}.\end{aligned} \tag{5.26}$$

Comparing the errors of (5.25) with the detection thresholds of (5.26), entries 1, 2 and 14 are highly likely to go undetected, while all other entries will, with high probability, be discarded. The example illustrates that some satellites have a relatively small impact on the horizontal position error and are, therefore, difficult to detect with the proposed method. On the other hand, the small impact on the horizontal position error also means that the integrity risk they represent is relatively small.

In the particular example of this section, no GNSS-based position solution can be computed. The system will have to resort to the APNT-based solution, which is less accurate, but more robust.

5.5.2 Single-constellation fault

Faults affecting the satellite signals of a single constellation are a situation that existing integrity monitors may struggle with. In the case of ARAIM, a disagreement between two constellations may result in a fault warning, but the response depends on the relative numbers of satellites visible for each constellation. Even in the more favorable case, ARAIM has no means available to resolve the disagreement or excluding the faulted satellites.

Given the range-based detection capability provided by the monitor $d^{(k)}$ of (5.12), our method has the ability to isolate the affected satellites. If sufficient satellites in view remain unflagged, a position solution can be computed. For this sanitized position solution it is again possible to define a position-domain consistency threshold $\mathbf{m}$, like in (5.4), based on the sanitized satellite geometry. From the sanitized geometry it is, again, possible to derive a detection threshold and a minimum detectable error.

If the number of approved satellites is insufficient to provide a position solution, or if the resulting sanitized geometry produces a position solution that is not consistent with $\mathbf{x}_A$, then the satellite-based navigation service will be declared unavailable.

For the example situation detailed above, it is conceivable that range errors are engineered to inject a position error $\Delta\mathbf{x}_E = [100\text{ m}, 0\text{ m}, 0\text{ m}]^T$, but do so only on GPS signals. The resulting range errors are now limited to GPS and result in two false negatives (mis-detections), six true negatives and seven true positives, which leads to nine available satellites of which two are actually faulted.

The resulting horizontal position error will have a bias of 5.70 m, which is an order of magnitude smaller than the 47.15 m error, if left unscreened.

The nature of the position-domain monitor $\mathbf{m}$ is such that faults become more easily detectable, the larger the disagreement between GNSS and APNT becomes. As an example, we simulate the behaviour of the monitor at $\mathbf{x}_{Ref}$ for a range of injected errors of different magnitudes $|\Delta\mathbf{x}_E|$, from zero to 400 m. We show the results of the simulation in Figure 5.6.

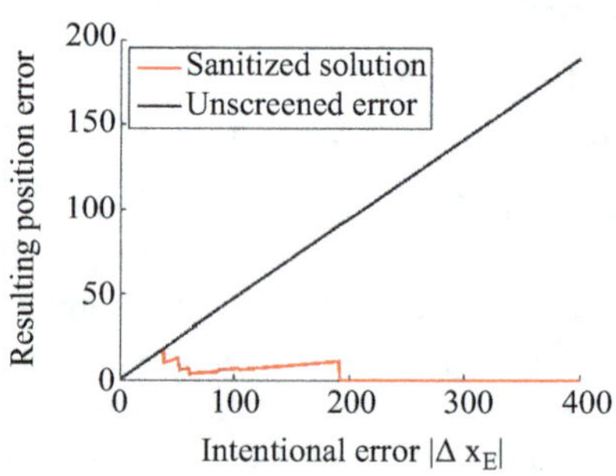

Figure 5.6 If only one constellation is affected by $\mathbf{x}_E$*, the position error because of spoofing can be controlled by using the proposed monitors* $\mathbf{m}$ *and* $d^{(k)}$

The plot in Figure 5.6 shows the benefit of screening GNSS measurements with the APNT position solution. There is a critical point at which GNSS-ranging errors are sufficiently large to inject errors, but the monitor is not sensitive to detect them. For particular situation considered for the plot in Figure 5.6 that maximum is at 36.6 m, but it depends on the geometry of the ranging sources, especially on those of the APNT system.

5.6 Discussion

Thus far, this chapter establishes how cross-checking GNSS and APNT information for consistency enables the observability of spoofing-induced position errors. In this section, we discuss some of the implications of the proposal, including the range-domain vs. position-domain monitoring and the protection of the user during a constellation or multi-constellation fault.

5.6.1 Range-domain vs. position-domain monitoring

The hybrid nature of consistency monitoring, in position and in range domain, takes advantage of the benefits of each method. For any given level of integrity, the position-domain monitor is sensitive to smaller faults than the combination of range-domain monitors as a whole. In addition, the position-domain monitor offers a computational benefit, as it provides a scalar test, as opposed to the satellite-by-satellite comparisons required for range-domain monitoring. The range-domain monitor, on the other hand, offers a more detailed insight into the potential errors, thus enabling detection and exclusion of affected satellites.

Monitoring in the position domain, using the monitor $\mathbf{m}$ of (5.4), gives a more sensitive detector for a given integrity risk than the range-domain monitor $d^{(k)}$ would. The advantage of range-domain monitoring, using $d^{(k)}$ of (5.12), is that it gives fault observability on individual PRNs, enabling fault exclusion at the measurement level.

With the aim of leveraging the benefits of both methods, our method is designed to permanently assert the consistency of solutions at position level, while range-domain consistency checks will only be invoked in the case of a detected fault. This concatenation of monitors provides an extra degree of freedom in allocating the false alert probability for each monitor. For the application in this chapter, it is sufficient to have both values at the same level, but future applications may benefit from this flexibility.

A position-domain monitor that errs on the side of false detections (higher p_{FA}) can be combined with an adequate range-domain-based FDE algorithm that again reduces the overall p_{FA} to the specified level, as indicated in (5.7) and (5.18). The required integrity risk will still prescribe a detection probability for faults of a given magnitude. The most appropriate trade-off in the false-alert probability between range-domain and position-domain monitoring, implicit in the definitions of the monitors $\mathbf{m}$ and $d^{(k0)}$, remains a topic for future research.

5.6.2 Protection during fault

In evaluating the benefits of our method, relative to the detection performance of existing methods such as ARAIM, we assume two fault scenarios without commenting on their likelihood. The first scenario is the all-out case, where all visible GNSS satellites have been spoofed; the second case is the single-constellation case, where all satellites of a particular constellation are in error.

The intent behind these threat scenarios is to test the behaviour of the method at the breaking points of ARAIM. It is not meant as a statement on the likelihood of either situation happening in reality. In fact, far more complex scenarios can be constructed by taking arbitrary combinations of satellites as being spoofed. The main benefit of the proposed method lies in its ability to detect any of these combinations, no matter how complex, and provide a confidence interval on the fault-free solution.

As described thus far, the method of checking the consistency between GNSS measurements and an APNT solution provides a confidence bound on the position estimate, even during a detected fault. The method provides integrity guarantees, as long as sufficient redundancy is available after removing all satellite measurements for which spoofing errors are detected. The fault exclusion mechanism itself, as described in (5.19), is a key tool in providing resilience to spoofing-induced outages of GNSS: by virtue of using APNT as a reference system, the new method provides higher continuity of service.

The two faults discussed above were not chosen based on the likelihood of occurrence, as it is difficult to assess the likelihood of a particular RFI fault occurring. Instead, the two scenarios were chosen as they represent the biggest threats to the integrity on ARAIM-based navigation systems.

5.7 Conclusion

We propose a method that uses terrestrial ranging to detect GNSS navigation errors because of radio interference or spoofing. The method compares the range measurements obtained from the GNSS receiver with the position fix from the APNT system and checks their consistency. We define a position-domain monitor and a range-domain monitor. The method was designed under the assumption that, at some point in the future, an APNT system will exist, capable of delivering RNP 0.3 or equivalent performance. Consequently, the monitor will work at a similar performance level, as the detection is driven by the least-performing system in the processing chain.

An interesting application of this research that will be left for further work is the possibility of monitoring GNSS for multi-satellite or constellation-wide faults. In fact, the method itself is able to detect even multi-constellation faults. The practical limitations of real-world and realistic future APNT systems do not yet provide the ranging or positioning accuracy that would support an integrity benefit to ARAIM services more demanding than RNP 0.3. If ever an APNT system were to boast meter or sub-meter level ranging errors, this method may translate into significant availability and integrity benefits.

References

[1] Osechas O, Nossek E, Belabbas B, *et al.* A Modular Approach to Integrity for APNT. In: Proc. of the Global Nav. Sat. Systems Conf. of the ION. Portland, OR; 2016.

[2] Berz G, Vitan V, Skyrda I, *et al.* Can Current DME Support PBN Operations with Integrity? In: Proceedings of the 26th International Technical Meeting of the ION Satellite Division, ION GNSS+; 2013.

[3] Berz G. ICAO GNSS RFI Mitigation Plan and associated EUROCONTROL Efforts. at UN ICG; 2016.

[4] Nossek E, Osechas O and Meurer M. Error Analysis for the Combination of Angular, Ranging, and Barometric Measurements. In: Proc. of the Global Nav. Sat. Systems Conf. of the ION. Portland, OR; 2016.

[5] Phelts E, Walter T and Enge P. Characterizing Nominal Analog Signal Deformation on GNSS Signals. In: Proceedings of the 22nd International Technical Meeting of The Satellite Division of the Institute of Navigation (ION GNSS 2009), Savannah, GA; 2009. p. 1343–1350.

[6] Osechas O, Misra P and Rife J. Carrier-Phase Acceleration RAIM for GNSS Satellite Clock Fault Detection. *Navigation*. 2012;59(3):221–235.

[7] Osechas O, Narayanan S, Crespillo OG, *et al.* Feasibility Demonstration of Terrestrial RNP with LDACS. In: Proc. of the Global Nav. Sat. Systems Conf. of the ION. Miami, FL; 2019.

[8] Tanil C, Khanafseh S, Joerger M, *et al.* An INS Monitor to Detect GNSS Spoofers Capable of Tracking Vehicle Position. *IEEE Transactions on Aerospace and Electronic Systems*. 2018;54(1):131–143.

[9] Swaszek P, Hartnett R and Seals K. GNSS Spoof Detection using Independent Range Information. In: Proc. of the International Technical Meeting of the ION. Monterey, CA; 2016.

[10] Osechas O, Perea S, Belabbas B, *et al.* Use of APNT to Protect GNSS-Based RNP Services from Intentional and Unintentional RF Interference. In: Proc. of the International Technical Meeting of the ION. Monterey, CA; 2017.

[11] Pullen S, Gao G, Tedeschi C, *et al.* The Impact of Uninformed RF Interference on GBAS and Potential Mitigations. In: Proceedings of the 2012 International Technical Meeting of the Institute of Navigation, Newport Beach, CA; 2012.

[12] Misra P and Enge P. *Global Positioning System: Signals, Measurements and Performance*, Second Edition. Massachusetts: Ganga-Jamuna Press; 2006.

[13] SC-159. Minimum Operational Performance Standards for Global Positioning System/Wide Area Augmentation System Airborne Equipment.

[14] Osechas O and Berz G. Improving the Availability of LDACS-Based APNT with Air-to-Air Ranging. In: Proc. of the Position, Localization and Navigation Symposium. Savannah, GA; 2016.

[15] Swaszek PF, Hartnett RJ and Seals KC. APNT for GNSS Spoof Detection. In: Proc. of the International Technical Meeting of the ION. Monterey, CA; 2017.

[16] Kumar R, Battista G and Osechas O. Optimized placement of LDACS stations to achieve RNP APCH with APNT. In: Proc. of the International Technical Meeting of the ION. Monterey, CA; 2017.
[17] Lawrence D. Alternative Positioning, Navigation, and Timing (APNT). Federal Aviation Administration; 2014.
[18] Schneckenburger N and Schnell M. Precise Aeronautical Ground Based Navigation Using LDACS1. In: Proc. of Integ. Comm., Nav., and Surveilance Conf. (ICNS). Herndon, VA; 2012.
[19] DO-189 Minimum Operational Performance Standards for Airborne Distance Measuring Equipment (DME) Operating within the Radio Frequency Range of 960–1215 MHz. RTCA; 1985.
[20] Thiasiriphet T, Schneckenburger N and Schnell M. Impact of the DME interference on the LDACS1 ranging performance. In: Proc. of the Global Nav. Sat. Systems Conf. of the ION. Tampa, FL; 2015.

Chapter 6

Security risk assessments and countermeasures for future aeronautical communication network architecture

Muhammad Ali[1], *Yim-Fun Hu*[1], *Rameez Asif*[1], *and Jiang-Ping Li*[1]

Cybersecurity vulnerabilities are relatively new regarding airworthiness in comparison with those encountered in terrestrial networks. The latest generation of aircraft such as Boeing and Airbus have included onboard aircraft data networks, which, on the one hand, offer innovative and robust ways of communications but, on the other hand, also introduce cybersecurity vulnerabilities to avionic communications. If these vulnerabilities are exploited, they may lead to irreparable damage to the information system that may cause a catastrophic failure. With this in mind, protection to the onboard data network and the data links between the air and ground communication infrastructure are of paramount importance. This chapter focuses on defining the future avionics communication system security architecture with security risk assessment and risk mitigation. It identifies the security requirements that future avionics communication system needs to comply with. A risk analysis is carried out to identify security assets and their security perimeters, vulnerability points, threats and attacks, the impacts of such threats, and the risk levels of threats with reference to the target future avionics communication network architecture. According to the risk analysis, security countermeasures are identified, which leads to the target network security architecture design. The method in the identification of the target network security architecture primarily follows the guidelines specified in EUROCAE ED-202 and CESG IAS standards.

The future avionics communication network security architecture is directly related to airworthiness security, which is defined as the protection of the airworthiness of an aircraft from information security threats. EUROCAE and ARINC define frameworks in assessing security threats and the risks that they impose on aircraft safety. Examples of security countermeasures to detect and prevent malicious attacks on information security will also be provided. This chapter mainly serves as a

[1]Department of Biomedical and Electronics Engineering, University of Bradford, Bradford, UK

recommendation for defining security processes and their implementation concerning the target network architecture.

6.1 Introduction

The need for adequate future communication capability to support the continuous growth in air traffic and evolving air traffic management (ATM) requirements within the global aeronautical communications operational framework is set to transform the communication paradigm from today's voice communications to digital data exchanges. Current VHF aeronautical communications systems used for analogue voice and data link can be used as required to cover the short- and medium-term needs [1] but will be insufficient to handle the vast amount of information transfer created by advanced aeronautical communications systems. Under the auspices of AP17, data link technologies for new communication infrastructures required to manage future aeronautic communication traffic demand and to provide an always-connected infrastructure for air-ground communication have been identified below:

1. A ground-based, high-capacity, airport surface data link system, referred to as the aeronautical mobile airport communications system (AeroMACS);
2. A ground-based data link system for continental airspace in general, referred to as the L-band digital aeronautical communications system (LDACS);
3. A satellite-based data link system for the oceanic and remote continental environment. It can also be used to complement the ground-based systems.

In addition, aircraft-to-aircraft communications or multi-hop/ad-hoc aircraft communications are also considered in future studies of air traffic management systems. Figure 6.1 shown below presents an overall aeronautical communications concept.

While these data link technologies are promising and efficient means to support future aeronautical communications systems, concerns have been raised in their vulnerability to malicious attacks that may lead to serious consequences in aircraft safety. Several analyses and studies have already been carried out by ARINC, ICAO, EUROCAE and others for aircraft information security and airworthiness security. State-of-the-art security techniques used for ground networks have been considered for securing aviation communications.

6.1.1 Fundamentals of cybersecurity concept

Computer security is defined as the protection afforded to an automated information system in order to preserve the integrity, availability and confidentiality of information system resources (includes hardware, software, firmware, information/data and telecommunications) [2]. This definition introduces three key objectives for information and communications systems security:

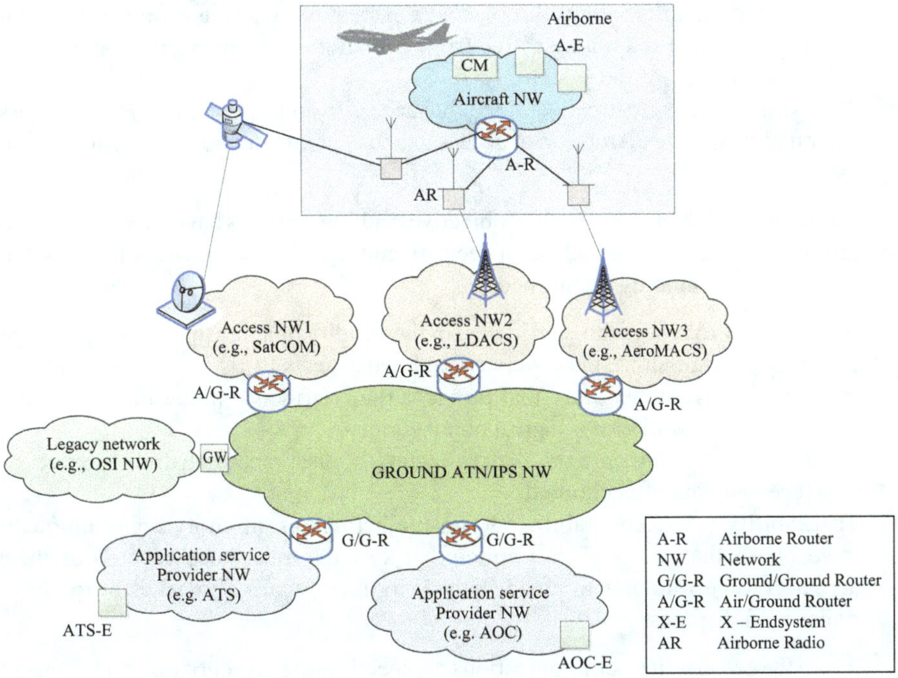

Figure 6.1 Future aeronautical communication concept

- **Confidentiality** to preserve authorized restrictions on information access and disclosure, including means for protecting personal privacy and proprietary information. A loss of confidentiality is the disclosure of information to unauthorized individuals, entities or processes. Confidentiality covers two areas:
  - Data Confidentiality which assures that private or confidential information is not made available or disclosed to unauthorized individuals.
  - Privacy which assures that individuals control or influence what information related to them may be collected and stored and by whom and to whom that information may be disclosed.
- **Integrity** to guard against improper information modification or destruction, including ensuring information nonrepudiation and authenticity. A loss of integrity is the
  - Failure to detect, prevent or guard against unauthorized modification or destruction of information.
  - Failure to recover from or accommodate the adverse effects of unauthorized modification or destruction of information.

  Integrity covers two related concepts [3]:
  - Data Integrity which assures that information and programs are altered or destroyed in an authorized manner.

  - System Integrity which assures that a system performs its intended function in an unimpaired manner, free from deliberate or inadvertent unauthorized manipulation of the system.
- **Availability** to ensure timely and reliable access to and use of information. A loss of availability is the disruption of access to or use of information or an information system.

In addition to the above three key objectives, additional ones have been introduced to present a more complete picture of security concepts including the following two which are most commonly mentioned:

- **Authenticity** to assure the confidence in the validity of a transmission, a message or message originator. This covers two related aspects [4]:
  - Peer Entity Authenticity that provides the confidence in the identity of the connected entities in a logical connection.
  - Data-origin Authenticity that provides the assurances that the source of received data is as claimed.
- **Traceability** (or Accountability) to ensure actions of an entity being uniquely traceable to that entity. This implies that systems must keep records of their activities to permit later forensic analysis to trace security breaches or to aid in transaction disputes.

From the aeronautics communications perspective, these security objectives need to be stringently fulfilled using cryptography and network security techniques to ensure that the safety of aircraft will not be hampered by any breach of security into the aircraft network and computing systems.

A breach of security will impact on organizations or individuals. FIPS 199 defines three levels of impact:

- **Low**: The loss (of confidentiality, integrity or availability) could be expected to have a limited adverse effect on organizational operations, organizational assets or individuals. This means that the loss may (i) cause degradation in mission capability to an extent and duration that the organization is able to perform its primary functions, but the effectiveness of the functions is noticeably reduced; (ii) result in minor damage to organizational assets; (iii) result in minor financial loss; or (iv) result in minor harm to individuals.
- **Moderate**: The loss could be expected to have a serious adverse effect on organizational operations, organizational assets or individuals that may (i) cause a significant degradation in mission capability to an extent and duration that the organization is able to perform its primary functions, but the effectiveness of the functions is significantly reduced; (ii) result in significant damage to organizational assets; (iii) result in significant financial loss; or (iv) result in significant harm to individuals that does not involve loss of life or serious, life-threatening injuries.
- **High**: The loss could be expected to have a severe or catastrophic adverse effect on organizational operations, organizational assets or individuals. The loss may (i) cause a severe degradation in mission capability to an extent and duration

that the organization is able to perform one of more primary functions, but the effectiveness of the functions is significantly reduced; (ii) result in major damage to organizational assets; (iii) result in major financial loss; or (iv) result in severe or catastrophic harm to individuals that does not involve loss of life or serious, life-threatening injuries.

6.1.2 Security attacks

ITU X.800 [5] and RFC 4949 [6] categorize security attacks into passive and active attacks. A passive attack attempts to learn or make use of information from the system but does not affect system resources. An active attack attempts to alter system resources or affects their operation.

6.1.2.1 Passive attacks

The goal of passive attacks is to obtain information that is being transmitted by eavesdropping or monitoring the transmission. Eavesdropping can result in the release of message content whereby an opponent is able to obtain the content of information, which may be sensitive, during transmission and subsequently release the sensitive information to other parties. A normal practice is to encrypt the message content to prevent information disclosure.

Malicious monitoring of transmissions is a form of traffic analysis attack. Even if a message is encrypted, by malicious monitoring the transmission, an attacker may still be able to observe the pattern of these messages to determine the location and identity of communication hosts and could observe the frequency and length of messages being exchanged, which in turn will enable the attacker to guess the nature of communication that was taking place.

Passive attacks are very difficult to detect since they will not disrupt the system or its resources. Thus the emphasis is on the prevention of such attacks rather than detection. Encryption is normally used to prevent passive attacks.

6.1.2.2 Active attacks

Active attacks involve modifications of the data stream or the creation of a false stream. There are four categories of active attacks:

Masquerade: This takes place when one entity pretends to be a different entity. An example is an authorized entity with few privileges to obtain extra privileges by impersonating an entity that has these extra privileges.
Replay: This involves the passive capture of information and its subsequent retransmission to produce an unauthorized effect.
Modification of messages: This involves altering some parts of a legitimate message or delaying or reordering that message to produce an unauthorized effect.
Denial of Service: This involves the prevention or inhibition of the normal use or management of communication facilities. This includes the suppression of all messages directed to a particular destination or the disruption of an entire network by disabling the network or by overloading it with messages so as to degrade the network performance.

Man-in-the-Middle Attack: The attacker makes independent connections with the victims and relays messages between them, making them believe that they are talking directly to each other over a private connection, when in fact the entire conversation is controlled by the attacker.

Some active attacks can be difficult to prevent but measures are available to mitigate the damage caused by them. Although 100% prevention will be impossible due to the wide variety of possibly unidentified potential physical, software and network vulnerabilities.

6.1.3 Security mechanisms

To counteract security attacks, security mechanisms and policies need to be enforced. ITU X.800 defines a list of security mechanisms. These including

Encipherment (encryption): The use of mathematical algorithms and zero or more encryption keys to transform data into an unintelligible form. Subsequent recovery of the original data requires corresponding mathematical algorithms and decryption keys.

- Digital signature: Data appended to or a cryptographic transformation of a data unit that allows a recipient of the data unit to prove the source and integrity of the data unit and protection against forgery.
- Access Control: Mechanisms that enforce access rights to resources.
- Data Integrity: Mechanisms that assure the integrity of a data unit or stream of data units
- Authentication Exchange: Mechanisms to ensure the identity of an entity by means of information exchange.
- Access Control: Mechanisms that enforce access rights to resources.
- Traffic Padding: The insertion of bits into gaps in a data stream to frustrate traffic analysis attempts.
- Routing Control: Enabling selection of particular physically secure routes for certain data and allows routing changes, especially when a breach of security is suspected.
- Notarization: The use of a trusted third party to assure certain properties of a data exchange.
- X.800 also defines pervasive security mechanisms that are not specific to any particular OSI security services or protocol layer. These include
- Trusted Functionality: Functionality that is perceived to be correct with respect to some criteria (e.g. the establishment of a security policy).
- Security Label: The marking bound to a resource (e.g. a data unit) that names or designates the security attributes of that resource (e.g. the use of VLAN tag).
- Event Detection: Detection of security-relevant events.
- Security Audit Trail: Data collected and potentially used to facilitate a security audit, which is an independent review and examination of system records and activities.

- Security Recovery: Ways to handle requests from mechanisms, such as event handling and management functions, and take recovery actions.

The above mechanisms involve directly or indirectly the application of cryptographic techniques. The most commonly used cryptography and network security techniques are listed as follows:

- Cryptographic Techniques
  - Symmetric key cryptography
  - Asymmetric key cryptography
  - Digital certificates
  - Hash functions
  - Digital signatures
- Network security techniques
  - IPSec
  - Point-to-point tunnelling protocol (PPTP)
  - Layer 2 tunnelling protocol (L2TP)
  - Transport Layer security (TLS)
  - Firewall
    - Packet filtering firewall
    - Circuit proxy firewall
    - Application proxy firewall
    - Stateful inspection packet filtering firewall
  - Intrusion detection and prevention system (IDPS)
  - Authentication and authorization
    - PPP authentication
    - IEEE 802.1X
    - EAP
    - RADIUS
    - DIAMETER
    - Biometric authentication

6.2 Security for data link communications in aviation

Present avionics standards are more focused on the safety assessment in terms of onboard systems failure or malfunction but any intended malicious actions that will sabotage the safety of the aircraft and its passengers have only been considered recently.

Although some security assessments of aircraft systems may suggest safety implications, there is a gap between the safety-oriented threat assessment and security-oriented threat assessment. The security-oriented threat assessment will have a separate and special impact on the system design which may not be completely covered by any safety-oriented threat assessment.

Government and industrial initiatives on security-related threat assessment concentrate mainly on air-ground (A/G) and ground-ground (G/G) communications.

Attacks in the ground networks include radio frequency spoofing of air traffic control, unauthorized access to computer systems, virus attacks to power and telecommunication infrastructural problems, whereas for A/G communications, numerous attacks on ACARS messages exchanged between the pilot and the ground entities have been launched in the past with the objective to obtain flight information.

Threat assessment of onboard aircraft communications networks will be dealt with thoroughly in Section 1.3 of this chapter within the target network architecture context.

6.2.1 Government and industrial initiatives on civil aviation cyber security

6.2.1.1 ICAO

ICAO is an agency of the United Nations formed in 1944 and its council adopts Standard and Recommended Practices (SARPs) following the Air Navigation Commission's (ANC) recommendations [7]. ICAO Aeronautical Communication Panels (ACPs) develop guidance material, recommendations and standards for air-ground and ground-ground voice and data aeronautical communications. ACP Working Group-I (WG-I) has considered the use of Internet Key Exchange (IKE) [8] and IPSec [9] in "Manual for the ATN using IPS Standards and Protocols (Doc 9896)" [10] to provide security in Aeronautical Telecommunications Networks (ATN) using Internet Protocol Suite (IPS) standards.

ICAO9896 [10] provides the requirements for the use of IPSec and IKE for both Ground-to-Ground and Air-to-Ground communication. The following are some of the key requirements presented in [10]:

(a) Ground-to-Ground Security using IPSec and IKE
- Compliance to Security Architecture as in RFC 4301 [10].
- Implementation of Encapsulating Security Protocol (ESP) as in RFC 4303 [11].
- Implementation of IP Authentication Header (AH) as in RFC 4302 [12].
- Implementation of IKEv2 as in RFC 4306 [13].
- Implementation of Cryptographic Algorithms for ESP and AH as in RFC 4835 [14]
- Implementation of Null Encryption Algorithm for IPSec security association as in RFC4835 but not the Null Authentication Algorithm.
- Implementation of Cryptographic Algorithms for IKEv2 as in RFC 4307
- Use of Internet X.509 Public Key Infrastructure (PKI) as in RFC 5280 [15] for authentication through digital signatures.

(b) Air-to-Ground Security Using IPSec and IKE
- Compliance to Security Architecture as in RFC 4301 [9].
- Implementation of Encapsulating Security Protocol (ESP) as in RFC 4303 [11].
- Implementation of AUTH-HMAC-SHA2-256-128 integrity algorithm as in RFC 4868[16] for ESP authentication while establishing IPSec association.

- Implementation of IKEv2 as in RFC 4306 [13] with the inclusion of RFC 4868[16], RFC 4753 [17], RFC 4754 [18] and RFC 3602 for the integrity, authentication and encryption methods.
- Use of Internet X.509 PKI as in RFC 5280[15] and RFC 3647[19] for authentication through digital signatures.
- Implementation of Mobile IPv6 operation with IKEv6 as in RFC 4877 [20].
- Optional implementation of TLS as in RFC 5246[21] in mobile and correspondent nodes of IPS using Cipher Suite TLS-ECDH-ECDSA_with_AES-128-CBC-SHA as in RFC 4492.
- Optional implementation of application layer security in mobile and correspondent nodes of IPS using Hashed Message Authentication Code (HMAC) as in RFC 2104 [22].

ICAO 9896 [10] also provides guidance for the fulfilment of the above-mentioned requirements through scenarios describing ground-ground, air-ground and end-to-end secure communications. It also recommends the exploration of other security mechanisms at data link layer, transport layer and application layer as an alternative to network layer security provided by IPSec.

6.2.1.2 ECAC

European Civil Aviation Conference (ECAC) [23] is an inter-governmental organization consisting of 44 European member states aiming to harmonize civil aviation practices and policies. One of ECAC's working group studies the cyber threats to aviation security and provides guidance material for countermeasures for such security threats. The ECAC work program for 2013-2015 has the following aims for its security-related activities:

- To promote a threat/risk-based approach to aviation security.
- To promote the development and use of security technology and equipment to address current and emerging threats.
- Based on ECAC Doc 30, to pursue the development of efficient, effective and uniform aviation security measures on a Pan-European basis.
- To promote cooperation with non-ECAC member states and regional/sub-regional organizations
- To support member states in achieving full compliance with European standards on aviation security

6.2.1.3 EUROCONTROL

EUROCONTROL is an international organization with an objective to achieve an efficient, safe and environment-friendly air traffic operations across whole of the European region through collaboration among 39 member states. It manages crisis situations and daily European ATM operations. EUROCONTROL has an advisory body known as the Security Team (SET) which provides consultation between the members for civil aviation security matters. SET includes the representatives from regulatory bodies, EUROCONTROL employees and representatives from national air navigation services. EUROCONROL plays a major role in the Single European

Sky ATM Research (SESAR) [24] programme which aims to define the future ATM systems. SESAR covers three areas of civil aviation security, namely ATM security standards, Communication Navigation and Surveillance (CNS) security and airspace security. An overview of security management in SESAR is provided in [25]. The "Manual for National ATM Security Oversight" [1] provides an overview on the role of cyber threats in ATM security and looks into the roles and responsibilities of various national and international actors like regulators, governments, operators, etc.

6.2.1.4 FAA

The Federal Aviation Authority (FAA) is an agency of the US Department Of Transportation (DOT) which manages all aspects of civil aviation in the US. Similar to SESAR, Next Generation Air Transportation System (NextGen) is undertaken by the FAA for the improvement of American National Air Space (NAS). NextGen considers the integration of security countermeasures in future ATM systems in contemplation of the added civil aviation security risks that NextGen will bring.

6.2.1.5 EUROCAE

European Organization for Civil Aviation Equipment (EUROCAE) develops specifications for performance and other documents devoted to the aviation community. It defines test specifications; Minimum Operational Performance Standards (MOPS), Minimum Aviation Systems Performance Standards (MASPS) and guidance documents in collaboration with RTCA Inc. Aeronautical Information System Security (AISS) WG-72 of EUROCAE recommends and provides guidance to ensure safe, secure and efficient operations amid the growing use of highly integrated electronic systems and network technologies onboard aircraft for CNS/ATM systems. WG-72 has published "Airworthiness Security Process Specification (ED-202)" for aeronautical network security [26] targeting the following areas:

(a) Aircraft and System Security Risk Management: ED202 [26] defines the guidelines for the management of risks in avionics system security and these guidelines include:
    - Definition of the security boundaries for assets of the system with their external dependencies. External dependencies are defined by the boundary of asset that needs protection against vulnerabilities and the asset could be an aircraft, a system within the aircraft or an item with the system.
    - Identification of threats to the asset.
    - Development of security architecture to address the threats and vulnerabilities.
    - Assessment of security risk through likelihood and impact analysis.
    - Addressing the security risks through security countermeasures and security effectiveness.
    - Assured implementation of the security countermeasures to address the threats and vulnerabilities.

(b) Specifications for the Airworthiness Security Process: ED202 [26] also defines the specifications for airworthiness security processes by defining the security perimeter, threat identification, risk assessment and risk countermeasures for:

- Preliminary Aircraft Processes
- Preliminary System Processes
- Preliminary Item Processes

The above three lead to security integration and implementation, security verification and final security assessment for the final System Processes and the final Aircraft Processes. These processes lead to the demonstration of compliance to certification needs.

6.2.1.6 RTCA

The Radio Technical Commission for Aeronautics (RTCA) is a volunteer organization. FAA has sponsored RTCA as a Federal Advisory Committee (FAC). RTCA follows FAA requests to develop comprehensive, industry-vetted and endorsed recommendations for the Federal government on issues ranging from technical performance standards to operational concepts for air transportation. RTCA has a Special Committee (SC) which develops guidance through a consensus among the SC members. The Aeronautical System Security committee (SC-216) [27] is a special committee that works jointly with EUROCAE WG-72 and has drafted security guidance including Continuing Air Worthiness Guidance for the Security of Aircraft Information System and Data Network (ED-204), Airworthiness Security Process Specification (ED-202A) and Airworthiness Security Methods and Considerations (ED-203).ED-202A and ED-203 provide further guidance for risk management through threat identification, threat assessment in terms of its likelihood and impact, security countermeasure and their effectiveness to address the threats and vulnerabilities. ED-204 provides guidelines for the security incident management and response in case of any security breaches.

6.2.1.7 A4A

Airlines for America (A4A) [28] is the only trade organization of principle US airlines. It was formerly known as Air Transport Association (ATA) of America. ATA e-business administers and publishes the specification established by A4A. To address the complex matter of digital security and its standardization in the aerospace industry, A4A created a Digital Security Working Group (DSWG) in July 1999. DSWG has developed specifications for digital security as "ATA Specification 42" [29] and these specifications are incorporated into the relevant aircraft system standards of ARINC. ATA Spec-42 is the civil aviation version of PKI of Internet Engineering Task Force (IETF) and is used as the PKI standard in RTCA and EUOROCAE documentation.

6.2.1.8 ARINC

In 1920 ARINC was established in the US with an aim of managing the radio communications for the Federal Communication Commission. In order to address the rapid growth and development of onboard avionics, the Airlines Electronics Engineering Committee (AEEC) was formed in 1949. AEEC drafts the technical solutions and standards for onboard systems. AEEC adopts the ARINC standards.

The ARINC 811 "Commercial Aircraft Information Security Concept of Operation and Process Framework" [30, 31] document provides guidance for security in aeronautical communication networks. It provides an overall security process framework along with the information security objectives and needs. It presents a detailed description of the various types of security assets in an aeronautical communication network and defines the procedure to be followed for proper asset identification. The proposed framework also looks into examples of threat identification, threat description, threat consequences that are collectively used for carrying out a detailed threat analysis. It also presents various examples of security policies which help in defining the security objectives of the system under consideration. These security objectives and needs initiate and facilitate the selection and implementation of appropriate security controls. A detailed guidance for the definition, operations and maintenance of these security controls is also provided.

6.2.1.9 CEN

The European Committee for Standardization (CEN) cooperates and collaborates with EUROCAE, EUROCONTROL and the European Aviation Safety Agency (EASA) in the field of ATM. Preparation and development of a standard for European ATM Networks is the main objective of CEN TC-377 [32]. Working group 1 of this committee is looking at "Information security for ATM". This WG is currently working on a draft standard titled "Air Traffic Management – Information Security for Organizations Supporting Civil Aviation Operations" which specifies the requirements and measures of an information security management system for civil aviation and presents the extended ISO 27001 standard for adoption in civil aviation.

6.2.1.10 JCG

In order to better understand the technical and business impacts of security on the civil aviation industry, AEEC and A4A DSWG formed a new group of members from EUROCAE, FAA, US DOT, A4A and AEEC [32]. This new group is called the Joint Coordination Group (JCG), and aims to drive the development of security-related standards. Following the suggestions from ICAO, JCG formed a Task Force with the participation from industry for an assisted improvement towards security in civil aviation. Although there are a multitude of activities taking place in order to address the security issues in civil aviation, there is still a need for the harmonization of these activities within the civil aviation industry and also between the civil aviation industry and IETF to come to some globally harmonized solutions of the security issues.

6.2.1.11 HMG Information Assurance Standards

To support the necessary requirements, referred as Mandatory Requirements, outlined in HMG Security Policy Framework (SPF), HMG Information Assurance (IA) Standards are issued jointly by CESG and Cabinet Office. These IA Standards set the minimum implementation requirements on measures to meet the Mandatory Requirements of HMG SPF. There are two IA Standards jointly issued by CESG and Cabinet Office namely IA Standard [33]. The two IA Standards complement one another in identifying, assessing, treating and managing risks to Information and Communication Technology (ICT) systems and services that handle, store and process government information. The CESG IAS describes risk assessment and risk treatment for ICT systems.

6.3 Target avionics network architecture

Cyber security vulnerabilities are relatively new regarding airworthiness in comparison with that encountered in terrestrial networks. The latest generation of aircraft such as Boeing and Airbus have included onboard aircraft data networks, which on the one hand offer innovative and robust ways of communications but on the other hand also introduce cyber security vulnerabilities to avionic communications [34]. If these vulnerabilities are exploited, they may lead to irreparable damage to the information system that may cause a catastrophic failure. With this in mind, protection to the onboard data network and to the data links between the air and ground communication infrastructure are of paramount importance.

Central to the target network architecture development are two major components known as Router subsystem (RoS) and Radio subsystem (RaS). Collectively the RoS and RaS offer the onboard communication infrastructure real-time delivery of avionic data and information between the aircraft and the ground over a set of heterogeneous radio links. Figure 6.2 represents the overall target network architecture for aircraft-ground communications. The network is divided into three segments namely Aircraft segment, Transport segment and the Ground segment. The Aircraft segment consists of the RoS and RaS. The RoS on the one hand interfaces with the onboard avionic data network that supports numerous applications of different criticality levels and on the other with the RaS. The RaS consists of a software-defined terminal capable of supporting heterogeneous waveforms, a waveform control and adaptation function that provides a uniform interface between the RoS and the different waveforms, an antenna system composed of multiple antennas that interface with the Transport segment for data and information delivery to and from the ground networks. Three radio access technologies are considered in the Transport segment, namely, VDL2, BGAN and Iridium. The Ground segment is composed of three Radio Access Networks (RANs) and their corresponding core networks, the Internet and the Aeronautical Telecommunication Network (ATN).

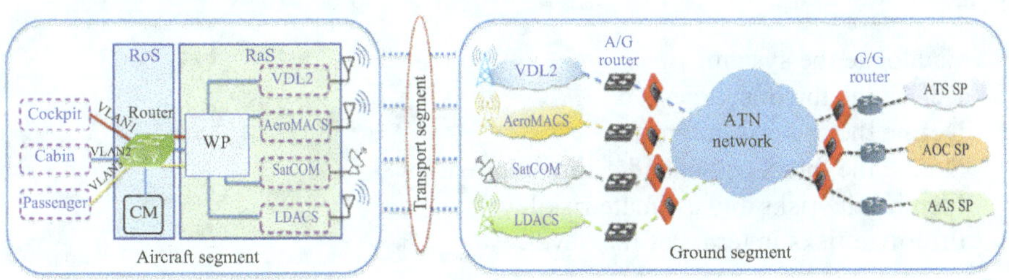

Figure 6.2 Target vision of air-ground communication infrastructure

6.3.1 Security requirements

For safety and security of the aircraft and its operations, all possible threats to the aircraft communication systems and its operations must be mitigated through efficient implementation of security mechanisms. These security mechanisms support different security features of the system and will ensure that the system meets the security requirements. First, the top-level general security requirements applicable to any aeronautical communication system needs to be carefully defined. While it is imperative that these general security requirements are met by the target network system, there may be additional specific security requirements that may also be applicable.

This section first aims to define the top-level general security requirements that are applicable to any aeronautical communication system. Based on these, the specific security requirements for the target network system are also presented. These requirements set the security objectives against target network system failures and for intentional or unintentional security attacks on the target network system.

6.4 Security risk assessment and treatment

Prior to defining the security architecture of target network, a risk assessment of the target network architecture needs to be carried out in order to identify threats, vulnerabilities and potential attacks such that suitable countermeasures can be developed and implemented. The target network security risk management combines the security risk management process defined in the EUROCAE ED202 [1] and risk assessment and treatment process defined in CESG IAS standards [33, 35]. The EUROCAE ED202 is defined specifically for aircraft environment, whereas the CESG IAS Standards provide a generic set of specifications to information security.

6.4.1 Risk assessment and treatment approach in target network

A security breach may have a direct impact on aircraft safety. As such, safety assessment is often considered hand in hand with security risk assessment. The risk assessment method in [33] specifies the six steps to allow security analysts to understand the system under consideration, define applicable threats and determine risks to the system with associated risk levels. The considered six steps in risk assessment are as follows:

1 Catalogue the system;
2 Define the threat sources;
3 Define the focus of interest;
4 Define the threat actors and estimate threat level;
5 Identify the risks and estimate risk levels;
6 Prioritize risks in terms of risk level.

Following the risk assessments, the CESG IAS also defines the risk treatment process, which includes the following four steps:

7 Apply baseline controls;
8 Risk treatment;
9 Develop an assurance plan;
10 Produce the security case.

A brief description on each step is extracted from [36] as below. Readers can refer to the standard [35–37] for further details.

6.4.1.1 Catalogue the system

The main objective of this step is to describe the system and agree with all interested parties what the system consists of, what the high-level business information exchange requirements are and the scope of the project's responsibility. This step is equivalent to identifying the security context as specified in ED202, which identifies the security environment, security perimeter of a piece of equipment or a system that are exposed to attacks through its network interfaces, logical data flow and other external dependencies.

6.4.1.2 Define and assess the threat source

This step aims to identify threat sources, based on professional judgment of who might have intentions to deliberately attack the system. A threat source is a person or organization that desires to breach security and ultimately will benefit from breach in some way. The CSG IAS distinguishes between threat sources and threat actors, although one person or organization may be both a source and an actor.

This step also defines the threat levels for threat sources (particularly the major threat sources relating to national security). Threat sources should be identified as whether they can act as threat actors to carry out their own attacks or would have to coerce or subvert another threat actor to act on their behalf.

6.4.1.3 Define the focus of interest

The FoI is a collection of assets with associated features that are the subject of a given risk assessment. In other words, the FoI conveniently groups the related assets so that the process of risk assessment can be accomplished for a group of assets instead of assessing each individual component of a system. However, it is important to determine the optimum grouping of assets into FoIs in order to apply a pragmatic, appropriate and cost-effective control at the right point in the system.

6.4.1.4 Define threat actors

The purpose of this step is to define the threat actor groups and assess the threat level posed by each threat actor or group. A threat actor is defined as a person who actually performs the attack or in the case of accidents, will cause the accident. For example, a criminal may wish to breach the confidentiality of some HMG data. The criminal in this case is the threat source. If the criminal persuades a system user to release the desired information to them, then the user is actually carrying out the attack and s/he is the threat actor. A threat actor group is a group of people who can reasonably be considered to have the same characteristics in terms of capability, motivation and opportunity to perform an attack.

Threat actors are specific to each FoI, therefore separate entries for each FoI must be generated in this step. A threat actor is someone who can actually carry out an attack on the FoI, so they must have an opportunity to do so. That opportunity may be very brief, but as attack can be planned and prepared in advance even a brief opportunity may be adequate.

The CSG IAS presents a key concept of threat actor type, which defines the types of attack that a threat actor can mount. Each threat actor belongs to one or more threat actor types according to the degree and type of access to an asset. Threat actor types include the following list of actors as shown in Table 6.1:

Table 6.1 Threat levels for threat actors [33]

Bystander (BY)	Normal User (NU)	Service Provider (SP)
Handler (HAN)	Person within Range (PWR)	Service Consumer (SC)
Indirectly Connected (IC)	Physical Intruder (PI) Subscriber (SSS)	Shared Service
Information Exchange Partner (IEP)	Privileged User (PU)	Supplier (SUP)

Readers can refer to [33] for a more detailed description of these threat actor types.

6.4.1.5 Identify specific risk and estimate the risk levels

This step is the key part of the security risk assessment procedure. It produces a list of identified risk in the system and also generates the list of risk levels for each identified risk.

6.4.1.6 Prioritize and present the risks

Step six is the final step in the risk assessment procedure. The main purpose of this step is to prioritize the list of generated risks and present it in a format that is easily understandable. An information risk can be thought of as the likelihood that a threat will exploit a vulnerability leading to a business impact. Within CSG IAS, a risk can be thought of as consisting of a number of components:

- Threat actor and threat actor type;
- Threat source;
- Compromise method;
- Property (C – Confidentiality, I – Integrity or A – Availability) of an asset or FoI and business impact level (IL) associated with the compromise of that property.

A compromise method is the broad type of attack by which a threat actor may attempt to compromise the C, I or A of an asset. It can be thought of as a generalization of vulnerability.

The CESG IAS defines business impact for different sectors of the UK including:

- Defence, International Relations, Security and Intelligence
- Public Order, Public Safety and Law Enforcement
- Trade, Economics and Public Finance
- Public Services
- Critical National Infrastructure (CNI)
- Personal/Citizen

The target network is assumed to belong to the sector of Public Order, Public Safety and Law Enforcement.

Business Impact Level (BIL) tables have been defined for each sector. With each BIL table, the business impacts are categorized and business impact statements are described for each business impact category according to its BIL. The CSG IAS defines seven levels of BIL, ranging from 0 (no impact) to 6 (extreme impact):

BIL0: No Impact.
BIL1: No Impact.
BIL2: Inconvenience or discomfort to an individual.
BIL3: Risk to an individual's personal safety and liberty.
BIL4: Risk to a group of individual's safety and liberty.
BIL5: Life-threatening and directly leading to limited loss of life.
BIL6: Leading directly to widespread loss of life.

Readers can refer to Appendix A of [33] for a more detailed description of BIL.

The risk level for an IAS risk is a combination of threat level and business impact level. A risk level is an indicative assessment of risk. In IAS, risk level is defined on a six-point scale: Very Low; Low; Medium; Medium High; High; Very high.

6.4.1.7 Apply baseline controls

All information systems must apply the full set of baseline controls except in the following defined situations:

- Where the information system has an impact of no more than Impact Level 1 (IL1) for each property of confidentiality, integrity or availability. In this case, appropriate baseline controls proportionate to AWARE level of the Segmentation Model (see Risk treatment below) can be applied at the discretion of the security analysts.
- Individual controls may be omitted from the baseline control set where it is not appropriate or reasonable to apply a particular control, for example, the baseline control will introduce extra overhead without adding any benefit to protecting the system.

6.4.1.8 Risk treatment

A Segmentation Model is presented in CESG IAS for the risk treatment process, which categories the risks based on the impact level imposed by each risk/threat in the system. The purpose of the Segmentation Model aims to segment responses to information risk based upon the types and capability of applicable threat. It is used to provide a framework to ensure that controls are both appropriate and proportionate to manage the risks to an ICT system. The model has four segments:

- AWARE aims to offer basic precautions against common nontargeted and opportunistic attacks by unsophisticated threat actors (Very Little and Little capability).
- DETER aims to protect against targeted attacks from the Internet by skilled threat actors (Limited Capability) using freely available tools with some bespoke modification.
- DETECT and RESIST aims to prevent targeted attacks from skilled threat actors (Significant capability) using bespoke tools that exploit known vulnerabilities.
- DEFEND aims to prevent all attacks against a system from the most sophisticated and highly capable threat actors (Formidable capability).

Except for situations indicated in (7) above, all information systems must apply a Baseline Control Set. The Baseline Control Set discussed above is a set of control aligned with the DETER level of the Segmentation Model is mandatory for HMG ICT systems and these controls will typically manage risks to information up to IL3 for Confidentiality, Integrity and Availability. Security controls for IL3 and IL4 fall under the category of DETECT and RESIST level, whereas the DEFEND level will treat all risks to information up to IL5 and IL6.

6.4.1.9 Develop an assurance plan

It is important that all security measures perform the functions that are expected of them and that the assessed baseline controls are maintained throughout the lifecycle of the system. A plan must be developed that describes how appropriate assurance will be gained in all controls applied to manage information risk. This includes the baseline controls as well as controls derived at higher levels of the Segmentation Model. Additionally, the plan must describe any assurance gaps (i.e. a gap between the required and delivered assurance activities).

6.4.1.10 Security case

The security case describes how all of the identified risks have been satisfactorily treated. It should document all significant decisions and their justification. The security case should contain the prioritized list of risks from CESG IAS, statement of conformance to baseline controls including any exceptions, assurance plan and any gaps in the application of controls as well as the plan to address these gaps. Figure 6.3 depicts the CESG IAS risk assessment and treatment methods. The target network security architecture is derived at the "Apply Basic Controls" step and also holds the capability to execute step 8 "Determine Control Measures" as shown in Figure 6.3.

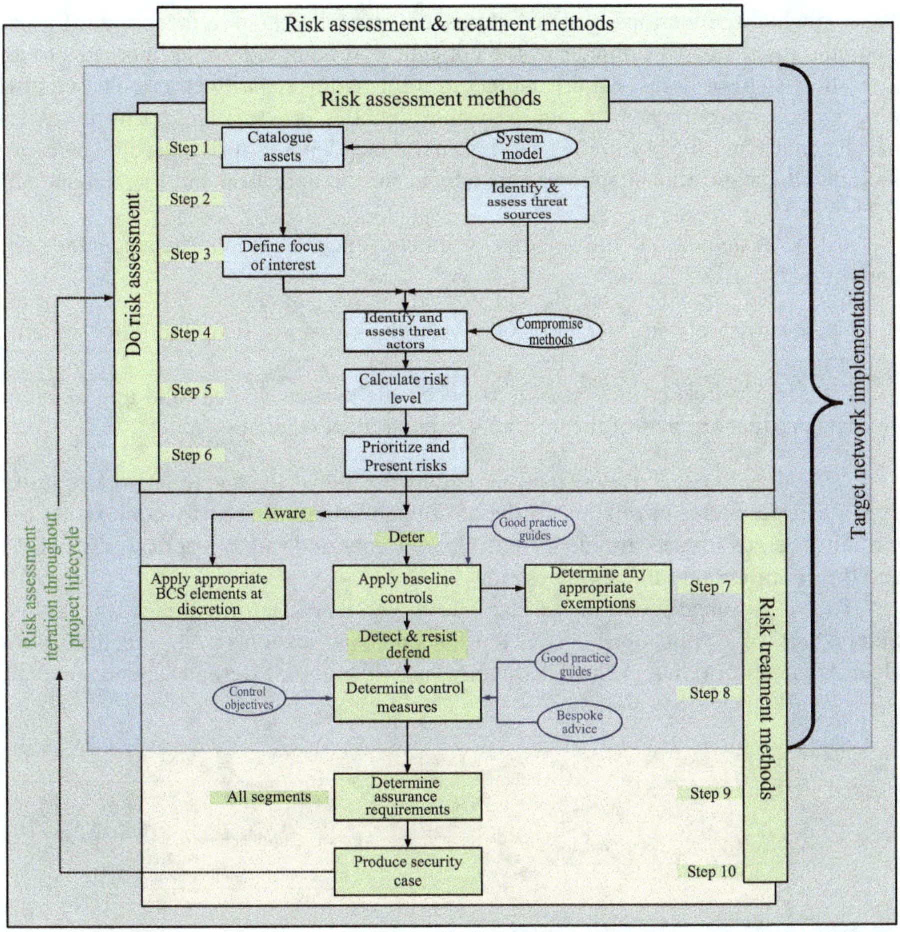

Figure 6.3 HMG IAS risk assessment and treatment method

6.4.2 Identification of target network security context

The CSG IAS standard suggested a domain-based modelling technique to catalogue the system in order to identify the information assets that need protection. An asset as defined in CSG IAS is "anything, which has value to an organization, its business operations and its continuity."

The adoption of the domain-based approach, however, is not mandatory in CSG IAS. In this chapter, the approach defined in ED-202 is adopted in cataloguing the system since the standard is catered towards aircraft systems. ED-202 used the term "security context of a system" to describe the security environment, the security perimeter and the assets of the system. Assets in ED-202 is defined as resources of the aircraft and systems which are subject to attack or may be used as part of an attack, including functions, systems, items, equipment, data, interfaces and information, and the security perimeter as the collection of boundaries and interfaces of an aircraft, system or items that are subject to information security attack. The context or external environment in which a system operates is characterized by its external dependencies. For example, the external dependencies at the aircraft level can include all organizational support procedures for the operation and maintenance of the aircraft.

At the system level, the security perimeter and its external dependencies are determined by ED-202:

- The security-related interactions between the system or item and other systems or items.
- The security-related interactions between the system or item and any external systems that are part of the aircraft's external dependencies.

According to the above definitions, Figure 6.4 shows the target network security assets and perimeter in relation to the aircraft information security context. At the aircraft level, two assets are identified: The Avionics and Cabin Applications (ACA) and the aircraft communication system.

The ACA consists of four assets: The Supervisor Control Application (SCA), Safety Service Application (SSA), the Cockpit Application (CA), and the Cabin Data Application (CDA). These assets are considered as external dependencies of

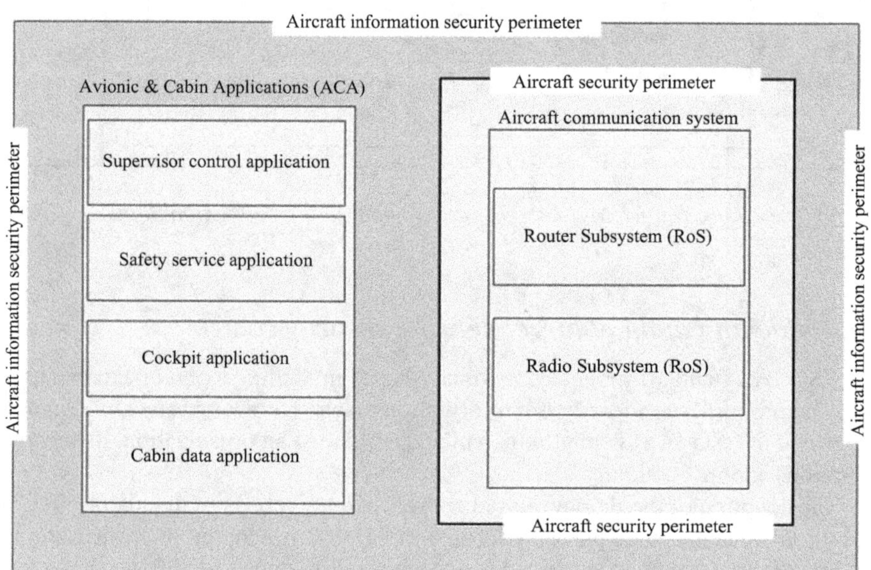

Figure 6.4 Network security assets at the aircraft level

the aircraft communications system. They interface directly with the aircraft communications system to enable control messages and application data/information generated by these assets will be passed through the aircraft communications system for transmission outside of the aircraft.

In this chapter, it is assumed that the four assets at the ACA level are physically segregated and secure. The main security objective of target network is to ensure that control messages and application data/information from/to the SCA, SSA, CA and the CDA are also securely protected within the target network security perimeter.

At the aircraft communications system level, two assets are identified: the RoS and the Radio subsystem (RaS). In the sections that follow, the assets and security perimeter, the external dependencies of the RoS and RaS are described in detail.

6.4.2.1 Assets and security perimeter of the RoS

At the RoS level, two assets are identified: The Router and the Communication Manager.

Router

The equipment, its network interfaces, logical data flow, external dependencies of the Router is shown in Figure 6.5.

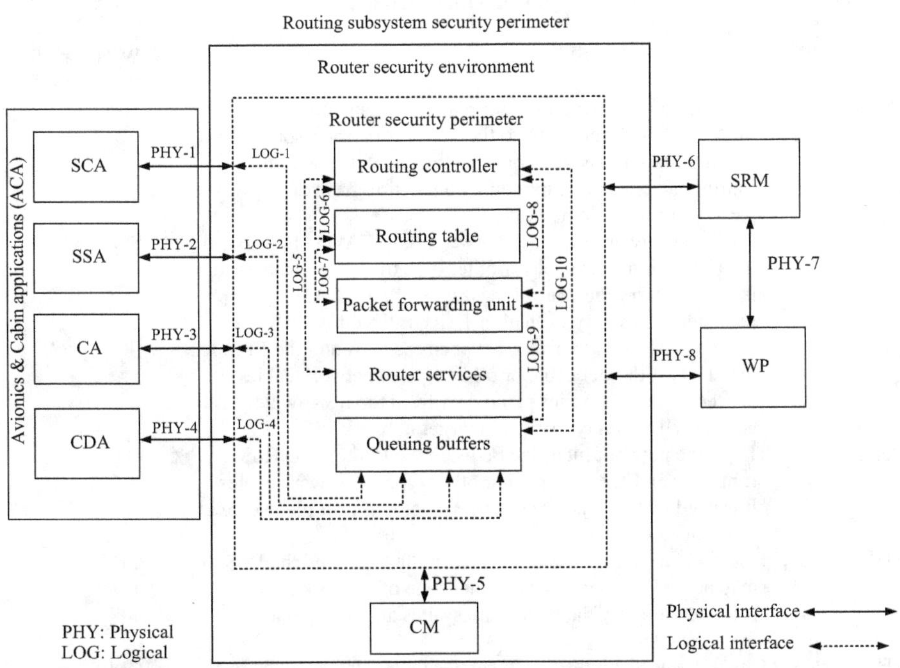

Figure 6.5 Security perimeter, assets and external dependencies of the Router in the RoS

Table 6.2 presents an overview of the functions of the assets in the RoS/Router system.

Table 6.2 Router assets descriptions

Assets at router level	Functions	External dependencies
Routing Controller	The routing controller performs important intelligent functions such as executing operating system instructions which includes system initialization, routing and switching functions, path computation, and routing table maintenance.	ACA Applications CM
Routing Table	This is used to store the latest routing information (source and destination addresses) to enable the router to route the packets generated by the SCA, SSA, CA, CDA to the correct radio link for transmission over the transport segment. The routing table is managed by the router operating system (IOS). Communication manager is the entity responsible for creating, deleting and updating entries in the router table.	N/A
Routing Services	These are built-in services supported by the router which includes SNMP, BootP and DHCP	CM
Queuing Buffer	This is used to store the user plane and control data from the SCA, SA, CA and CDA in a buffer for processing.	ACA CM RM WP (Waveform Processor)
Packet Forwarder	This is used to forward the user plane data generated by the SCA, SSA, CA and CDA to the waveform processors once the radio link is assigned or once the entry is made in the routing table for corresponding connection from any particular traffic domain.	N/A
PHY-1	This is the network interface between the SCA and the RoS. It may be an Ethernet cable or other type of network cables. It is used to carry the control messages of the SCA, which are considered as safety-critical in TARGET NETWORK.	SCA
PHY-2	This is the network interface between the SSA and the RoS. It may be an Ethernet cable or other type of network cables. It is used to carry the user plane data from the SSA, which are considered as safety-critical in target network.	SCA, SSA
PHY-3	This is the network interface between the CA and the RoS. It may be an Ethernet cable or other type of network cables. It is used to carry the control messages and user plane data from the CA.	CA
PHY-4	This is the network interface between the CDA and the RoS. It may be an Ethernet cable or other type of network cables. It is used to carry the control messages and user plane data from the CDA.	CDA
PHY-5	This is a physical interface. It carries the control/management plane messages from the CM related to QoS support and radio resource management, mobility management, security management, network management.	CM

Table 6.2 Continued

PHY-6	This is a physical interface for forwarding SCA, CDA and CM control messages from the router to the RM for adaptation.	RM
PHY-7	This is a physical interface between the RM and WP to carry out adaptation and configuration operations.	RM, WP
PHY-8	This is a physical interface for routing the SCA, SA, CA and CDA user plane data from the router to the WP.	WP

Communication Manager (CM)

The Communication Manager is a part of the RoS and is a piece of software hosted in a computer that interfaces directly with the Router. Figure 6.6 shows the network interface, logical data flow and the external dependency of the CM.

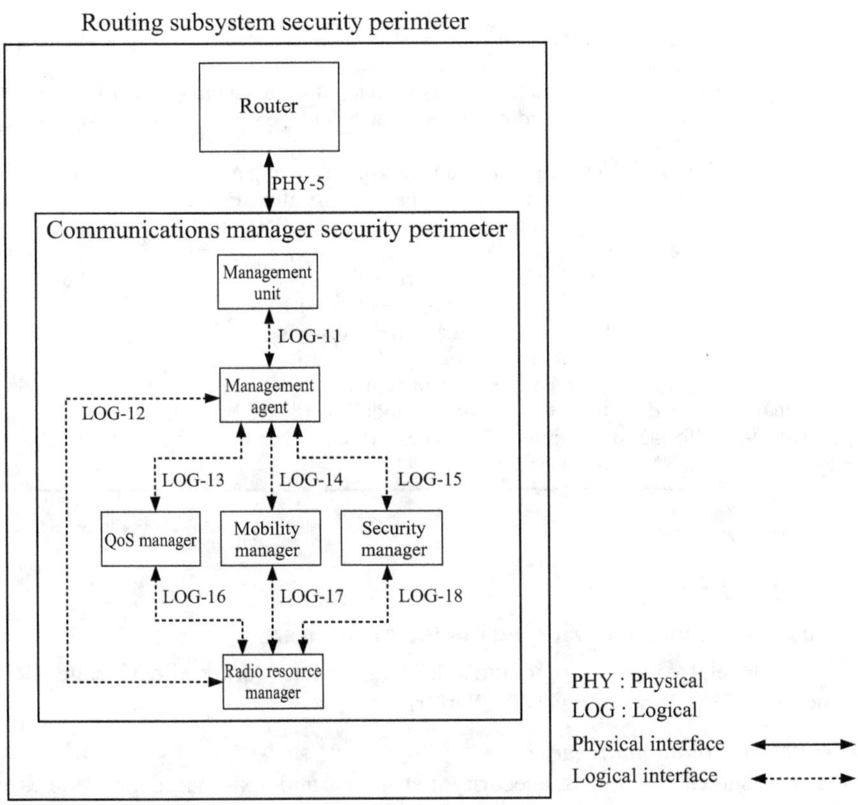

Figure 6.6 Security perimeter, assets and external dependencies of the CM in the RoS

Table 6.3 presents an overview of the functions of the assets in the CM.

Table 6.3 CM assets description

Assets at CM level	Functions	External dependencies
NMU	This is a logical function that interacts with the management agents to perform management plane functions and is responsible for logging all network management-related data including fault management, performance management, security management and configuration. It also communicates with the management agents in the RM and the WP to perform management plane functions.	Router RM WP
SM	This is a logical function that is used to set up the security policies for assets in ACA and for the communication between router and other target network components such as CM and RM.	Router
QM	This is a logical function that is used to set up the QoS policies	Router
RRM	This is a logical function that is used to allocate the radio link resource according to the security and QoS policies set for the individual applications.	ACA RM, Router
MM	This is a logical function that is used for carrying out all mobility-related functions such as address allocation and handover management. It interacts with the RRM for handover execution.	Router
PHY-5	This is a physical interface. It carries the control/management plane messages from the CM related to QoS support and radio resource management, mobility management, security management and network management.	Router
Management Agent (MA)	MA is defined in the functional model. It collects FCAPS information for internal modules, which is then passed to the NMU for processing.	N/A

6.4.2.2 Assets and security perimeter of the RaS

At the RaS level, two assets are identified: The target network Radio Manager (RM) and the Waveform Processing (WP) Platform.

Target network radio manager

Figure 6.7 shows the assets, security perimeter and external dependencies of the RM.

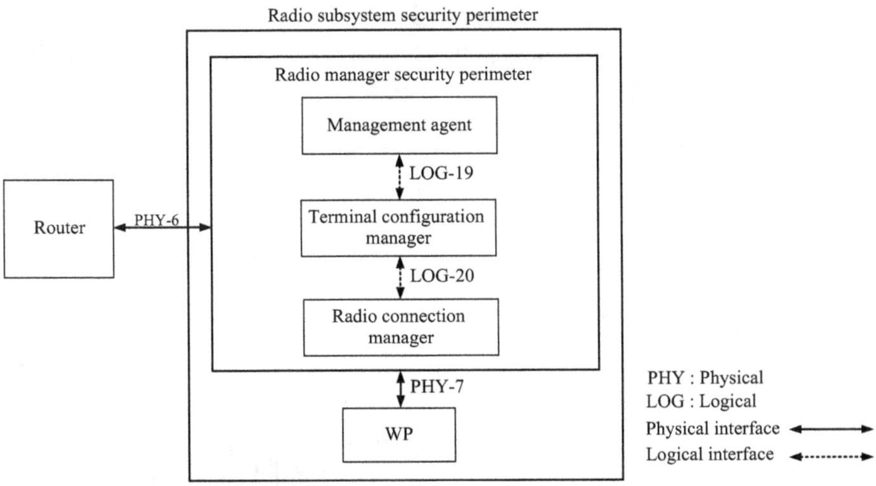

Figure 6.7 Security perimeter, assets and external dependencies of the RM in the RaS

Table 6.4 presents an overview of the functions of the assets in the RM.

Table 6.4 RM assets, their functions and dependencies

Assets at RM level	Functions	External dependencies
Management Agent (MA)	MA is defined in the functional model. It collects FCAPS information, which is then passed to the NMU for processing.	Router CM
Terminal Configuration Manager	The Terminal Configuration Manager performs the installation/uninstallation of the radio applications on the RS. It is also responsible for the management of the radio parameters of the radio applications.	WP Router CM
Radio Connection Manager	The RCM in the target network Radio Manager (RM) provides the functions for activating/deactivating unified radio applications according to user requests, and for the overall management of application data flows, which can also be switched from one radio application to another. The protocol adaptation is required in the RCM in order to provide a unified control interface to the higher-level control functional entities. The RCM acts as the MIHF to provide media independent command, information and event services.	WP Router CM
PHY-6	This is a physical interface for forwarding SCA, CDA and CM control messages from the router to the RM for adaptation.	Router
PHY-7	This is a physical interface between the RM and WP to carry out adaptation and configuration operations.	WP

Waveform processor platform

Figure 6.8 presents the assets, security perimeter and external dependencies of the RM.

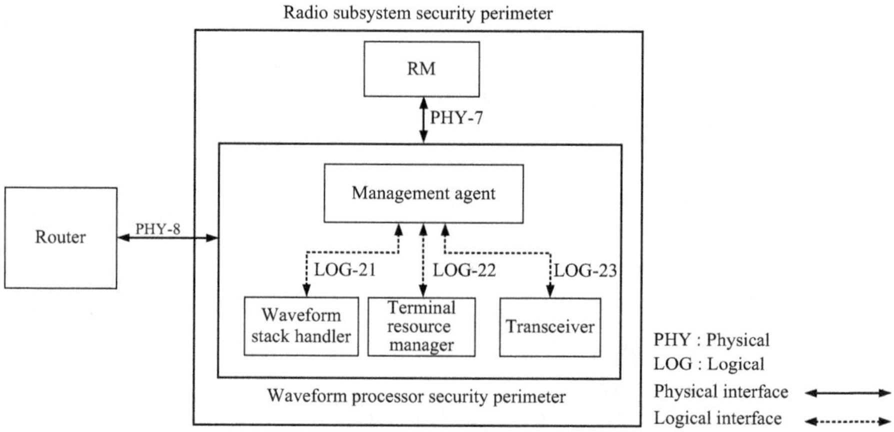

Figure 6.8 Security perimeter, assets and external dependencies of the WP in the RaS

Table 6.5 presents an overview of the functions of the assets in the RM.

Table 6.5 WP assets descriptions

Assets at WP level	Functions	External dependencies
Management Agent	MA is defined in the functional model. It collects FCAPS information, which is then passed to the NMU for processing.	CM Router
Terminal Resource Manager	The Terminal Resource Manager provides the functions to manage the hardware resources (such as processor, memory and storage etc.) of the WP in order to share them simultaneously among the active radio waveform applications, while guaranteeing their real-time requirements.	RM
Waveform Stack Handler	The Waveform Stack Handler sets up the lower layer connections of the radio waveforms through the waveform library that contains the software codes of the protocol stacks of the standardized radio technology, which can be dynamically executed and configured in the RS.	RM Router
Transceiver	The Transceiver is a reconfigurable device comprising both a transmitter and a receiver for a SDR.	N/A
PHY-7	This is a physical interface between the RM and WP to carry out adaptation and configuration operations.	RM
PHY-8	This is a physical interface for routing the SCA, SA, CA and CDA user plane data from the packet forwarder to the WP.	Router

6.5 Target network security risk assessment using CESG IAS

Table 6.6 presents step 1 of target network risk assessment according to the CESG IAS standard. The first and second columns represent the list of identified assets identifier and their names, the third column presents the description of each identified asset in the system.

Table 6.6 TARGET network assets list

Target network asset list		
Asset identifier	**Asset name**	**Description/notes**
1	Routing Controller	The routing controller performs important intelligent functions such as executing operating system instructions which includes system initialization, routing and switching functions, path computation, and routing table maintenance.
2	Routing Services	These are built-in services supported by the router which includes SNMP, BootP and DHCP
3	Queuing Buffer	This is used to store the user plane and control data from the SCA, SA, CA and CDA in a buffer for processing.
4	Packet Forwarder	This is used to forward the user plane data generated by the SCA, SSA, CA, CDA to the waveform processors once the radio link is assigned or once the entry is made in the routing table for corresponding connection from any particular traffic domain.
5	PHY-1	This is the network interface between the SCA and the RoS. It may be an Ethernet cable or other type of network cables. It is used to carry the control messages of the SCA, which are considered as safety-critical in target network.
6	PHY-2	This is the network interface between the SSA and the RoS may be an Ethernet cable or other type of network cables. It is used to carry the user plane data from the SSA, which are considered as safety-critical in target network.
7	PHY-3	This is the network interface between the CA and the RoS may be an Ethernet cable or other type of network cables. It is used to carry the control messages and user plane data from the CA.
8	PHY-4	This is the network interface between the CDA and the RoS may be an Ethernet cable or other type of network cables. It is used to carry the control messages and user plane data from the CDA.
9	PHY-5	This is a physical interface. It carries the control/management plane messages from the CM related to QoS support and radio resource management, mobility management, security management, network management.
10	PHY-6	This is a physical interface for forwarding SCA, CDA and CM control messages from the router to the RM for adaptation.
11	PHY-7	This is a physical interface between the RM and WP to carry out adaptation and configuration operations.
12	PHY-8	This is a physical interface for routing the SCA, SA, CA and CDA user plane data from the router to the WP.
13	NMU	This is a logical function that interacts with management agent to perform management plane functions and is responsible for logging all network management-related data including fault management, performance management, security management and configuration. It also communicates with the management agents in RM and WP to perform management plane functions.

(*Continues*)

Table 6.6 (Continued)

Target network asset list		
Asset identifier	**Asset name**	**Description/notes**
14	SM	This is a logical function that is used to set up the security policies for assets in ACA and for the communication between router and other target network components such as CM and RM.
15	QM	This is a logical function that is used to set up the QoS policies.
16	RRM	This is a logical function that is used to allocate the radio link resource according to the security and QoS policies set upon the individual applications.
17	MM	This is a logical function that is used for carrying out all mobility-related functions such as address allocations and handover management. It interacts with the RRM for handover execution.
18	PHY-5	This is a physical interface. It carries the control/management plane messages from the CM related to QoS support and radio resource management, mobility management, security management, network management.
19	Management Agent_RoS	MA is defined in the functional model. It collects FCAPS information for internal modules, which is then passed to the NMU for processing.
20	Management Agent_RaS	MA is defined in the functional model. It collects FCAPS information, which is then passed to the NMU for processing.
21	Terminal Configura-tion	The Terminal Configuration Manager performs the installation/uninstallation of the radio applications on the RS. It is also responsible for the management of the radio parameters of the radio applications.
22	Radio Connection Man	The RCM in the target network Radio Manager (RM) provides the functions for activating/deactivating unified radio applications according to user requests, and overall management of application data flows, which can also be switched from one radio application to another. The protocol adaptation is required in the RCM in order to provide a unified control interface to the higher-level control functional entities. The RCM acts as the MIHF to provide media-independent command, information and event services.
23	Management Agent_WP	MA is defined in the functional model. It collects FCAPS information, which is then passed to the NMU for processing.
24	Terminal Resource Manager	The Terminal Resource Manager provides the functions to manage the hardware resources (such as processor, memory and storage, etc.) of the WP in order to share them simultaneously among the active radio waveform applications, while guaranteeing their real-time requirements.
25	Waveform Stack Handler	The Waveform Stack Handler sets up the lower layer connections of the radio waveforms through the waveform library that contains the software codes of the protocol stacks of the standardized radio technology, which can be dynamically executed and configured in the RS.
26	Transceiver	The Transceiver is a reconfigurable device comprising both a transmitter and a receiver for a SDR.

6.6 Define and assess threat sources

Step 2 of risk assessment process produces a detailed list of threat sources and also provides the attributes associated with each threat source. The attributes of a threat source include Capability of a threat source, Priority of a threat source and Threat Level associated with the capability and priority of a threat source.

CESG IAS has defined five categories of threat sources in Table 1 of [33] based on the capabilities of the threat sources and these categories range from 1 to 5, reflecting the capability of a threat source on cause an impact on a system [33]:

1. Very Little: The threat source has almost no capabilities or resources.
2. Little: The threat source has very modest capabilities and resources.
3. Limited: The threat source has modest capabilities and resources.
4. Significant: The threat source is capable and has significant resources.
5. Formidable: The threat source is extremely capable and well resourced.

The priority reflects the urgency level or the prime objective of a threat source and CESG IAS has defined five levels of these priorities for threat sources in Table 2 of [33] ranging from 1 to 5:

1. Very Low (Indifferent): The threat source is very unlikely to attempt any attack on the system.
2. Low (Curious): The threat source will attempt to attack the system on an occasional basis and is willing to devote very few resources to the attack(s).
3. Medium (Interested): The threat source will attempt to attack the system on a frequent basis and is willing to devote few resources to the attack(s).
4. High (Committed): The threat source will attempt to attack the system on a persistent and frequent basis and is willing to devote several resources to the attack(s).
5. Very High (Focused): The threat source has a primary aim to attack the system.

Threat levels are based on the capability level and the priority of threat sources as shown in Table 6.7

Table 6.8 presents the outcomes of step 2 for the risk assessment process. The "Source of Threat Assessment column" indicates whether the threat assessment is carried out internally by an organization (In-House) or the threat assessment is carried by an accredited external organization. CESG, CPNI and Defence Intelligence Staff (DIS) for Ministry of Defence (MOD) are the examples of accredited external organizations.

Table 6.7 Threat levels for threat sources [33]

		Capability level of threat source				
		1 Very little	2 Little	3 Limited	4 Significant	5 Formidable
Priority of threat source	1: Indifferent	Negligible	Negligible	Low	Low	Moderate
	2: Curious	Negligible	Negligible	Low	Moderate	Substantial
	3: Interested	Negligible	Low	Moderate	Substantial	Severe
	4: Comitted	Low	Low	Moderate	Severe	Severe
	5: Focused	Low	Moderate	Substantial	Severe	Critical

The "Influencer" column indicates a threat source who wishes to have an occurrence of compromise of a system security. An influencer may launch an attack directly or may use others to carry out the desired action. An assessment should be made as to whether it is believed that the influencer will attempt to influence threat actors (such as Normal Users) through coercion or bribery. A simple Yes or No should be recorded in this column.

Table 6.8 Threat source list

Threat sources					
Source name	**Description (and rationale)**	**Threat level**	**Source of threat assessment**	**Influencer (Y/N)**	**Threat actor (Y/N)**
1 - Terrorists	The goal of this threat source group is to spread terror throughout the world by causing casualties, weakening the economy and deploying attacks on public places and or transport systems. In case of target network, possible attacks can include making the communication system unavailable using DoS/DDoS attacks, modifying or disrupting the communications through message modification attacks and stealing classified information from the communication links or devices.	Critical	In-House	Y	Y
		Critical	In-House	Y	Y
		Critical	In-House	Y	Y
2 - Industrial spies	The goals of attack from this group of threat sources are mostly motivated by profits. Attacks can include assaults on competitor's communication infrastructure, theft of trade secrets, and gain access and blackmail affected industry using potential public exposure as a threat.	Critical	In-House	Y	Y
		Critical	In-House	Y	Y
		Critical	In-House	Y	Y
3 - Hackers	This group of threat sources can pose a relatively high threat through an isolated or brief disruption of the system, causing serious damage, including extensive hardware equipment and or software damage or loss of life. The likelihood of (an exceptionally skilled and malicious) a hacker attempting and succeeding in launching harmful	Severe	In-House	Y	Y
		Severe	In-House	Y	Y

(Continues)

Table 6.8 (*Continued*)

	attacks is increasing with the increasing numbers of hackers' population. The goals of this class of threat sources are to get gain unauthorized access to the restricted resources, disruption of networks and associated computer systems, weakening or breaking the security infrastructure, earning money and achieving recognition with an exploit.	Severe	In-House	Y	Y
		Severe	In-House	Y	Y
		Severe	In-House	Y	Y
4 - Criminal groups	This group of threat sources seeks to attack systems for financial profit. Specifically, organized crime groups use spam, phishing and spyware/malware to commit identity theft and online fraud to gain unauthorized access to the system and then blackmail the responsible person for public exposure of information which they might have stolen. The criminal groups can also hand over the sensitive information to any other third party for the sake of money.	Moderate	In-House	Y	Y
		Moderate	In-House	Y	Y
		Moderate	In-House	Y	Y
5- Phishers	The threat source group who executes phishing schemes in an attempt to steal identities or information for financial profit. Phishers may also use spam and spyware/malware to accomplish their objectives. In target network, either the phishers or another threat actor used by the phishers execute the phishing schemes on the target network network with direct or indirect access to the network.	Severe	In-House	Y	Y
		Low	In-House	Y	Y
		Moderate	In-House	Y	Y
6 -Spyware/malware authors	This group of threat sources write software programs with malicious commitment. There have been several cases of destructive computer viruses and worms harming data on the computers on the networks. Common examples are explore.zip worm, Chernobyl virus, code red and slammer virus etc. The likelihood of spreading spyware/malware is higher from the cabin data domain (CDA) of target network as the actors in this domain is highly untrusted; however, actors from other domains may also be involved in spreading malicious viruses.	Substantial	In-House	Y	Y
		Low	In-House	Y	Y
		Substantial	In-House	Y	Y

A threat source can also be a threat actor. The Analyst should make a judgement as to whether they believe this to be the case and enter a Yes or No in the Threat Actor column.

6.7 Define focus of interest (FoI)

Table 6.9 presents the outcomes of step 3 for the risk assessment process. The values in column 1, 2 and 3 are a high-level entities, set of assets associated with those entities and rational behind grouping the set of assets together

Table 6.9 Network focus of interest

Target network focus of interest		
FoI name	**Assets**	**Rationale**
CM	9, 13, 14, 15, 16, 17, 18	This includes the physical links between the router and the CM and between the sub-components of CM such as SM, QM, MM, RRM and NMU. An attacker may intend to target any component of CM for malicious activities like the disruption of communication (Availability attack).
Router	1, 2, 3, 4, 5, 6, 7, 8, 9, 10, 12, 18	This includes all the features in the router along with the interface of router to different traffic domains, CM, RM and WP. An adversary may launch confidentiality, integrity and availability attacks on the router from any interfaces.
RM	10, 11, 19, 20, 21, 22	This includes the interface between router and RM and the sub-components of RM such as management agent, TCM and RCM. An attacker may target RM via router to malfunction the RM's sub-components such as TCM and RCM etc.
WP	11, 12, 23, 24, 25, 26	This includes the interface between the WP and the router and between sub-components of the WP such as management agent, waveform stack handler, TRM and transceiver etc. An attacker may launch attacks on WP via router or from the ground to malfunction it and disrupt or monitor the communication traffic.

6.8 Define threat actors

Similar to threat source, CESG IAS has defined five categories of threat actors based on the native and enhanced capabilities of the threat actors in Table 4 of [33] and these categories range from 1 to 5, reflecting the native and enhanced capability of a threat actor to cause an impact on a system. These five categories of capability include: Very Little (1), little (2), Limited (3), Significant (4) and Formidable (5) in the increasing order of threat actor's native or enhanced capability.

The native or enhanced motivation reflects the urgency level or the prime objective of a threat actor and IAS has defined five levels of these native and enhanced motivation in Table 5 of [33] for threat actors ranging from 1 to 5. These levels are: Very Low:1 (Indifferent), Low:2 (Curious), Medium:3 (Interested), High:4 (Committed) and Very High:5 (Focused).

Threat levels of threat actors are based on the native/enhanced capability level and the native/enhanced threat sources and CESG IAS has defined six different threat levels for threat sources. These six different threat levels are shown below in Table 6.10.

Table 6.10 Treat levels for threat actors [33]

		CAPABILITY LEVEL OF THREAT ACTOR				
		1 VERY LITTLE	2 LITTLE	3 LIMITED	4 SIGNIFICANT	5 FORMIDABLE
PRIORITY OF THREAT ACTOR	1: INDIFFERENT	Negligible	Negligible	Low	Low	Moderate
	2: CURIOUS	Negligible	Negligible	Low	Moderate	Substantial
	3: INTERESTED	Negligible	Low	Moderate	Substantial	Severe
	4: COMITTED	Low	Low	Moderate	Severe	Severe
	5: FOCUSED	Low	Moderate	Substantial	Severe	Critical

6.9 Identify estimate and priorities risks

This step is one of the key parts of the risk assessment process as it results in the generation of a list of risks and an indication of a risk level for each identified risk. This step is repeated for each FoI/Threat actor group combination. The CESG IAS considers risk level as the combination of maximum Business Impact Level (BIL) and threat level. IAS has defined a six-point scale for risk level: Very Low, Low, Medium, Medium-High, High and Very High. Table 6.11 shows different risk levels for several combinations of Max BILs and threat levels.

Table 6.11 Risk levels [33]

		Threat level					
		Negligible	Low	Moderate	Substantial	Severe	Critical
Buisness impact level (BIL)	BIL0	Very Low	Very Low	Very Low	Very Low	Very Low	Very Low
	BIL1	Very Low	Very Low	V ery Low	Low	Low	Low
	BIL2	Very Low	Low	Low	Medium	Medium	Medium
	BIL3	Very Low	Low	Medium	Medium	Medium-High	Medium-High
	BIL4	Low	Medium	Medium	Medium-High	High	High
	BIL5	Medium	Medium	Medium-High	High	High	Very High
	BIL6	Medium	Medium	Medium-High	High	Very High	Very High

This is the final step which actually prioritizes identified risks and presents it in an easily understandable format. Table 6.12 represents the outcomes after applying step 6. Table 6.12 has three distinct columns (i.e. Risk ID, Description and Risk level). This is presenting maximum of three samples from each risk level identified in the

Table 6.12 Prioritized risk-reduced list in target network

Prioritized risk reduced list in target network					
Risk ID	**Description**	**Risk level**	**Risk ID**	**Description**	**Risk level**
R1	Passengers (influenced by Terrorists), as a Normal User, Deliberately disrupts Router, compromising its availability and having a potential Business Impact at BIL 5.	Very High	R8	Pilot/Copilot (influenced by Terrorists), as a Handler, Tampers with equipment in CM, compromising its integrity and having a potential Business Impact at BIL 5.	Medium-High
R2	Passengers (influenced by Terrorists), as a Bystander, Tampers with equipment in Router, compromising its availability and having a potential Business Impact at BIL 5.	Very High	R9	Crew Members (influenced by Criminal groups), as a Privileged User, Changes the configuration of Router, compromising its integrity and having a potential Business Impact at BIL 5.	Medium-High
R3	Passengers (influenced by Terrorists), as a Normal User, Changes the configuration of Router, compromising its availability and having a potential Business Impact at BIL 5.	Very High	R10	Maintenance staff (influenced by Criminal groups), as a Privileged User, Changes the configuration of CM, compromising its integrity and having a potential Business Impact at BIL 5.	Medium
R4	Passengers (influenced by Hackers), as a Bystander, Tampers with equipment in Router, compromising its integrity and having a potential Business Impact at BIL 5.	High	R11	Pilot/Copilot (influenced by Criminal groups), as a Privileged User, Changes the configuration of RM, compromising its integrity and having a potential Business Impact at BIL 5.	Medium
R5	Crew Members (influenced by Terrorists), as a Bystander, Tampers with equipment in Router, compromising its availability and having a potential Business Impact at BIL 5.	High	R12	Pilot/ Copilot, as a Privileged User, Accidentally disrupts Router, compromising its integrity and having a potential Business Impact at BIL 5.	Medium
R6	Crew Members (influenced by Terrorists), as a Bystander, Impersonates a user of Router, compromising its availability and having a potential Business Impact at BIL 5.	High	R13	Maintenance Staff, as a Privileged User, Accidentally disrupts RM, compromising its availability and having a potential Business Impact at BIL 5.	Low
R7	Passengers, as a Normal User, Accidentally disrupts Router, compromising its availability and having a potential Business Impact at BIL 5.	Medium-High	R14	Maintenance Staff, as a Privileged User, Accidentally disrupts WP, compromising its availability and having a potential Business Impact at BIL 5.	Low

target network architecture. The actual number of risks identified and their level are presented in the “Risk Treatment” section.

The following subsections demonstrate the steps involved in CESG IAS risk treatment. The process of risk treatment is a natural continuation of CESG IAS risk assessment. The method of risk treatment described in CESG IAS comprises four steps which are as follows:

- Applying the baseline controls
- Treat risk at different levels of segmentation model
- Develop assurance plan
- Produce the security case

The following subsections address the risk treatment starting from step 7 of CSG IAS standard.

6.10 Apply baseline countermeasure controls

This is the first step in risk treatment. In this step, a complete set of baseline controls must be applied to the entire information system, except for the contexts where the impact of the information system is lesser than equal to the IL1 for each security attribute such as confidentiality, integrity and availability. Other contexts are inappropriateness, where an individual control is omitted from the baseline control set. A set of countermeasures for each vulnerable asset in the system have been developed and applied to secure the asset from potential threats.

6.11 Risk treatment

In this step, the segmentation model with four different segmentations as described in [33] is utilized to align the impact level implied by any particular threat with the risk level. Four segments defined in CSG IAS risk treatment are as follows:

> AWARE: it aims to provide the basic defences against common nontargeted and opportunistic attacks by unsophisticated threat actors with the capability of Very Little or Little capability. The levels of capabilities for threat actors are described in IAS Table 4 of [33]. These threat actors may download and use publicly available hacking tools, but will not be capable of engaging in any personalized development or specific targeting of attacks.
>
> DETER: it aims to provide protection against targeted attacks from the Internet by skilled threat actors with Limited capability using freely available tools with some custom-made modification. The threat actor may be able to download a tool from the Internet, understand how that tool works and how it exploits vulnerability, so that they can modify the tool to function in a way that may make a successful compromise more likely or more likely to avoid simple detection systems.
>
> DETECT and RESIST: provide protection against targeted attacks from skilled threat actors with significant capability using personalized tools that exploit

known vulnerabilities in the target system. These threat actors may be funded and motivated by threat sources such as investigative journalists, criminal groups or terrorists. This segment also aims to resist attacks from Foreign Intelligence Services (FIS) and to detect such attempted attacks when they occur, although it may not stop them.
DEFEND: DEFEND aims to prevent all attacks against a system from the most sophisticated and highly capable threat actors with formidable capability, such as FIS.

Table 6.13 represents the relationship between segments in the segmentation model and the risk levels.

Table 6.13 Mappings between risk levels and segments [33]

Risk Levels	**Segments**
Low	AWARE
Medium	DETER
Medium-high	DETECT and RESIST
>= High	DEFEND

Application of the Baseline Control Set can be considered as an application of DETER level controls to all risks. This will typically treat all risks to IL2 and IL3 information, up to and including Medium risk level. All Good Practice Guides (GPGs) in CESG portfolio will provide advice that is aligned to DETER and will support the Baseline Control Set ...[33]. The identified risks which are categorized as DETECT and RESIST and DEFEND require both baseline controls as well as advance controls to avoid the potential hazards of these risks. Table 6.14 shows the classification of risks from Table 6.13 into different segments of the segmentation model defined in CESG IAS risk treatment. Table 6.14 comprises three distinct columns which are described as follows:

- **Risk ID:** It uniquely identifies each detected risk
- **Segment:** It classifies how the risk has been handled once detected
- **Risk Level:** The risk level classifies the severity of risk in the system

For the simplicity and ease of presentation, the results presented in Table 6.14 are reduced to maximum of three same level of risks samples from actual number of risks identified and treated. The statistical details of actual number of risks identified and treated are presented in Table 6.15.

The application of ED-202 [1] and CESG IAS [33] standard on the target future avionics network produced a large amount of tabular data which have been reduced to include in this chapter. Particularly the list of risks shown in Table 6.12 and the list of segmented risk levels presented in Table 6.14 are reduced to avoid capturing too much of space in this chapter. Table 6.15 presents the actual statistics of security risk assessment on targeted future avionics network which were obtained after applying CESG

Table 6.14 Risk segmentation after applying countermeasures for risk mitigation

Risk ID	Segment	Risk level	Risk ID	Segment	Risk level	Risk ID	Segment	Risk level
R1	DEFEND	Very High	R6	DEFEND	High	R11	DETER	Medium
R2	DEFEND	Very High	R7	DETECT and RESIST	Medium-High	R12	DETER	Medium
R3	DEFEND	Very High	R8	DETECT and RESIST	Medium-High	R13	AWARE	Low
R4	DEFEND	High	R9	DETECT and RESIST	Medium-High	R14	AWARE	Low
R5	DEFEND	High	R10	DETER	Medium	-	-	-

Table 6.15 Detailed statistics of security risk assessment and treatment using CESG IAS [33]

	Total identified	Defended	Detected and resisted	Deter	Aware
Very High	6	6	–	–	–
High	34	34	–	–	–
Medium high	44	–	44	–	–
Medium	32	–	–	32	–
Low	2	–	–	–	2

IAS [33]. The first column presents different types of risks identified during risk assessment process. The second column presents the total number of risks identified against each category of identified risks, the third column presents total number of defended risks from each category of identified risks, column four presents the total number of risks detected and resisted from "Medium High" category of identified risks. Finally, column 5 and column 6 presents the total number of risks "deterred" and the total number of risks which the system is "aware of" against each category of risks identified during risk assessment and treatment process.

6.12 Conclusion

The main purposes of this chapter are to review the ongoing standardization activities and state-of-the-art security mechanisms in relation to aircraft information security. The proposed work also produced and utilized an amalgamation of security risk assessment standards from avionics and traditional IP network domains. This chapter first gives a summary introduction on the evolution of aeronautical communications. It then provides a comprehensive review on activities carried out by different standardization bodies in relation to aircraft information security and airworthiness security. This is followed by a review of security challenges in aeronautical communications together with a review of the state-of-the-art security techniques to tackle those challenges. A risk analysis is carried out to identify security assets and their security perimeters, vulnerability points, threats and attacks, the impacts of such threats and the risk levels of threats regarding the target network architecture. According to the risk analysis, security countermeasures are identified, which leads to the target network security architecture design. The method in the identification of the target network security architecture primarily follows the guidelines specified in EUROCAE ED-202 and CESG IS1 standards. This chapter only serves as a recommendation for defining security processes and their implementation in relation to the target network architecture. The security mechanisms defined in this document should by no means be considered up to the aircraft certification level for handling information security threats which may ultimately affect aircraft safety.

References

[1] EUROCAE-WG72, "ED202 – Airworthiness Security Process Specification", June 2014.

[2] ICAO, "Cyber Security for Civil Aviation," in Thirteen Air Navigation Conference, Montreal, October 2018.

[3] T. W. a. K. S. Anoop Singhal, "Guide to Secure Web Services Recommendations of the National Institute of Standards and Technology," National Institute of Standards and Technology, U.S, Department of Commerce August, 2007.

[4] W. Stallings, Network Security Essentials: Applications and Standards, 4th Edition ed. Pearson Education, Inc., publishing as [Prentice-Hall, 1 Lake Street, Upper Saddle River, NJ 07458]: Prentice Hall, 2010.

[5] I. X.800, "Security Architecture for Open System Interconnection for CCITT Applications," in Data communication networks, Open system interconnections, security architecture and applications, ed, 1991.

[6] R. Shirey, "RFC 4949: Internet Security Glossary," IETF2007.

[7] ICAO, "Manual on the Aeronautical Telecommunications Network (ATN) using Internet Protocol Suite (IPS) Standards and Protocols," Doc 9896, 2011.

[8] D. Harkins and D. Carrel, "IETF RFC-2409: The Internet Key Exchange," RFC 2409, 1998.

[9] S. Kent and K. Seo, "IETF RFC-4301: Security Architecture for the Internet Protocol," RFC 4301, 2005.

[10] ICAO, "Manual on the Aeronautical Telecommunications Network (ATN) using Internet Protocol Suite (IPS) Standards and Protocols," Doc 9896, 2010.

[11] S. Kent, "IETF RFC-4303: IP Encapsulating Security Payload (ESP)," IETF RFC-4303, December 2005.

[12] S. Kent, "IETF RFC-4302: IP Authentication Header," IETF RFC-4302, December 2005.
[13] C. Kaufman, "IETF RFC-4306: Internet Key Exchange (IKEv2) Protocol," December 2005.
[14] V. Manral, "IETF RFC-4835: Cryptographic Algorithm Implementation Requirements for Encapsulating Security Payload (ESP) and Authentication Header (AH)," IETF RFC-4835 April 2007.
[15] D. Cooper, S. Santesson, S. Farrell, S. Boeyen, R. Housley, and W. Polk, "IETF RFC-5280: Internet X.509 Public Key Infrastructure Certificate and Certificate Revocation List (CRL) Profile," IETF RFC-5280, May 2008.
[16] S. F. S. Kelly, "IETF RFC-4868: Using HMAC-SHA-256, HMAC-SHA-384, and HMAC-SHA-512 with IPsec," IETF RFC-4868, May 2007.
[17] J. Solinas, "IETF RFC-4753: ECP Groups for IKE and IKEv2," IETF RFC-4753, January 2007.
[18] J. S. D. Fu, "IETF RFC-4754: IKE and IKEv2 Authentication Using the Elliptic Curve Digital Signature Algorithm (ECDSA)," IETF RFC-4754 January 2007.
[19] W. Ford, R. Sabett, C. Merrill, and S. Wu, "IETF RFC-3647: Internet X.509 Public Key Infrastructure Certificate Policy and Certification Practices Framework," IETF RFC 3647, November 2003.
[20] A. N. V. Devarapalli, F. Dupont, "IETF RFC-4877: Mobile IPv6 Operation with IKEv2 and the Revised IPsec Architecture," IETF RFC-4877, April 2007.
[21] E. R. T. Dierks, "IETF RFC-5246: The Transport Layer Security (TLS) Protocol Version 1.2," IETF RFC-5246, August 2008.
[22] M. B. H. Krawczyk, R. Canetti, "IETF RFC-2104: HMAC: Keyed-Hashing for Message Authentication," IETF RFC-2104 February 1997.
[23] ECAC, https://www.ecac-ceac.org, Accessed 16 March 16, 2021.
[24] SEASAR, http://www.sesarju.eu, Accessed 16 March 16, 2021.
[25] R. Koelle and M. Hawley, "Sesar security 2020: How to Embed and Assure Security in System-of-systems Engineering?," in Integrated Communications, Navigation and Surveillance Conference (ICNS), 2012.
[26] WG-72, http://www.eurocae.net/working-groups/wg-list/41-wg-72.html, Accessed 16 March 16, 2021.
[27] SC216, http://www.rtca.org/content.asp?pl=108&sl=33&contentid=82, Accessed 16 March 16, 2021.
[28] R. L. Rivest, A. Shamir, and L. Adleman, "A method for obtaining digital signatures and public-key cryptosystems," Communications of the ACM, vol. 21, pp. 120–126, 1978.
[29] ATA, "Aviation Industry Standards for Digital Information Security" 2009.
[30] M. Olive, R. Oishi, and S. Arentz, "Commercial Aircraft Information Security An Overview of ARINC Report 811," 2006.
[31] ARINC, "ARINC Report 811: 811 Commercial Aircraft Information Security Concepts of Operation and Process Framework," ed: ~Aeronautical radio, INC., 20th December, 2005.
[32] CPNI, "Cyber Security in Civil Aviation," Presented at the Centre for the Protection of National Infrastructure, August 2012.

[33] CESG, “HMG IA Standard No. 1 Technical Risk Assessment,” Risk Assessment and Risk Treatment, 2009.
[34] CAAP, “Administration of Aircraft and Related Ground Support Network Security Programs,” CAAP 232A-1(0), March 2013.
[35] C. Office, “HMG Security Policy Framework,” October 2013.
[36] K. a. Martin, “Aviation Communication Infrastructure Security,” Presented at the Integrated Communications, Navigation and Surveillance Conference (ICNS), 2012.
[37] CAAP, “Administration of Aircraft and Related Ground Support Network Security Programs,” CAAP 232A-1(0), March 2013.

Chapter 7
Physical layer security in wireless HetNets avionics and satellite communication systems

Rameez Asif[1], *Muhammad Ali*[1], *Yim-Fun Hu*[1], *and George Oguntala*[2]

Today's aeronautical communications increasingly rely on wireless communication to exchange information within an aircraft, between aircrafts, between aircraft and the ground network, and between aircraft and the satellite. The future avionics communications industry foresees an amalgamation of different radio access technologies to provide sufficient bandwidth for the growing demand for avionics applications onboard the aircraft as well as on the ground. This amalgamation of different types of radio access technologies, on one hand, brings much-needed resources, but, on the other hand, it also leads to the growing numbers of cybersecurity attacks which point at different domains and attract a variety of cybersecurity attackers. Aeronautical networks consist of several interconnected devices that enable data management and ensure that the messages using the aircraft applications are exchanged within a given timeframe. This constraint makes the design of the communication system highly challenging because of the requirement of a guaranteed bandwidth and a pre-defined quality of service. These issues raise several security concerns as to the security of both personal and sensitive data. Almost within all wireless communication systems, the initial countermeasures for vulnerabilities such as authentication, integrity and confidentiality are employed in the top layers within the protocol stack using different encryption schemes that are based on mathematical operations and the security provided by the techniques is usually referred to as the computational security. Even though these techniques have shown to be very effective over the years but the research has shown that when the aim is to design a secure communication system, then the attributes of the physical layer such as its imperfections must also be considered, which can then result in additional security, e.g., in physical layer, white noise is considered an imperfection but this noise can be used to hide secure messages. If these techniques are employed in addition to the traditional security measures without the loss of data rate, then they can complement the traditional security techniques.

[1]Department of Biomedical and Electronics Engineering, University of Bradford, Bradford, UK
[2]Department of Health, Education and Life Sciences, Birmingham City University, Birmingham, UK

This chapter aims to highlight some of the secrecy mechanisms and provide some insight into the physical layer techniques that can be employed for a secure and reliable transmission. The chapter will also highlight the performance metrics that are employed to evaluate the channel secrecy capacity and secrecy outage probability. Finally, to sum up, an example of a physical layer attack is presented, and it is shown how the PHY-layer security technique can be employed as a countermeasure.

7.1 Introduction

Air traffic management (ATM) systems are going through a major modernisation process throughout the world, most notably the Single European Sky (SES) in Europe and the NextGen in the US. As the technological pillar of SES, SESAR (Single European Sky ATM Research) coordinates and concentrates all EU research and development (R&D) activities in ATM. As part of the ATM modernization process in Europe, new datalinks were proposed for aeronautical communications [1].

1. A ground-based, high-capacity, airport surface datalink system, referred to as the aeronautical mobile airport communications system (AeroMACS).
2. A ground-based datalink system for continental airspace in general, referred to as the L-band digital aeronautical communications system (LDACS).
3. A satellite-based datalink system for the oceanic and remote continental environment. It can also be used to complement the ground-based systems.

In addition, aircraft-to-aircraft communications or multihop/ad hoc aircraft communications are also considered in future ATM systems.

Figure 7.1 [2] shows the use of heterogeneous networks (HetNets) and radio technologies for air traffic control to provide an always-connected infrastructure for air-ground (A/G) and air-air (A/A) communications.

Figure 7.2 presents an overall high-level architecture of the avionics network consisting of both the onboard networks and air-to-ground networks. It also highlights the potential threats and vulnerabilities in the network at different locations.

To secure the communications between the aircraft segment and ground segment several techniques can be employed such as

- Firewalls
- IDPS on the aircraft and on the ground network
- Encryption and Decryption (DTLS)
- Traffic Domain Segregation
- Authentication Servers (As Third-Party Servers)
- Certification Servers (As Third-Party Servers)
- Anti-jamming on the radio links

In terms of traffic domain segregation, safety-critical flight control communication domain and non-safety-critical airline/passenger communication domain are physically interconnected and may even share the same datalinks for exchanging

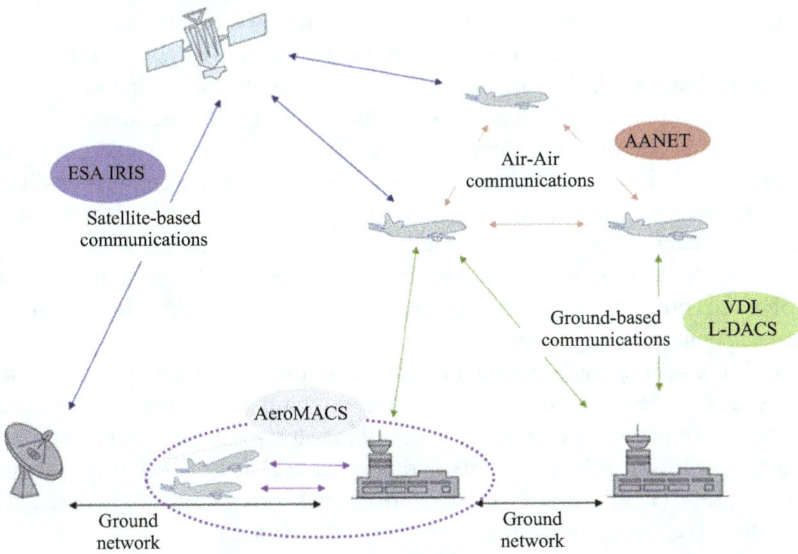

Figure 7.1 Wireless network in the sky [2]

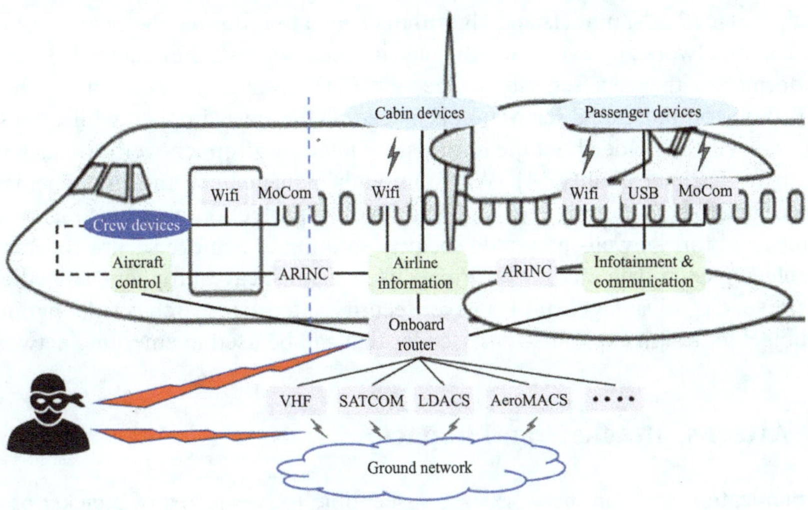

Figure 7.2 Avionics network

information with the ground network. While physical segregation of different communication domains may offer the best way to prevent a cross-domain security breach, this would increase the costs and weight of an aircraft. Logical segregation can provide a more efficient solution. However, logical segregation of different communication domains requires dedicated security measures such as VLANs, DTLS and gateway firewalls to ensure onboard and end-to-end communication security.

Aircraft or airport vehicles broadcast their surveillance information including their identity, position and other information to the nearby aircraft and ground control to facilitate airborne traffic situation awareness, spacing, separation and self-separation operations. The advancement in technology and with the advent of quantum computing it is becoming easier for malicious adversaries such as hackers or terrorists to launch virtual aircraft in the sky for a denial-of-service attack by overwhelming traffic control capacity or causing aircraft to take dangerous evasive manoeuvres. To counter such attacks, it is important to ensure aircraft applications message integrity and authenticity, to implement access control at all external communication interfaces of an aircraft and anti-jamming techniques to protect the datalink. In addition, aeronautical communication technologies that do not have include encryption requirements to secure the messages before they are sent out will be prone to security attacks, e.g., possible espionage of important personnel onboard the aircraft or preparation of other attacks. To avoid such exploits, security measures can be enhanced through a combination of physical layer security and/or other higher-layer security techniques.

The physical (PHY) layer is not only concerned with network hardware but also contains useful resources including definitions of hardware specifications, data transmission and reception, encoding and signalling, topology, and physical design to improve the security of the wireless networks.

PHY layer security (PLS) has been an emerging and dynamic paradigm that investigates methods, models and algorithms aimed to reinforce the security of communication networks by exploiting the physical characteristics of channels [3]. From the information-theoretic security perspective, PHY layer security exhibits the advantage that no computational restriction is placed on the eavesdropper, whilst the same expression can be made about the information the eavesdropper overheard as a function of the channel quality [4]. With thorough management and implementation, physical layer security provides an added level of protection over existing security schemes and form a well-integrated security solution to achieve secure coverage by guaranteeing extra data rate for legitimate users against eavesdroppers. Nonetheless, with the several advantages of PHY layer security, intense research is still ongoing on the subject to design explicit security codes that can be used in emerging networks.

7.2 Attacks, attackers and impact

Communication and data networks are susceptible to two forms of attacks: passive and active attacks. In passive attacks, the intruder tries to obtain information that is transmitted via eavesdropping on or monitoring transmission. However, in active attacks, the intruder employs some form of modification on the data stream, which results in a false data stream [5]. The characteristics of these security attacks are summarised in Table 7.1.

The detection of passive attacks is often arduous since these attacks do not involve any alteration of the data. For example, the eavesdropper aims to intercept data transmission by remaining in the transmit coverage area of the source node to overhear the communication session [6]. The prevention of these attacks is feasible

Table 7.1 Security attacks

Security attacks	Characteristic of attack
Passive attacks	
• Eavesdropping	Interception of confidential messages
• Traffic analysis	Adversaries determine locations and identities of network resources by intercepting transmitted messages
Active attacks	
• Information disclosure	Deliberate disclosure of information to unauthorised nodes
• Message modification	Compromised node modifies by performing additions or deletion of the data stream
• Replay attack	Adversary captures data stream and subsequently retransmit to produce an unauthorised effect
• Masquerade attack	Compromised node deceives authentication system by acting as a legitimate node to usurp network resources
• Denial of service attack	Breach of legitimate transmission by inhibiting normal management of communication facilities

by employing instruction detection techniques. In contrast, it is arduous to prevent active attacks because of the variety of software, physical and network vulnerabilities. For example, in DoS attack or jamming, the adversary aims to prevent legitimate users from accessing network resources by generating intentional interference to disrupt communications between authorised users, such that by disrupting the wireless communication of legitimate users, the jammer impairs the availability of network resources to legitimate users [7]. To this end, the emphasis in passive attacks is on attack prevention rather than detection, whilst in active attacks, the focus is to detect the attack and recover from disruptions or effects of the attack.

The impact of security attacks on the aircraft information and communication system depends on the attack objectives and the type of attackers. Attackers can be classified into five categories:

Military forces: Military forces can launch an attack on an enemy plane from proximity. There have been reported incidents of unmanned aerial vehicles intruding an attack on aircraft communication system using the onboard equipment.

Intelligence services: Intelligence services exploit security vulnerabilities for threat preventions and can alter information based on different developments such as personal, business or political.

Terrorists: Terrorists are usually either politically or religiously motivated and their intent is destruction and harm, and they try to sabotage the whole aircraft, hijack an aircraft or attack the aircraft control system to destabilise the establishment and cause loss of public confidence. Their threats are credible, and actions have a devastating impact.

Business competitors: They are usually after the business insider information such as business model (e.g., information using AAC) to either outperform the competition or ruin their credibility and reputation and cause business espionage.

Hackers: Hackers usually have the specialist knowledge to the malicious apps. Hackers can be in a group or a single person who carry out the attacks either for thrill and amusement or with an intent to abuse such as crashing an aeroplane for the purpose of suicide or take revenge on a certain known passenger. This attack group has the resources to destruct the aircraft security systems as well as privacy systems.

The following identifies potential impacts when an aircraft or ground communication network is compromised [8]:

- Potential loss of human life
- Population at risk – the population can be the passengers onboard an aircraft or population on ground (e.g., a plane crash into a motorway)
- Economic or financial loss
- National security impact
- Loss of or disruption to operations or service
- Loss of public confidence

7.3 Fundamental security requirements of aeronautical communications

Security is defined as the protection afforded to an automated information system in order to preserve the integrity, availability and confidentiality of information system resources (includes hardware, software, firmware, information/data and telecommunications) [9]. This definition introduces some key objectives for information and communications systems security:

Confidentiality: Confidentiality refers to the requirement that the information is only read and interpreted by the intended recipients and it ensures the privacy of the communication between the sender and the receiver. Confidentiality is required for all safety and mission-critical data communication to avoid disclosure of critical information. Suitable mechanisms are required to prevent any accidental or intentional disclosure of confidential information to unauthorised entities.

ARINC811 describes confidentiality as the protection against the unauthorised disclosure of information and defines the level of confidentiality required for different aeronautical information. While it would be expected that the confidentiality requirements for ATC data would be high, ARINC811 specifies that it requires low confidentiality. On the other hand, it specifies that the confidentiality requirements for AOC, AAC and APC data are high.

ARINC823 mandates that data confidentiality should be provided for all ACARS messages. Hence the ACARS Messaging security (AMS) should provide the communicating entities an assurance that the content of an AMS message is protected from unauthorised disclosure during message transmission

Integrity: Integrity refers to the requirement that the information remains unchanged while traversing from the authentic source to the destined recipient. It is essential to devise and implement mechanisms that prevent unauthorised destruction or modification of information. For safety and mission-critical services it is necessary to maintain the integrity of the information from source to destination to ensure safe and secure operation of the aircraft.

ARINC811 describes integrity as the protection against unauthorised modification or destruction of information and defines the need of integrity for different aeronautical information. It specifies that the integrity requirement for ATC and APC data is high, while that for AOC and AAC data is medium.

ED202 describes integrity as safeguarding the accuracy and completeness of information and processing methods.

ARINC823 mandates that data integrity should be provided for all ACARS messages. Hence the AMS system should provide the communicating entities with the assurance that an AMS message is protected against modification after creation and transmission by an authorised source. Modification includes insertion, substitution or deletion of message content, whether accidental (e.g., errors caused by a noisy transmission channel) or intentional (e.g., errors introduced by an unauthorised entity).

Authentication: Authentication refers to the requirement that the identity of system entities or users is only used by the same entities or users who these claims to be. For safe and secure operation of the aircraft and its communication, it is essential to have robust mechanisms against unauthorised access to system resources and protection against information sent from malicious entities masquerading as authorised users.

ARINC 823 mandates that data authentication should be provided for all ACARS messages. The AMS message authentication (also known as data origin authentication) security service mitigates spoofing threats and provides a message recipient with the assurance that the source of an AMS message is as claimed. In other words, message authentication provides aircraft crew or an automated aircraft information system with confidence that messages originate from the claimed ground source (e.g., DSP or airline operations). Conversely, it provides a dispatcher or automated ground information system with confidence that messages originate from the claimed airborne source (e.g., an aircraft information system).

Authorisation and access control: Authorisation refers to the requirement to allocate different privileges to different users. Different entities may be able to access a given system but may have different access rights in terms of what they can access or change in the system. Access control refers to the prevention of unauthorised use of a resource, including the prevention of the use of a resource in an unauthorised manner. This helps in limiting the flow of information from the resources of a system only to authorised persons, programs, processes or other system resources on a network. To ensure the safety and security of the aircraft it is important to define an access boundary for the authorised entities/users and define different access profiles for the different groups of system users, e.g., a separate access profile for the pilots and a separate access profile for the cabin crew.

Non-repudiation: Non-repudiation refers to the requirement that a user or a system entity cannot deny sending or receiving a message. This is to have verifiable evidence of the source of any malicious activity. Authentication, integrity and digital signatures provide powerful means for non-repudiation.

Intrusion detection: Intrusion detection refers to the requirement that there exists a robust technique to detect any residual or potential security breaches. The efficiency of detection procedures will help ensure the safe and secure operation of the system and will also trigger the timely execution of countermeasure strategies against the detected malicious activity. Such countermeasure techniques will remove or decrease the exploitation probability of system vulnerabilities by detecting and reacting to security attacks. The countermeasures are taken as a response to some malicious activity and responses should be developed with service impacts in mind. The response may vary from shutting down a specific service to shutting down an entire domain of data communication.

Traceability: Traceability refers to the requirement for a mechanism to facilitate the full audit of any actions performed. It facilitates prompt event investigation in case of any occurrence of security or safety breach events. It also facilitates the development of enhanced security and safety procedures. Traceability generally encompasses the time stamp for each of the sequential events along with the identity of involved user or system entity.

From the aeronautics communications perspective, these security objectives need to be stringently fulfilled using cryptography and network security techniques to ensure that the safety of aircraft will not be hampered by any breach of security into the aircraft network and computing systems.

A breach of security will impact on organisations or individuals. Federal Information Processing Standards (FIPS) Publication 199, Standards for Security Categorization of Federal Information, and Information Systems (FIPS 199) define three levels of impact:

- **Low:** The loss (of confidentiality, integrity or availability) could be expected to have a limited adverse effect on organisational operations, organisational assets or individuals. This means that the loss may (i) cause degradation in mission capability to an extent and duration that the organisation is able to perform its primary functions, but the effectiveness of the functions is noticeably reduced; (ii) result in minor damage to organisational assets; (iii) result in minor financial loss; or (iv) result in minor harm to individuals.
- **Moderate:** The loss could be expected to have a serious adverse effect on organisational operations, organisational assets or individuals that may (i) cause a significant degradation in mission capability to an extent and duration that the organisation is able to perform its primary functions, but the effectiveness of the functions is significantly reduced; (ii) result in significant damage to organisational assets; (iii) result in significant financial loss; or (iv) result in significant harm to individuals that does not involve loss of life or serious, life-threatening injuries.
- **High:** The loss could be expected to have a severe or catastrophic adverse effect on organisational operations, organisational assets or individuals. The loss may (i) cause a severe degradation in mission capability to an extent and duration that the organisation is able to perform one of more primary functions, but the

effectiveness of the functions is significantly reduced; (ii) result in major damage to organisational assets; (iii) result in major financial loss; or (iv) result in severe or catastrophic harm to individuals that does not involve loss of life or serious, life-threatening injuries.

7.4 Security mechanisms

To counteract security attacks, security mechanisms and policies need to be enforced. ITU X.800 defines a list of security mechanisms. These including

- Encipherment (encryption): The use of mathematical algorithms and zero or more encryption keys to transform data into an unintelligible form. Subsequent recovery of the original data requires corresponding mathematical algorithms and decryption keys.
- Digital signature: Data appended to or a cryptographic transformation of a data unit that allows a recipient of the data unit to prove the source and integrity of the data unit and protection against forgery.
- Access Control: Mechanisms that enforce access rights to resources.
- Data Integrity: Mechanisms that assure the integrity of a data unit or stream of data units.
- Authentication Exchange: Mechanisms to ensure the identity of an entity by means of information exchange.
- Traffic Padding: The insertion of bits into gaps in a data stream to frustrate traffic analysis attempts.
- Routing Control: Enabling selection of physically secure routes for certain data and allows routing changes, especially when a breach of security is suspected.
- Notarisation: The use of a trusted third party to assure certain properties of data exchange.

X.800 also defines pervasive security mechanisms that are not specific to any OSI security services or protocol layer. These include

- Trusted Functionality: Functionality that is perceived to be correct with respect to some criteria (e.g., the establishment of a security policy).
- Security Label: The marking bound to a resource (e.g., a data unit) that names or designates the security attributes of that resource (e.g., the use of VLAN tag).
- Event Detection: Detection of security-relevant events.
- Security Audit Trail: Data collected and potentially used to facilitate a security audit, which is an independent review and examination of system records and activities.
- Security Recovery: Ways to handle requests from mechanisms, such as event handling and management functions, and take recovery actions.

7.5 Physical layer security

Claude Shannon in his work detailed in [10] proposed to use his mathematical theory regarding communication to quantitatively measure the level of secrecy in an encryption system. However, Shanon's model has very stringent security requirements. In 1975, Wyner introduced the wiretap channel (WTC) model [11], which generalises the problem of secure and reliable communication in a noisy medium and this became the basis for the physical layer security, following his earlier work with Gray [12], Wyner model proposes a more relaxed security constraint, which suggested that if the codeword length is sufficiently large then the rate of information leakage asymptotically vanishes. Wyner's WTC is illustrated in Figure 7.3.

In Figure 7.3, the objective of the WTC is twofold as Alice must encode the message M into a codeword X^n of length n such that Bob upon receiving Y^n can reliably recover the message, i.e.,

$$\mathbb{P}^{\{M \neq \widehat{M}\}} \overrightarrow{n \to \infty} \, 0 \tag{7.1}$$

It is worth noting that the codeword X^n makes use of the channel n, i.e., $X^n = \{X_1 \ldots X_n\}$ where X_i is sent in the ith channel. The same holds for $Y^n = \{Y_1 \ldots Y_n\}$ and $Z^n = \{Z_1 \ldots Z_n\}$, which describes the corresponding channel outputs of Bob and Eve, respectively [13]. However, the message must be kept secret from Eve. To achieve secrecy in the presence of a noisy channel, which makes Shannon secrecy criterion of statistical independence between the confidential message and the channel output Z^n at Eve hard to realise, Wyner defined secrecy in terms of conditional entropy or equivocation [14]. This criterion is often referred to as weak secrecy and is expressed in terms of mutual information as

$$\frac{1}{n}(M; Z^n) \overrightarrow{n \to \infty} \, 0 \tag{7.2}$$

Specifically, the conditional entropy $\frac{1}{n}H(M|Z^n) \approx \frac{1}{n}H(M)$ is required so that the knowledge of Z^n does not decrease the uncertainty rate about M.

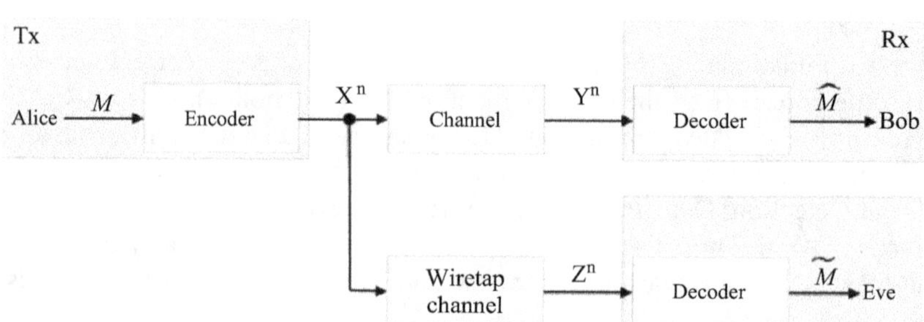

Figure 7.3 Wyner's wiretap system

Nonetheless, the weak secrecy criterion expressed in (7.5), which describe the information leaked about M to Eve in terms of normalisation rate by the block length has its vulnerabilities, which is ameliorated by dropping the division by n to

$$I(M;Z^n)\,\overrightarrow{n \to \infty}\,0 \tag{7.3}$$

Equation (7.3) expressed the strong secrecy condition with the intuition to have the total amount of information leaked to Eve vanish as $n \to \infty$. The strong secrecy criterion for the WTC was first considered by Csiszár [15] and Maurer and Wolf [16]. Readers interested in studies of further interesting approaches to achieving strong secrecy are referred to the following articles [17–22]. Further, recent works strengthen the strong secrecy criterion by considering the semantic security, where Eve is unable to decode the message but also unable to obtain any information about it all [23].

In general, the two secrecy requirements by Shannon and Wyner differs and seemingly look conflicting, as, on the one hand, the codeword is required to encode the message for transmission and recovery by Bob (reliability) whilst, on the other hand, the codeword is impractical for Eve (security). However, both requirements are possible simultaneously and the so-called secrecy capacity characterises the maximum rate at which both criteria are met. Ulterior then further generalised Wyner's work to different channel types such as Gaussian and fading channels.

It is important to understand that for physical layer security the requirement for perfect secrecy is much greater than trying to prevent the eavesdropper to be able to decode properly.

From the works of Shannon and Wyner, one concludes that the ruling cryptographic paradigm can never provide information-theoretic security because the communication channel between the friendly parties and the eavesdropper is noiseless and the secrecy capacity is zero. Moreover, existing key-distribution schemes based on the computational model require a trusted third party as well as complex protocols and system architectures. If multiple keys are to be generated, it is usually possible to do so only from a single shared secret and at the price of reduced data protection.

Moreover, since broadcast and superposition, which are two fundamental characteristics of any wireless medium to ensure reliable and/or secure communication in the presence of adversarial attacks, the issue of security in future 5G wireless networks becomes a critical concern [24]. Generally, most of today's security architecture rely on bit-level cryptographic methods and associated protocols at various level of data processing stack [25]. However, most bit-level cryptographic techniques suffer from several drawbacks including inadequate standardised security protection within public wireless networks, strong constraints and high additional cost for public network users even in the presence of enhanced ciphering and authentication protocols. Therefore, the development of new data security approaches from information theory fundamentals that focuses on secrecy capacity of the propagating channel, often referred to physical layer security, has been receiving intense attention in the literature [26–32].

Physical later security is a dynamic research field that investigates methods, models and algorithms aimed to reinforce the security of commination networks by exploiting the properties of the physical layer protocol. The subject of physical layer

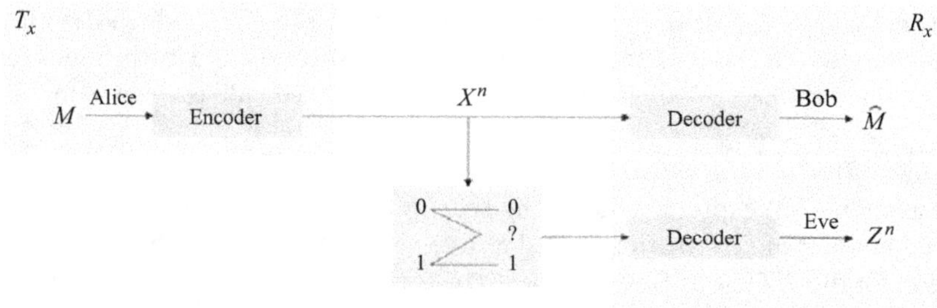

Figure 7.4 Binary erasure wiretap system [4]

security is particularly critical with the proliferation of 5G networks. Physical layer security techniques possess desirable edges over cryptographic methods for 5G networks in two distinct folds: Firstly, physical layer security does not require complex computations, such that even in the presence of eavesdroppers equipped with powerful computational devices in the 5G network, secure communication would still be achieved. On the other hand, the security of computation-based cryptography methods is usually compromised if the eavesdropper possesses sufficient computational capabilities to solve the hard-mathematical security formulation.

To derive the performance metrics, it is imperative to highlight the role of noise using an illustrative model presented in Figure 7.4.

In Figure 7.4, Alice is transmitting a legitimate message to Bob over a noiseless channel in the form of binary codewords of length n while Eve the eavesdropper receives a corrupted message that has the erasure probability $\epsilon \in (0, 1)$. The transmitted messages $M \in [\![1, M]\!]$ with uniform random probability. Codewords $X^n \in \{0, 1\}^n$. The corrupted message at Eve can be written as $Z^n \in \{0, 1, ?\}^n$. Every transmitted message employs a different codeword for encoding, thus making the rate of reliable transmission equal to $\frac{1}{n}\,\mathbb{H}(M) = (1/n)\log M$. Achieving perfect secrecy is a very stringent tasks along with ensuring that the message M and codeword X^n are statistically independent, rather we assume that both the transmission and the coding is secure if $lim_{n\to\infty}\mathbb{I}(M; Z^n) = 0$ is guaranteed. Thus, the information leaked starts to vanish as n approaches infinity. The main design hurdle is to choose an encoder that can satisfy the last statement.

In summary, the information theoretic basis to secrecy provides a measure of the average-information. Based on these fundamentals secrecy requirements can be defined when designing a wireless communication system however the designed system will not be able to guarantee a fully secured system because of the channel state information requirement that cannot be accurately predicted.

7.6 Secrecy performance metrics

The issue of secrecy and confidentiality arise in virtually all communications networks and secrecy capacity is an important aspect of this issue. Secrecy capacity plays a vital role in the PLS as it defines the bounds of communication security over different channels. Secrecy capacity and channel capacity can be considered as counterparts when there are both reliability and secrecy constraints.

7.6.1 Secrecy capacity

To evaluate the physical layer security the most widely adapted performance metric is the secrecy capacity S_c [33]. It is defined as the difference in capacity between two channels (i.e., the main channel and the WTC). In other words, it distinguishes the max secrecy rate at which the receiver can reliably recover the information but the eavesdropper cannot [34]. For a fading channel that is static over a period t (Quasi-Static) as shown in Figure 7.5, the secrecy capacity S_c can be written as:

$$\begin{aligned} S_C &= \max\{S_M - S_W, 0\} \\ &= \max\{U \log_2(1+\gamma_M) - U \log_2(1+\gamma_W), 0\} \end{aligned} \tag{7.4}$$

In (7.4), $\gamma_Z = |H_{TZ}|^2 T_p/N_0$ for $Z \in \{\text{M, W}\}$ represents the signal-to-noise ratio (SNR) and the channel coefficient for the main channel is H_{TM} and for the WTC is H_{TW}, respectively. The transmit power is shown by T_p, and the average noise power is represented by N_0. The channel capacity for the main channel is S_M whereas for the WTC it is represented by S_W. Without introducing any new constraints and maintaining the generality the normalised bandwidth can be considered as BW $= 1$.

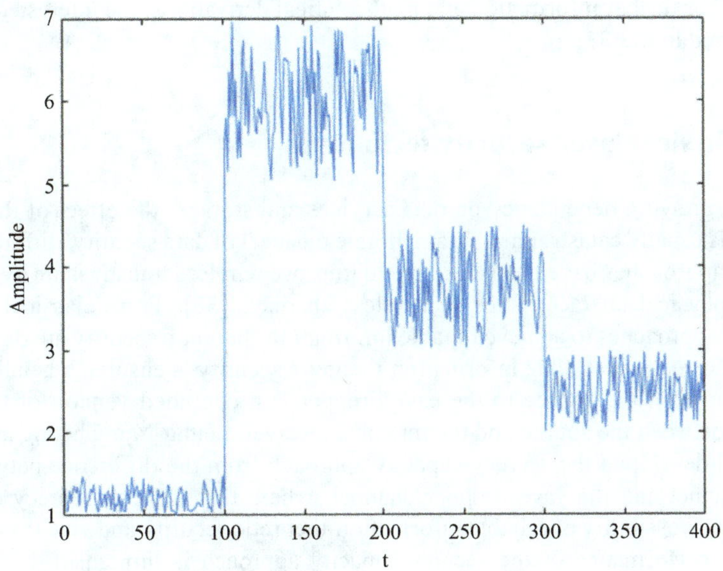

Figure 7.5 Quasi-static channel

Using the above conditions if the SNR of the legitimate link is better than that of the eavesdropper then the secure transmission can be achieved i.e.,

$$S_C = \begin{cases} \log_2\left(\dfrac{1+\gamma_M}{1+\gamma_W}\right), & \gamma_M > \gamma_W \\ 0, & \text{if } \gamma_M \leq \gamma_W \end{cases} \tag{7.5}$$

Equation (7.5) has been widely extended to derive the secrecy outage probability (SOP).

7.6.2 Secrecy outage probability

The definition of SOP is the probability of the secrecy capacity falling below the secrecy rate S_R, which means that the security requirement cannot be guaranteed by the secrecy rate. In [35] the mathematical model is formulated as below:

$$\begin{aligned} SOP &= p_r\{S_C(\gamma_M, \gamma_W) < S_R \\ &\overset{(a)}{=} p_r\left(\frac{1+\gamma_M}{1+\gamma_W}\right) < 2^{S_R} \\ &\overset{(b)}{\geq} p_r\left(\frac{\gamma_M}{\gamma_W}\right) < 2^{S_R} \end{aligned} \tag{7.6}$$

The SOP in (a) in (7.6) shows that if $S_R < S_C$, the legitimate channel will be better than the WTC, which in turn guarantees secure communication as stated in [36]. Most of the recent research on the issue of PLS focuses on (b) in (7.6) because of several reasons. For further information and mathematical derivations, the interested readers are referred to [33,37].

7.7 Physical layer security techniques

With the growing dependency on data services and storage, the effect of the eavesdropper is usually catastrophic. To ameliorate the level of data security, different PHY security approaches have been designed to improve wireless transmission by exploiting the physical characteristics of wireless channels [38]. From classical studies, different approaches to achieve reliable information-theoretic security are developed. Wyner showed that reliable information-theoretic security is ensured when the WTC stretching from the source to the eavesdropper is a degraded version of the main channel between the source and the intended receiver. Leung-Yan-Cheong and Hellman [39] developed the secrecy capacity approach from the difference between the main channel and the eavesdropper channel, where that positive secrecy capacity implies the possibility of reliable information-theoretic security and vice versa. However, the performance of the secrecy capacity approach is limited in the wireless channel, when deep fading occurs in the main channel because of shadowing from obstacles resulting in significant degradation of the wireless secrecy capacity [40].

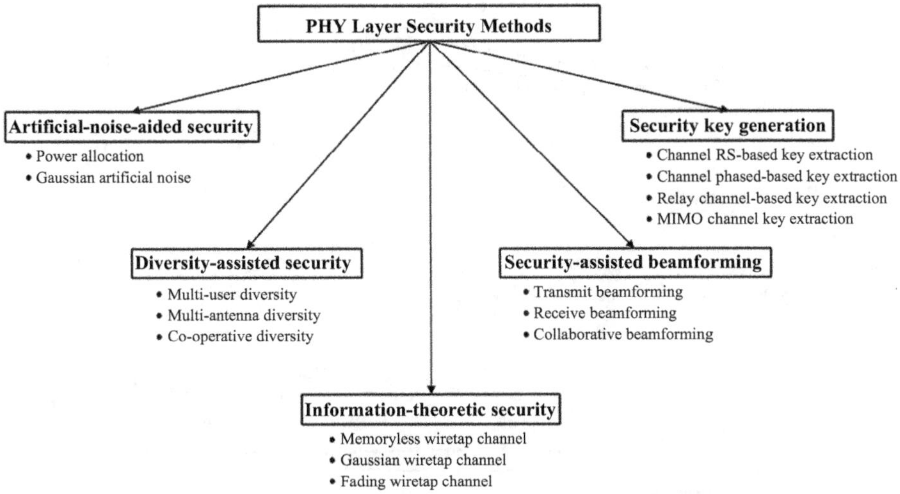

Figure 7.6 Taxonomy of PHY layer security methodologies

Therefore, intense research efforts about wireless PHY layer security have been contributed to the literature resulting in several novel PHY layer security methodologies summarised in Figure 7.6 as:

i. Information-theoretic security method [41–43]
ii. Artificial-noise-aided security method [44–48]
iii. Diversity-assisted security method [49,50]
iv. Security-oriented beamforming method [51–57]
v. Security key generation method [58–71]

7.7.1 Artificial noise-aided security

The key idea behind the artificial-noise-aided security approach is to allow the transmitter to generate specific interfering signals or artificial noise such that Eve is adversely affected by the artificial noise (AN), whilst Bob remains unaffected. The AN-aided approach in the presence of an eavesdropper is illustrated in Figure 7.7.

7.7.2 Diversity-assisted security

The diversity security approach is invoked to improve the PHY layer security of wireless transmission. The diversity-aided security approach enhances wireless security without additional power in contrast to the artificial-noise-aided security paradigm, which dissipates additional power assigned to AN generation. The antenna diversity or multiple-antenna-aided transmit diversity constitutes an effective means to combat the fading effect, which invariably increases the secret capacity of the wireless transmission. The diversity-assisted security using MIMO is illustrated in Figure 7.8.

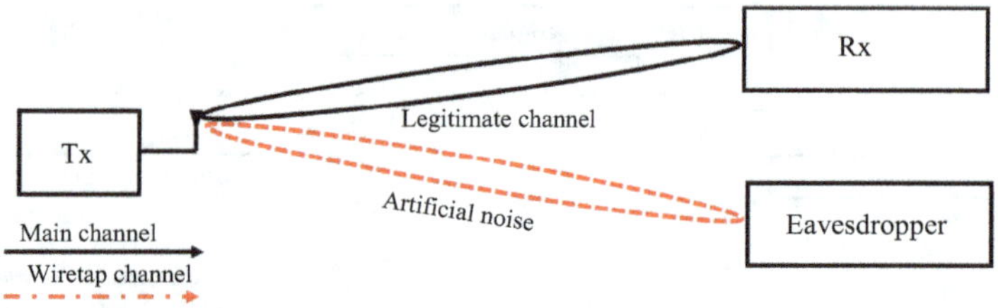

Figure 7.7 Legitimate transmitter and receiver nodes with eavesdropper for a wiretap channel with AN

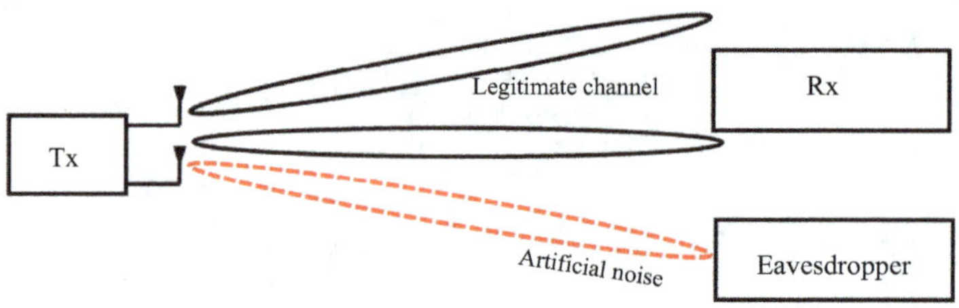

Figure 7.8 MIMO system nulls in the direction of an eavesdropper

The advantage of using spatial diversity is that the nulls can be directed towards the MITM using beamforming.

7.7.3 Cooperative diversity for security

Cooperative diversity is used to enhance the PHY layer security by enhancing the reliability of the transmitted message and extending coverage. The cooperative-diversity approach makes use of devices such as relays as illustrated in Figure 7.9. The relay receives the information from the transmitter and then retransmit it to the receiver. By placing the relays between the main transmitter and the receiver it helps extend the coverage area. Also, the relays help in the diverse configuration of the system into different architecture according to requirements e.g., to degrade the channel of the eavesdropper so that they can act as jammers and send AN. Different protocols are used to achieve cooperative diversity such as amplify and forward and if any decoding is required then decode and forward [73].

It is worth noting that the untrusted relays can act as eavesdropper [74] as shown in Figure 7.10. However, maximum diversity gain can be achieved by using the cooperating networks when the relay is untrusted [75].

7.7.4 Channel coding for security

With the introduction of information-theoretic security and a further extension via the introduction of the wiretap model where different coding techniques can be employed

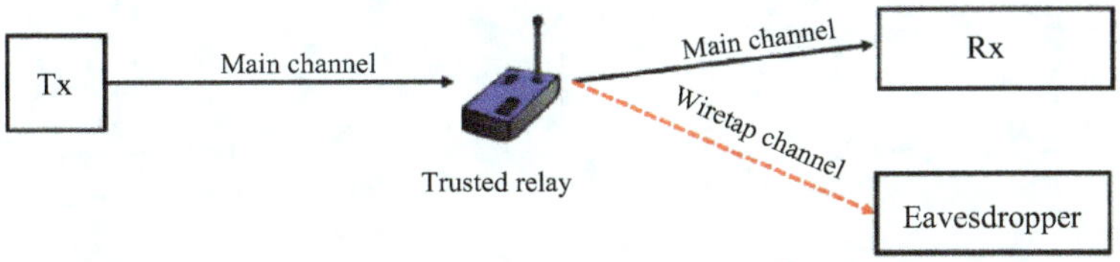

Figure 7.9 Wiretap channel relay network with trusted relay

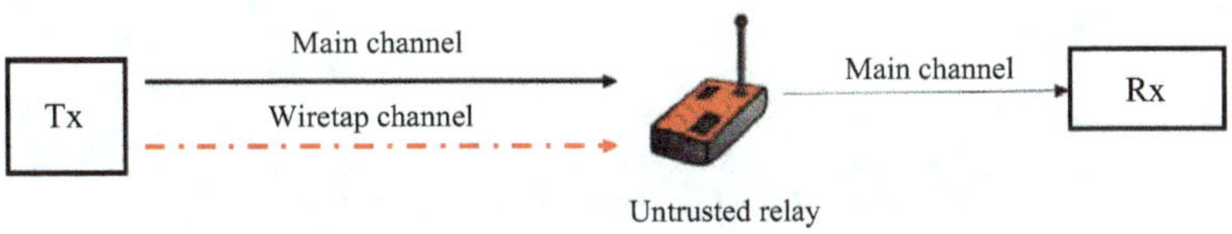

Figure 7.10 Wiretap channel relay network with untrusted relay

for both reliable and secure communication, information-theoretic security has gained research attention however the design of codes to maintain security in a WTC is a very difficult task and this field still has a long way to go to achieve perfect secrecy. Since secure communication cannot be guaranteed primarily by the exploitation of the noisy channel, different coding approaches including polar coding, lattice coding and low-density-parity-check (LDPC) coding have been proposed in the literature to implement information-theoretic security to countermeasure to the eavesdropper attack and secure the communication data. However, readers are referred to these works for further study on physical layer security from the information-theoretic perspective [76–82].

7.7.4.1 Polar coding

Based on the previous works of Shannon and Wyner and a further generalisation of Wyner model by Csiszar and Korner in [83] to broadcast channels that are not necessarily degraded, Arikan introduced the channel polarisation method for binary memoryless symmetric channels and generalised for arbitrary q-ary alphabets [84] and Galois field [85] to achieve the secrecy capacity of the WTC [86]. In channel polarisation, several independent realisations, N of a binary-input discrete memoryless channel is converted to a new set of N binary-input channels. Polar codes have also been extended to lossy and lossless source coding [87–90] and in several multi-user problems [91].

However, we begin a basic description of polar codes in a precise manner, as illustrated in Figure 7.4. A transmitter wants to send a confidential message, X to an authorised receiver through the main channel. The transmission is also perceived by an eavesdropper via the WTC. Both channels, (N and O) are block fading erasure channels or binary symmetric channels with n states $\{n_c|i = 1, \ldots, n\}$. The channel transmission is initially established as: $q_i = 1 - 2p_i + x_i$ where x_i is noise. Bits of p_i in P are encoded from the frozen bits and source message bits of m_i in M and the

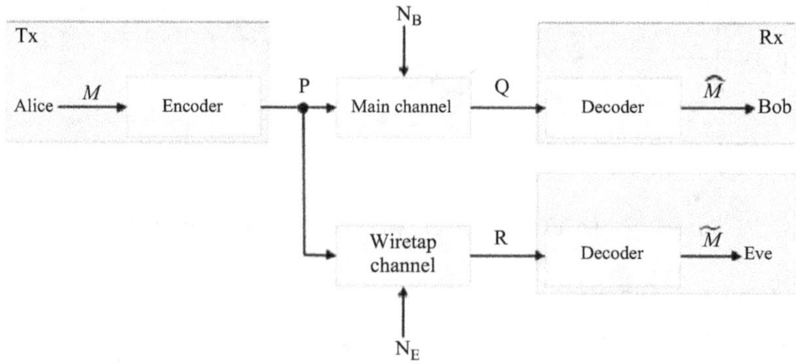

Figure 7.11 Gaussian wiretap channel

received bits of q_i in Q will be decoded as m_i in M by the successive cancellation decoder. To this end, polar coding schemes under weak and strong secrecy criteria have been proposed in the literature. Under the weak secrecy criterion, Wei and Ulukus developed polar codes for the general WTC by relaxing the symmetric and degraded constraints under the weak secrecy condition [92]. Fakoorian and Swindlehurst further extend the work by investigating multiple access WTC (MA-WTC), the broadcast channel with a confidential message (BC-CM) and the interference channel with a confidential message (IC-CM) [93]. Andersson *et al.* proposed the polar coding scheme and decode-and-forward protocol for bi-directional relay networks with common and confidential message [94]. From other works, it is shown that polar codes can achieve secrecy capacity, capacity-equivocation region and key agreement of binary-input symmetric memoryless degraded WTC [95–98].

For the strong secrecy criterion, Şaşoğlu and Vardy proposed a multi-block polar coding scheme for the WTC model to achieve both data reliability and security [99]. Gulcu and Barg proposed a simple polar coding scheme for asymmetric general WTC and BC-CM to achieve secrecy capacity [100]. Ekrem and Ulukus studied the MIMO WTC to obtain maximum capacity-equivocation [101]. The study establishes the accuracy of jointly Gaussian auxiliary random variables and channel input to further evaluate the work. However, the polar coding scheme is further extended for discrete memoryless BC-CM under a more general condition [102]. Recently, Wang *et al.* proposed a low-complexity polar code scheme using superposition coding to improve the secrecy capacity of a WTC with a shared key between the transmitter and the authorised receiver [103]. Zhao *et al.* proposed a new hash chaining structure to evaluate the good bits sacrificing problem and obtain a strong encryption polar coding scheme for a two-way WTC [104]. A summary of existing works on polar code is presented in Table 7.2.

7.7.4.2 Lattice coding

Lattice codes are codes whose codewords are represented as points in a lattice. Lattice code provides the significant solution to diverse communication problems because of

Table 7.2 Existing survey of polar codes for PHY layer security

Channel	Criterion	Key contributions	References
General WTC	Weak secrecy	Secrecy capacity and secrecy rate regions	[92]
Deterministic WTC	Weak secrecy	Secrecy capacity	[93]
Bi-directional relay networks with CMs	Weak secrecy	Capacity-equivocation region	[94]
Binary symmetric memoryless degraded WTC	Weak secrecy	Secrecy capacity	[95]
Binary symmetric memoryless degraded WTC	Weak secrecy	Capacity-equivocation region	[96,97]
Binary symmetric memoryless degraded WTC	Weak secrecy	Establish key agreement	[98]
Binary symmetric memoryless degraded WTC	Strong secrecy	Security and reliability	[99]
General WTC-BC with CM	Strong secrecy	Secrecy capacity	[100]
Discrete memoryless BC-CM	Strong secrecy	Secrecy capacity	[102]
WTC with a shared key	Strong secrecy	Secrecy capacity with low complexity	[103]
Symmetric binary-input general WTC	Strong secrecy	Secrecy rate with improved upper bound	[104]

its group structure and is employed for two fundamental reasons: on the one hand, as a codeword space used in coset coding for the Gaussian WTCs, whilst, on the other hand, as a structured framework to achieve PHY layer network coding [105]. Table 7.3 provides a list of works employing lattice codes. Mathematically, a lattice is a discrete additive subgroup of $\mathbb{R}^n$, such that any point in a lattice $L \subset \mathbb{R}^n$, is expressed as a generator matrix $M \in \mathbb{R}^{nxd}$, as:

$$L = \{x \in \mathbb{R}^n | x = M_{\omega,\omega} \in \mathbb{Z}^d \tag{7.7}$$

where d is the dimension of the lattice and ω is the lattice coordinates.

It is worth stating that channel models including Gaussian, Rayleigh block fading and Rayleigh fast fading channels can be regarded as points in $\mathbb{R}^n$. From (7.1), if the column of M is linearly independent over $\mathbb{Z}$, then ω of the lattice points are unique. Moreover, assume , $d = n$, then the lattice is of full rank. However, the sublattice of a lattice of dimension d in $\mathbb{R}^n$ is an additive subgroup and has a generator matrix MZ such that:

$$\left|\frac{L_a}{L_b}\right| = |\det Z| \tag{7.8}$$

Table 7.3 Lattice codes for PHY layer security

Channel	Criterion	Key contributions	References
Gaussian WTC	Secrecy gain	Definition of secrecy gain established	[108,109]
Gaussian WTC	Secrecy gain	Method to investigate secrecy gain with results is established	[110,111]
Gaussian WTC	Secrecy gain	Construction of best lattice code for dimensions $8 < n \leq 23$	[112]
Rayleigh WTC	Secrecy gain	Construction of wiretap lattice code	[113]
Gaussian WTC	Weak Secrecy	Construction of nested lattice code	[114]
Gaussian WTC	Strong secrecy	Design of wiretap lattice code	[115]
Gaussian broadcast channel with CM	Strong secrecy	Superposition lattice code is proposed	[116]

where $Z \in \mathbb{Z}^{dxk}$ and k represents the dimension of the sublattice and for a square matrix, Z [106,107]. The key parallelotope P_L of a lattice L is defined as:

$$P_L = \{x \in \mathbb{R}^n \mid x = M_{\omega,\omega} \in 0, 1^n\} \tag{7.9}$$

Moreover, the volume of the lattice Vol(L) for full-rank lattice is the volume of the key parallelotope and is given by

$$Vol(\mathrm{L}) = |\det M| \tag{7.10}$$

For a general rank, d,

$$Vol(\mathrm{L}) = \left(\det\left(v_i.v_j\right)_{i,j=1}^{d}\right)^{1/2} \tag{7.11}$$

where v_i are the generator vectors and $(v_i.v_j)_{i,j-1}^{d}$ is the Gram matrix, which determines the geometry of the lattice basis. However, it is worth noting that the volume of any lattice is independent of the type of key parallelotope since if L has two different generator matrices M and M', then M' would generate a sublattice,

$$M' = MZ \tag{7.12}$$

with

$$\left|\frac{L}{L}\right| = 1 = |\det Z| \tag{7.13}$$

Hence,

$$|\det M| = \left|\det M'\right| \tag{7.14}$$

Further, the dual lattice L of a full rank L generated by M is obtained by

$$M^{-T} := \left(M^{T}\right)^{-1} \tag{7.15}$$

The dual lattice is well-defined as it does not depend on the type of generator matrix

Then can be written as:

$$M'^{-T} = M^{-T} Z^{-T} \tag{7.16}$$

where Z^{-T} is an integer matrix and as shown in (7.16), M^{-T} and M^{-T} would generate the same lattice.

7.7.4.3 Low-density-parity-check (LDPC) coding

Low-density-parity-check (LDPC) codes are linear codes obtained from sparse bipartite graphical representations. The graphical representation is equivalent to matrix representation by examining the adjacency matrix of the graph. The graph with n left nodes (message nodes) and r right nodes (check nodes) results in a linear code of block length n and dimension at least $n - r$. This implies that the n coordinates of the codeword are associated with the message nodes. The codewords are those vectors $c_1 \ldots\ldots c_n$ such that for all r the sum of the neighboring positions among r is zero [116]. LDPC codes are useful for secrecy when the main channel between Alice and Bob is better than the WTC between Alice and Eve. The applicability of LDPC codes comes from their capability to make the channel error catastrophic to Eve, so that negligible to no information is leaked [117]. The efficiency of LDPC codes for PHY layer security can be improved through dummy bits insertion and puncturing under the criterion of BER where the secrecy information bits are hidden in the punctured bit [118]. A summary of some interesting works on LDPC codes for PHY layer security is presented in Table 7.4.

Table 7.4 LDPC codes for PHY layer security

Key channel	Paradigm	Eve channel	Constituent codes	References
Noiseless	Weak secrecy	BEC	Duals of LDPC	[119]
Noiseless	Strong secrecy	BEC	Duals of LDPC	[120]
BEC	Weak secrecy	BEC	Two-edge LDPC	[121]
Gaussian	BER	Gaussian	Punctured LDPC	[122]
Gaussian	BER	Gaussian	Non-punctured LDPC	[123]
Gaussian	BER	Gaussian	Non-punctured LDPC	[124]
Parallel Rayleigh	Equivocation rate of Eve	Parallel Rayleigh	Irregular LDPC	[125]

7.8 Physical layer – attack and countermeasures

Radio jamming is one of the most direct forms to launch a disruptive DoS attack. As most of the aviation services use the wireless spectrum and there is no physical boundary this raises several security issues such as intrusion, interception, eavesdropping, spoofing and jamming. The reason jamming is the first choice for most attackers is that these attacks are trivial to carry out and because the sensor nodes are restricted to a standard frequency, they can easily be deprived of any countermeasures such as frequency hopping, security key transmission, etc. Jamming can be categorised as an active destructive attack launched by a malicious adversary to cause DoS this makes jamming a much more serious threat as compared to any passive attack such as eavesdropping. If the frequency channel used for the transmission of security and safety-critical information from the satellite such as troop's location behind the enemy lines or the sensory data from aircraft or the location information from the GPS to the aircraft is jammed by a malicious adversary by broadcasting noise or any interference signal that can disrupt the reception of the legitimate signal it can lead to severe consequences.

Jamming has become a topic of great concern in GPS even with the spread spectrum (SS) capability of the GPS receivers. Jamming is one of the major threats to the datalinks as it is very hard to detect and countermeasure and when the information travels from air to ground or especially from satellite to ground as in the case of GPS satellites the received signal strength is often below –110 dB whereas the jammer is on ground and within its range it can transmit a high-powered white noise to block the spectrum. Jamming can be categorised as active/ persistent or reactive. The main challenge faced by the wireless communication system designers for aircrafts and satellites is to avoid the active jamming attacks because the reactive jammer only transmits the jamming signal when it senses the transmission, and the transmitter can avoid that by increasing its low probability detection. With such a low received power from the GPS satellite conventional secrecy techniques such as encryption are not enough and some form of physical layer technique is required to counter any attack on the GPS signal such as antenna diversity, beamforming or channel coding. Below is a scenario of malicious jamming attack and its various types along with physical layer security-based SS coding technique frequency hopping SS that can be employed as an anti-jamming technique. Jamming techniques are also being studied as a technique of physical layer security [126,127].

7.8.1 Jamming techniques

Jamming signals can be characterised in three different classes: (1) band Jamming, (2) tone jamming and (3) partial time jamming. Most of the existing literature on the topic of jamming assumes that the jamming pattern is known and is also constant for the time of signal transmission, however with advancement in technology and several commercial off the shelf jamming devices the attacker may launch a cognitive jamming attack, which is adaptive and varies with time.

7.8.2 System model

Using the basics of wireless communication perspective, we assume that $y(t)$ is our desired signal, which has now either convolved with the channel in the frequency domain or multiplied in the time domain. Figure 7.12 shows a simplified setup of a spoofer. Mathematically we can write:

$$y(t) = x(t) + n \tag{7.17}$$

After the jammer introduces added noise then at the receiver the received signal before equalisation can be written as:

$$y(t) = x(t) + n_{ch} + n_J \tag{7.18}$$

So, it can be now seen from (7.18) how the SNR will be reduced at the receiver. This concept can be extended a little further and, in more detail, using three receiving elements.

Assume a receiver with a linear antenna array of three isotopically radiating elements that are separated at a distance of $\frac{\lambda}{2}$, where λ represents the wavelength of the incident signals. The reason for selecting $\frac{\lambda}{2}$ as the separating distance is to prevent the inter-element interreference also known as mutual coupling as shown in Figure 7.13.

Consider that the incident signal that is arriving at the array is from a source located in the far-field zone and is a narrow band signal. Also, the central antenna is the reference point. The antennas are located on the 3D cartesian coordinate system at O (0,0,0), A(d,0,0) and B(-d,0,0). As these antennas are separated by some distance $\frac{\lambda}{2}$, there will be a time delay between the incident signal arriving at each element, the reason for assuming the distance as $\frac{\lambda}{2}$ is to prevent any mutual coupling between the

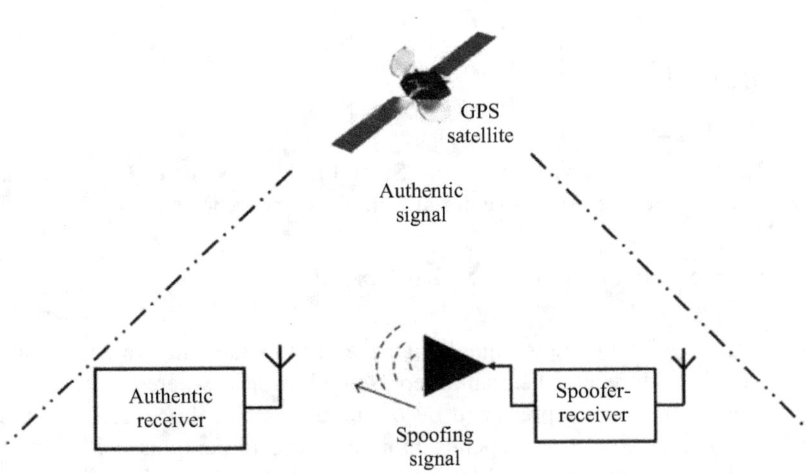

Figure 7.12 GPS satellite signal and spoofer

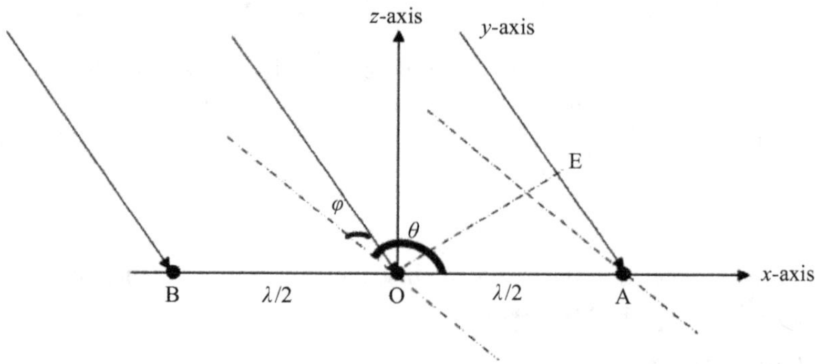

Figure 7.13 Linear antenna array with three elements

elements of the array. Time lag is represented by τ. Then the delay from the reference antenna can be calculated mathematically as:

$$\tau_A = \frac{|AE|}{c} = \frac{d}{c}\cos\theta\cos\varphi \tag{7.19}$$

$$\tau_B = \frac{d}{c}\cos\theta\cos\varphi \tag{7.20}$$

where Phi (ϑ) is the azimuth angle and elevation is represented by Theta (θ), c is the speed of light. As the antennas are isotropic in nature the scanning range of $\vartheta \in (0, 359°)$ and $\theta \in (0, 90°)$, speed of light $c = 3 \times 10^8\ m/s$.

The steering vector $a(\theta, \varphi)$ of the signal received by each element from (θ, φ), respectively, can be mathematically described as:

$$a(\theta, \varphi) = \begin{bmatrix} 1 \\ e^{-j\omega_c\tau_A} \\ e^{-j\omega_c\tau_B} \end{bmatrix} = \begin{bmatrix} 1 \\ e^{j\beta\cos\theta\cos\varphi} \\ e^{-j\beta\cos\theta\cos\varphi} \end{bmatrix} \tag{7.21}$$

In (7.21), $\omega_c = 2\pi f$, where $f = c/\lambda$. Substituting the value of f with c/λ, $\omega_c = 2\pi c/\lambda$. Now the received signal vector at time t_n can be expressed as:

$$y(n) = \sum_{i=1}^{I} a(\theta_i, \varphi_i)\, s_i(n) + \sum_{k=I+1}^{K} a(\theta_k, \varphi_k)\, j_k(n) + e(n) \tag{7.22}$$

In (7.22), $e(n)$ is a complex number and represents the additive white gaussian noise vector. The Ith received signal is represented by $s_i(n)$ where $i = 1, 2, 3, \ldots, I$. The jamming signals are represented by k where $(k = I + 1, I + 2, I + 3, \ldots, K)$, respectively. The value of $s_i(n)$ depends upon the protocol being used whereas k can be independent and indefinite. If $K(t)$ is the jamming interference at time (t), which is either stationary or not and the autocorrelation function for $K(t)$ can be written as:

$$C_k(t, \tau) = \mathrm{E}\{K(t, \tau)K(t)\} \tag{7.23}$$

Then the power spectral density (PSD) can be calculated as:

$$Q_k(t,f) = \mathcal{F}\{C_k(t,\tau)\} = \int_{-\infty}^{\infty} C_k(t,\tau)e^{-j2\pi f\tau}d\tau \tag{7.24}$$

where time-varying power of the jamming interference signal is

$$P_k(t) = \int_{-\infty}^{\infty} Q_k(t,f)df \tag{7.25}$$

Assume that the value of $P_k\ (t)$ is finite, then the maximum power of the jamming interference signal can be expressed as:

$$P_k(t) \leq P_{k,\max} \tag{7.26}$$

However, if the jamming interference signal is wide sense stationary, then the autocorrelation function, power spectral density and the jamming interference signal power all will be time-invariant.

To obtain the values for the three different jamming types assume that the specific frequency band for any communication and the starting frequency is f_0 and the ending frequency is f_1 and the time period is from t_0 to t_1 and $P(t)$ is constant then.

1. If $Q_k(t,f) = \frac{P_k}{f_1-f_0} \triangleq B_j, \forall f \in [f_0,f_1], \forall t\ [t_0,t_1]$, then the jamming interference signal $K(t)$ with power spectral density of $Q_k(t,f)$ returns full band jamming.
2. If $Q_k(t,f) = P_k\delta\,(f - f_l)\,, \forall t\ [t_0,t_1]$, and $f_l \in [f_0,f_1]$ will return the tone jamming with single tone. To obtain multi tone jamming,

$$Q_k(t,f) = \sum_{l=1}^{L} P_k(t,l)\delta\,(f - f_l), \qquad q.t. \sum_{l=1}^{L} P_k(t,l) = P_k \tag{7.27}$$

In (7.27), L represents multiple tones and $P_k(t,l)$ is the power of the lth tone f_l.

3. Finally, for the partial-time method, let us assume we have two time instant t_a and t_b, where $(t_b \leq t_a) \in [t_0,t_1]$ then,

If $\forall f \in [f_0, f_1]$,

$$Q_k(t,f) = \begin{cases} 0, & if\, t \in [t_0,t_b)\, or\ t \in (t_b,t_1], \\ \dfrac{P_k}{f_1 - f_0}, & if\ t \in [t_a,t_b], \end{cases} \tag{7.28}$$

7.8.3 Countermeasures

The fight against illegal jamming has attracted a great deal of attention for the avionics future communications infrastructure. The reason behind this attention is the adverse effects that the jamming can have on the security of the growing air traffic as well as the on the quality of service to the passengers. A decade ago, these frequency disturbances were ignored or dealt with passively. But with the advent of technologies and the experience of different key players in the aeronautical industry has since then changed and thus the mindset has shifted from passive to aggressive. There are now different anti-jamming techniques that are available to both the aircraft operators and controllers. The key technique is described below.

7.8.3.1 Spread spectrum (SS)

This technique employs signalling and spreads the signal over a wide frequency range with bandwidth that is considerably large using a pseudo-noise. This is the most aggressive and effective physical layer security-based anti-jamming technique. The two most employed SS techniques are direct sequence SS (DSSS) where the transmitted data is spread over different frequencies and frequency hopping SS (FHSS) in which the central frequency is continuously changed during a bit duration thus making it extremely difficult for the eavesdropper to wiretap the signals of the SS. In a patent filed in 1903 Nikola Tesla presented the first ever concept of frequency hopping (FH).

One well-known DSSS-based technique that is mostly used to achieve secure communication is Code Division Multiple Access (CDMA). In DS-CDMA the same channel is shared among different users by altering the codes used for spreading. This allows the users to distinguish the signals. In DS-CDMA first a code sequence is used for the signal at the transmitter and then the PN sequence is added to spread signal to achieve scrambling. Thus, hiding the signal behind noise to evade wiretapping. To ensure physical layer security of this technique the size of the PN scrambling sequence is usually relatively very long to prevent any attacks.

These methods were initially developed for military communications and without a doubt these systems are very well suited for voice communications.

Security weaknesses against jamming

The security of these techniques highly relies on PN sequences and their randomness. If a sequence segment of $2n$-bit is intercepted, it then the whole sequence along with the characteristic polynomial can be reconstructed. Because the characteristic polynomial is publicly available even an interception of an n-bit sequence will allow the recovery of the complete PN sequence. In short with the computing powers and applications available today the PN sequence can be broken and recovered. To completely jam the authorised signal in CDMA the same constellation can be employed from the recreated spreading code from a legitimate user and a different signal can be transmitted. In the case of frequency hopping, once the jammer has recovered the hopping pattern, jamming can be launched, which will jam every available frequency hopping band, and thus destroying the effective communication. Because of this lack of security, these techniques fail under jamming attacks.

Low Spectral efficiency

Another limitation of SS-based techniques such as CDMA and frequency hopping are the low spectral efficiency caused by the multipath propagation, asynchronisation and collision. This results in the loss of the total available bandwidth, which is a scarce resource.

Self-jamming in multiple access CDMA (MA-CDMA) is another limitation. All the authorised users have a fixed frequency band for transmission. The receiver exploits the orthogonality present between the signature waveforms but if this orthogonality is lost because of multipath propagation and asynchronisation that will result in self jamming. Thus, the CDMA-based communication system that has a spreading vector S can only support $S/4$ number of users.

Similarly, for multiple access frequency hopping SS, every user hops separately using a unique PN sequence. Thus, when multiple users transmit using the same frequency band it results in collision, which increases the bit error rate and the probability of that happening is close to 0.5. To calculate the collision probability, assume that the number of available channels is S_c and the total active users are U, the channels have equal probability and user hopping is independent and each user is transmitting using a single carrier then,

$$Q_c = 1 - \left(1 - \frac{1}{S_c}\right)^{U-1} \tag{7.29}$$

If the total number of available channels is large, then $Q_c \approx U - 1/S_c$. Thus, theoretically the system can support very limited user provided the usable spectrum is large enough.

Finally, in FHSS, each hop is transmitting the same information and similarly in CDMA systems for each period same signals with phase shifted variants is transmitted. Thus, neither of these systems are very efficient because both these systems employ repeated coding, which is not a very suitable coding choice.

7.8.3.2 MIMO-based anti-jamming

The ability of the MIMO systems to provide the interference cancellation has been extended in massive MIMO where the total number of antennas have been increased but at the cost of size, but these systems are very suitable for AeroMACS base stations (BS) as they have no power constraints. Therefore because of these advantages these techniques have been proposed to prevent illegal frequency jamming. In [128] authors presented a design of a receiver to combat jamming attacks based on the idea of frame reservation of wiretapper channel estimations. In this technique part of the pilots from the frame were separated and then used for the wiretapper channel estimation and at the same time providing the legitimate user the ability to estimate its own channel using the law of large numbers. This allows the design of a linear spatial filter at the BS for the mitigation of the jamming signal and recovery of the legitimate signal.

7.8.3.3 Multiple base stations for anti-jamming

Another technique that has been proposed as a jamming countermeasure, which is closely related to the massive MIMO concept is the use of separate base stations to nullify the jamming signal using rerouting. If the serving base station gets attacked by frequency jamming the traffic is rerouted via the alternative base station. The key idea behind this technique is that if the user cannot acquire or is unable to decode the serving base stations MIB it starts searching for the MIB of the neighbouring base station to establish a new link.

7.9 Conclusion

Wireless networks have been continuously evolving and their applications are increasing as the technology is becoming more widely available and the cost is decreasing. With the advancement the threats to the wireless technologies are also on the rise as, on one hand, there are devices to provide the secure transmission, on the other hand, there are devices that can intercept and decode those secure transmissions. To provide both secure and reliable transmissions scientists and engineers have come up with different computational techniques and physical layer techniques to ensure close to perfect secrecy transmission. After Wyner's proposal of taking advantage of noise to degrade the wiretappers channel several new concepts have been presented on how to employ physical layer techniques in addition to the computational methods to improve secrecy. This chapter provides an insight into future aeronautical communications, their uses, requirements and threats. Depending on the use of the access network technology and its purpose the results can be both advantageous and devastating. On one hand, Jammers have their advantages whereas at the same time jammers in the hands of malicious adversaries can cause denial of service attacks on the critical infrastructure and safety critical networks such as avionics communications networks. Most of the computational techniques that are employed for secrecy are on the higher layer in the OSI protocol stack and they rely on strong mathematical models to maintain secrecy however with the advent of quantum computing now the malicious adversaries have the computing power to decode those transmissions. Physical layer techniques provide an additional security to those computational methods such as cryptography and encryption. Some of those techniques have been discussed and how they can improve both the security and reliability. PHY layer security techniques are not without limitation so in the context of anti-jamming the limitations of SS techniques are low spectral efficiency etc.

References

[1] N. Fistas, B. Phillips, A. Faa, and J. Budinger, "Future communication study – Action plan 17 final conclusions and recommendations report," 2007.

[2] Y. F. H. M. Ali, A. Ahmed and P. Pillai, "Technical note on secure routing," *Deliverable of Innovate U.K funded project: Secure Integrated Network Communications for Broadband and ATM Connectivity (SINCBAC),* vol. WP3, April 2014.

[3] M. Bloch, "Fundamentals of physical layer security," in *Physical Layer Security in Wireless Communications*, X. Zhou, L. Song, and Y. Zhang Eds., (Wireless Networks and Mobile Communications Series), Dr. Yan Zhang, Ed. Boca Raton, Florida, USA: CRC Press, Taylor & Francis Group, 2014, ch. 1, pp. 1–13.

[4] M. Bloch and J. Barros, *Physical Layer Security – From Information Theory to Security Engineering*. The Edinburgh Building, Cambridge, UK: Cambridge University Press, 2011, p. 345.

[5] W. Stallings, *Cryptography and Network Security: Principles and Practice*, 6th ed. Pearson Education Inc., 2014.

[6] A. Perrig, J. Stankovic, and D. Wagner, "Security in wireless sensor networks," *Commun. ACM,* vol. 47, no. 6, pp. 53–57, 2004, doi: 10.1145/990680.990707.

[7] A. Mpitziopoulos, D. Gavalas, C. Konstantopoulos, and G. Pantziou, "A survey on jamming attacks and countermeasures in WSNs," *IEEE Communications Surveys & Tutorials,* vol. 11, no. 4, pp. 42–56, 2009.

[8] M. Strohmeier, M. Schäfer, M. Smith, V. Lenders, and I. Martinovic, "Assessing the impact of aviation security on cyber power," in *2016 8th International Conference on Cyber Conflict (CyCon)*, 2016: IEEE, pp. 223–241.

[9] T. W. a. K. S. Anoop Singhal, "Guide to secure web services recommendations of the National Institute of Standards and Technology," in "Special Publication 800-95," National Institute of Standards and Technology, U.S, Department of Commerce, August, 2007.

[10] C. E. Shannon, "Communication theory of secrecy systems," *The Bell System Technical Journal,* vol. 28, no. 4, pp. 656–715, 1949, doi: 10.1002/j.1538-7305.1949.tb00928.x.

[11] A. D. Wyner, "The wire-tap channel," *The Bell System Technical Journal,* vol. 54, no. 8, pp. 1355–1387, 1975, doi: 10.1002/j.1538-7305.1975.tb02040.x.

[12] R. M. Gray and A. D. Wyner, "Source coding for a simple network," *The Bell System Technical Journal,* vol. 53, no. 9, pp. 1681–1721, 1974.

[13] H. V. Poor and R. F. Schaefer, "Wireless physical layer security," *Proceedings of the National Academy of Sciences,* vol. 114, no. 1, p. 19, 2017, doi: 10.1073/pnas.1618130114.

[14] A. Wyner, "The common information of two dependent random variables," *IEEE Transactions on Information Theory,* vol. 21, no. 2, pp. 163–179, 1975, doi: 10.1109/TIT.1975.1055346.

[15] I. Csiszár, "Almost independence and secrecy capacity," *Problems Inform. Transmission,* vol. 32, no. 1, pp. 40–47, 1996.

[16] U. Maurer and S. Wolf, "Information-theoretic key agreement: from weak to strong secrecy for free," in *Advances in Cryptology — EUROCRYPT 2000*, Berlin, Heidelberg, B. Preneel, Ed., 2000// 2000: Springer Berlin Heidelberg, pp. 351–368.

[17] M. R. Bloch and J. N. Laneman, "Strong secrecy from channel resolvability," *IEEE Transactions on Information Theory,* vol. 59, no. 12, pp. 8077–8098, 2013, doi: 10.1109/TIT.2013.2283722.

[18] I. Devetak, "The private classical capacity and quantum capacity of a quantum channel," *IEEE Transactions on Information Theory,* vol. 51, no. 1, pp. 44–55, 2005, doi: 10.1109/TIT.2004.839515.

[19] T. S. Han, H. Endo, and M. Sasaki, "Reliability and secrecy functions of the wiretap channel under cost constraint," *IEEE Transactions on Information*

Theory, vol. 60, no. 11, pp. 6819–6843, 2014, doi: 10.1109/TIT.2014.2355811.

[20] M. Hayashi, "General nonasymptotic and asymptotic formulas in channel resolvability and identification capacity and their application to the wiretap channel," *IEEE Transactions on Information Theory,* vol. 52, no. 4, pp. 1562–1575, 2006, doi: 10.1109/TIT.2006.871040.

[21] J. Hou and G. Kramer, "Effective secrecy: Reliability, confusion and stealth," in *2014 IEEE International Symposium on Information Theory*, 29 June–4 July 2014, pp. 601–605, doi: 10.1109/ISIT.2014.6874903.

[22] M. Bellare, S. Tessaro, and A. Vardy, "Semantic security for the wiretap channel," in *Advances in Cryptology – CRYPTO 2012*, Berlin, Heidelberg, R. Safavi-Naini and R. Canetti, Eds., 2012//2012: Springer Berlin Heidelberg, pp. 294–311.

[23] A. Mukherjee, S. A. A. Fakoorian, J. Huang, and A. L. Swindlehurst, "Principles of physical layer security in multiuser wireless networks: a survey," *IEEE Communications Surveys & Tutorials,* vol. 16, no. 3, pp. 1550–1573, 2014, doi: 10.1109/SURV.2014.012314.00178.

[24] 5G_PPP, "White paper: 5G vision," *European Commission,* pp. 1–16, 2015. [Online]. Available: https://5g-ppp.eu/wp-content/uploads/2015/02/5G-Vision-Brochure-v1.pdf.

[25] E. Tekin and A. Yener, "The general Gaussian multiple-access and two-way wiretap channels: achievable rates and cooperative jamming," *IEEE Transactions on Information Theory,* vol. 54, no. 6, pp. 2735–2751, 2008, doi: 10.1109/TIT.2008.921680.

[26] A. Khisti and G. W. Wornell, "Secure transmission with multiple antennas—Part I: The MISOME wiretap channel," *IEEE Transactions on Information Theory,* vol. 56, no. 7, pp. 3088–3104, 2010, doi: 10.1109/TIT.2010.2048445.

[27] A. Khisti and G. W. Wornell, "Secure transmission with multiple antennas—Part II: The MIMOME wiretap channel," *IEEE Transactions on Information Theory,* vol. 56, no. 11, pp. 5515–5532, 2010, doi: 10.1109/TIT.2010.2068852.

[28] X. He and A. Yener, "MIMO wiretap channels with unknown and varying eavesdropper channel states," *IEEE Transactions on Information Theory,* vol. 60, no. 11, pp. 6844–6869, 2014, doi: 10.1109/TIT.2014.2359192.

[29] N. Yang, P. L. Yeoh, M. Elkashlan, R. Schober, and I. B. Collings, "Transmit antenna selection for security enhancement in MIMO wiretap channels," *IEEE Transactions on Communications,* vol. 61, no. 1, pp. 144–154, 2013, doi: 10.1109/TCOMM.2012.12.110670.

[30] L. Dong, Z. Han, A. P. Petropulu, and H. V. Poor, "Improving wireless physical layer security via cooperating relays," *IEEE Transactions on Signal Processing,* vol. 58, no. 3, pp. 1875–1888, 2010, doi: 10.1109/TSP.2009.2038412.

[31] S. A. A. Fakoorian and A. L. Swindlehurst, "Solutions for the MIMO Gaussian wiretap channel with a cooperative jammer," *IEEE Transactions on Signal Processing,* vol. 59, no. 10, pp. 5013–5022, 2011, doi: 10.1109/TSP.2011.2161298.

[32] X. Yang and A. L. Swindlehurst, "Limited rate feedback in a MIMO wiretap channel with a cooperative jammer," *IEEE Transactions on Signal Processing,* vol. 64, no. 18, pp. 4695–4706, 2016, doi: 10.1109/TSP.2016.2572042.

[33] X. He and A. Yener, "Providing secrecy with structured codes: two-user Gaussian channels," *IEEE Transactions on Information Theory,* vol. 60, no. 4, pp. 2121–2138, 2014, doi: 10.1109/TIT.2014.2298132.

[34] J. D. V. Sánchez, L. Urquiza-Aguiar, M. C. P. Paredes, and D. P. M. Osorio, "Survey on physical layer security for 5G wireless networks," *Annals of Telecommunications,* pp. 1–20, 2020.

[35] P. K. Gopala, L. Lai, and H. El Gamal, "On the secrecy capacity of fading channels," *IEEE Transactions on Information Theory,* vol. 54, no. 10, pp. 4687–4698, 2008.

[36] V. U. Prabhu and M. R. Rodrigues, "On wireless channels with M-antenna eavesdroppers: characterization of the outage probability and ε-outage secrecy capacity," *IEEE Transactions on Information Forensics and Security,* vol. 6, no. 3, pp. 853–860, 2011.

[37] L. Wang, *Physical layer security in wireless cooperative networks*. Springer, 2018.

[38] K. Guo, K. An, B. Zhang, *et al.*, "Physical layer security for multiuser satellite communication systems with threshold-based scheduling scheme," *IEEE Transactions on Vehicular Technology,* vol. 69, no. 5, pp. 5129–5141, 2020.

[39] S. Leung-Yan-Cheong and M. Hellman, "The Gaussian wire-tap channel," *IEEE Transactions on Information Theory,* vol. 24, no. 4, pp. 451–456, 1978, doi: 10.1109/TIT.1978.1055917.

[40] S. Goel and R. Negi, "Secret communication in presence of colluding eavesdroppers," in *MILCOM 2005–2005 IEEE Military Communications Conference*, 17–20 Oct. 2005, pp. 1501–1506, vol. 3, doi: 10.1109/MILCOM.2005.1605889.

[41] A. Khisti, G. Wornell, A. Wiesel, and Y. Eldar, "On the Gaussian MIMO wiretap channel," in *2007 IEEE International Symposium on Information Theory*, 24–29 June 2007, pp. 2471–2475, doi: 10.1109/ISIT.2007.4557590.

[42] T. Chrysikos, T. Dagiuklas, and S. Kotsopoulos, "A closed-form expression for outage secrecy capacity in wireless information-theoretic security," Berlin, Heidelberg, 2010: Springer Berlin Heidelberg, in *Security in Emerging Wireless Communication and Networking Systems*, pp. 3–12.

[43] F. Oggier and B. Hassibi, "The secrecy capacity of the MIMO wiretap channel," *IEEE Transactions on Information Theory,* vol. 57, no. 8, pp. 4961–4972, 2011, doi: 10.1109/TIT.2011.2158487.

[44] X. He, A. Khisti, and A. Yener, "MIMO broadcast channel with arbitrarily varying eavesdropper channel: secrecy degrees of freedom," in *2011 IEEE Global Telecommunications Conference - GLOBECOM 2011*, 5–9 Dec. 2011, pp. 1–5, doi: 10.1109/GLOCOM.2011.6134247.

[45] S. Goel and R. Negi, "Guaranteeing secrecy using artificial noise," *IEEE Transactions on Wireless Communications,* vol. 7, no. 6, pp. 2180–2189, 2008, doi: 10.1109/TWC.2008.060848.

[46] X. Zhou and M. R. McKay, "Secure transmission with artificial noise over fading channels: achievable rate and optimal power allocation," *IEEE Transactions on Vehicular Technology,* vol. 59, no. 8, pp. 3831–3842, 2010, doi: 10.1109/TVT.2010.2059057.

[47] F. Wu, W. Wang, B. Yao, and Q. Yin, "Effective eavesdropping in the artificial noise aided security scheme," in *2013 IEEE/CIC International Conference on Communications in China (ICCC)*, 12–14 Aug. 2013, pp. 214–218, doi: 10.1109/ICCChina.2013.6671117.

[48] W. Liao, T. Chang, W. Ma, and C. Chi, "QoS-based transmit beamforming in the presence of eavesdroppers: an optimized artificial-noise-aided approach," *IEEE Transactions on Signal Processing,* vol. 59, no. 3, pp. 1202–1216, 2011, doi: 10.1109/TSP.2010.2094610.

[49] D. Goeckel, S. Vasudevan, D. Towsley, S. Adams, Z. Ding, and K. Leung, "Artificial noise generation from cooperative relays for everlasting secrecy in two-hop wireless networks," *IEEE Journal on Selected Areas in Communications,* vol. 29, no. 10, pp. 2067–2076, 2011, doi: 10.1109/JSAC.2011.111216.

[50] Y. Zou, X. Wang, and W. Shen, "Optimal relay selection for physical-layer security in cooperative wireless networks," *IEEE Journal on Selected Areas in Communications,* vol. 31, no. 10, pp. 2099–2111, 2013, doi: 10.1109/JSAC.2013.131011.

[51] Y. Zou, J. Zhu, X. Wang, and V. C. M. Leung, "Improving physical-layer security in wireless communications using diversity techniques," *IEEE Network,* vol. 29, no. 1, pp. 42–48, 2015, doi: 10.1109/MNET.2015.7018202.

[52] J. Zhang and M. C. Gursoy, "Collaborative relay beamforming for secrecy," in *2010 IEEE International Conference on Communications*, 23–27 May 2010, pp. 1–5, doi: 10.1109/ICC.2010.5501835.

[53] A. Mukherjee and A. L. Swindlehurst, "Robust beamforming for security in MIMO wiretap channels with imperfect CSI," *IEEE Transactions on Signal Processing,* vol. 59, no. 1, pp. 351–361, 2011, doi: 10.1109/TSP.2010.2078810.

[54] C. Jeong, I. Kim, and D. I. Kim, "Joint secure beamforming design at the source and the relay for an amplify-and-forward MIMO untrusted relay system," *IEEE Transactions on Signal Processing,* vol. 60, no. 1, pp. 310–325, 2012, doi: 10.1109/TSP.2011.2172433.

[55] H. Qin, X. Chen, Y. Sun, M. Zhao, and J. Wang, "Optimal power allocation for joint beamforming and artificial noise design in secure wireless communications," in *2011 IEEE International Conference on Communications Workshops (ICC)*, 5–9 June 2011, pp. 1–5, doi: 10.1109/iccw.2011.5963524.

[56] N. Romero-Zurita, M. Ghogho, and D. McLernon, "Outage probability based power distribution between data and artificial noise for physical layer

security," *IEEE Signal Processing Letters,* vol. 19, no. 2, pp. 71–74, 2012, doi: 10.1109/LSP.2011.2178067.

[57] M. Qian, C. Liu, and Y. Fu, "Distributed beamforming designs to improve physical layer security in wireless relay networks," *EURASIP Journal on Advances in Signal Processing,* vol. 2014, no. 1, p. 56, 2014/04/28 2014, doi: 10.1186/1687-6180-2014-56.

[58] T. Lv, H. Gao, and S. Yang, "Secrecy transmit beamforming for heterogeneous networks," *IEEE Journal on Selected Areas in Communications,* vol. 33, no. 6, pp. 1154–1170, 2015, doi: 10.1109/JSAC.2015.2416984.

[59] M. Steiner, G. Tsudik, and M. Waidner, "Diffie–Hellman key distribution extended to group communication," presented at the Proceedings of the 3rd ACM conference on Computer and communications security, New Delhi, India, 1996. [Online]. Available: https://doi.org/10.1145/238168.238182.

[60] J. E. Hershey, A. A. Hassan, and R. Yarlagadda, "Unconventional cryptographic keying variable management," *IEEE Transactions on Communications,* vol. 43, no. 1, pp. 3–6, 1995, doi: 10.1109/26.385951.

[61] Y. Li, L. J. Cimini, and N. R. Sollenberger, "Robust channel estimation for OFDM systems with rapid dispersive fading channels," *IEEE Transactions on Communications,* vol. 46, no. 7, pp. 902–915, 1998, doi: 10.1109/26.701317.

[62] B. Muquet, M. d. Courville, and P. Duhamel, "Subspace-based blind and semi-blind channel estimation for OFDM systems," *IEEE Transactions on Signal Processing,* vol. 50, no. 7, pp. 1699–1712, 2002, doi: 10.1109/TSP.2002.1011210.

[63] A. A. Hassan, W. E. Stark, J. E. Hershey, and S. Chennakeshu, "Cryptographic key agreement for mobile radio," *Digital Signal Processing,* vol. 6, no. 4, pp. 207–212, 1996/10/01/1996, doi: https://doi.org/10.1006/dspr.1996.0023.

[64] S. Mathur, W. Trappe, N. Mandayam, C. Ye, and A. Reznik, "Radio-telepathy: extracting a secret key from an unauthenticated wireless channel," presented at the *Proceedings of the 14th ACM International Conference on Mobile Computing and Networking*, San Francisco, California, USA, 2008. [Online]. Available: https://doi.org/10.1145/1409944.1409960.

[65] S. Jana, S. N. Premnath, M. Clark, S. K. Kasera, N. Patwari, and S. V. Krishnamurthy, "On the effectiveness of secret key extraction from wireless signal strength in real environments," presented at the Proceedings of the 15th annual international conference on Mobile computing and networking, Beijing, China, 2009. [Online]. Available: https://doi.org/10.1145/1614320.1614356.

[66] S. Gollakota and D. Katabi, "Physical layer wireless security made fast and channel independent," in *2011 Proceedings IEEE INFOCOM*, 10–15 April 2011, pp. 1125–1133, doi: 10.1109/INFCOM.2011.5934889.

[67] Y. E. H. Shehadeh, O. Alfandi, K. Tout, and D. Hogrefe, "Intelligent mechanisms for key generation from multipath wireless channels," in *2011 Wireless Telecommunications Symposium (WTS)*, 13–15 April 2011, pp. 1–6, doi: 10.1109/WTS.2011.5960848.

[68] Q. Wang, H. Su, K. Ren, and K. Kim, "Fast and scalable secret key generation exploiting channel phase randomness in wireless networks," in *2011 Proceedings IEEE INFOCOM*, 10–15 April 2011, pp. 1422–1430, doi: 10.1109/INFCOM.2011.5934929.

[69] J. W. Wallace, C. Chen, and M. A. Jensen, "Key generation exploiting MIMO channel evolution: Algorithms and theoretical limits," in *2009 3rd European Conference on Antennas and Propagation*, 23–27 March 2009, pp. 1499–1503.

[70] K. Zeng, D. Wu, A. Chan, and P. Mohapatra, "Exploiting multiple-antenna diversity for shared secret key generation in wireless networks," in *2010 Proceedings IEEE INFOCOM*, 14–19 March 2010, pp. 1–9, doi: 10.1109/INFCOM.2010.5462004.

[71] T. Shimizu, H. Iwai, and H. Sasaoka, "Physical-layer secret key agreement in two-way wireless relaying systems," *IEEE Transactions on Information Forensics and Security,* vol. 6, no. 3, pp. 650–660, 2011, doi: 10.1109/TIFS.2011.2147314.

[72] S. M. Alamouti, "A simple transmit diversity technique for wireless communications," *IEEE Journal on Selected Areas in Communications,* vol. 16, no. 8, pp. 1451–1458, 1998, doi: 10.1109/49.730453.

[73] M. Edman, A. Kiayias, and B. Yener, "On passive inference attacks against physical-layer key extraction?," presented at the *Proceedings of the Fourth European Workshop on System Security*, Salzburg, Austria, 2011. [Online]. Available: https://doi.org/10.1145/1972551.1972559.

[74] J. M. Hamamreh, H. M. Furqan, and H. Arslan, "Classifications and applications of physical layer security techniques for confidentiality: A comprehensive survey," *IEEE Communications Surveys & Tutorials,* vol. 21, no. 2, pp. 1773–1828, 2018.

[75] M. Chraiti, A. Ghrayeb, C. Assi, and M. O. Hasna, "On the achievable secrecy diversity of cooperative networks with untrusted relays," *IEEE Transactions on Communications,* vol. 66, no. 1, pp. 39–53, 2017.

[76] Y. Liang, H. V. Poor, and S. Shamai, "Secure communication over fading channels," *IEEE Transactions on Information Theory,* vol. 54, no. 6, pp. 2470–2492, 2008.

[77] R. Liu and W. Trappe, *Securing Wireless Communications at the Physical Layer*. Springer New York: Springer 2010.

[78] Y. Shiu, S. Y. Chang, H. Wu, S. C. Huang, and H. Chen, "Physical layer security in wireless networks: a tutorial," *IEEE Wireless Communications,* vol. 18, no. 2, pp. 66–74, 2011.

[79] Y. Zou, J. Zhu, X. Wang, and L. Hanzo, "A survey on wireless security: technical challenges, recent advances, and future trends," *Proceedings of the IEEE,* vol. 104, no. 9, pp. 1727–1765, 2016.

[80] X. Zhou, L. Song, and Y. Zhang, *Physical Layer Security in Wireless Communications*. Boca Raton, FL, USA: CRC Press – Taylor and Francis Group, LLC, 2014.

[81] I. Csiszar and J. Korner, “Broadcast channels with confidential messages,” *IEEE Transactions on Information Theory,* vol. 24, no. 3, pp. 339 - 348, 1978, doi: 10.1109/TIT.1978.1055892.

[82] E. Şaşoğlu, E. Telatar, and E. Arikan, “Polarization for arbitrary discrete memoryless channels,” in *2009 IEEE Information Theory Workshop*, 11–16 Oct. 2009, pp. 144–148.

[83] R. Mori and T. Tanaka, “Channel polarization on q-ary discrete memoryless channels by arbitrary kernels,” in *2010 IEEE International Symposium on Information Theory*, 13–18 June 2010, pp. 894–898.

[84] E. Arikan, “Channel polarization: a method for constructing capacity-achieving codes for symmetric binary-input memoryless channels,” *IEEE Transactions on Information Theory,* vol. 55, no. 7, pp. 3051–3073, 2009.

[85] H. S. Cronie and S. B. Korada, “Lossless source coding with polar codes,” in *2010 IEEE International Symposium on Information Theory*, 13–18 June 2010, pp. 904–908.

[86] S. Çayci and O. Arikan, “Lossless polar compression of q-ary sources,” in *2013 IEEE International Symposium on Information Theory*, 7–12 July 2013, pp. 1132–1136.

[87] L. Liu, J. Shi, and C. Ling, “Extracting Wyner’s common randomness using polar codes,” in *2016 8th International Conference on Wireless Communications & Signal Processing (WCSP)*, 13–15 Oct. 2016, pp. 1–5.

[88] S. B. Korada and R. L. Urbanke, “Polar codes are optimal for lossy source coding,” *IEEE Transactions on Information Theory,* vol. 56, no. 4, pp. 1751–1768, 2010.

[89] N. Hussami, S. B. Korada, and R. Urbanke, “Performance of polar codes for channel and source coding,” in *2009 IEEE International Symposium on Information Theory*, 28 June–3 July 2009, pp. 1488–1492.

[90] Y. Wei and S. Ulukus, “Polar coding for the general wiretap channel with extensions to multiuser scenarios,” *IEEE Journal on Selected Areas in Communications,* vol. 34, no. 2, pp. 278–291, 2016, doi: 10.1109/JSAC.2015.2504275.

[91] S. A. Fakoorian and A. L. Swindlehurst, “On the optimality of polar codes for the deterministic wiretap channel,” in *2013 Asilomar Conference on Signals, Systems and Computers*, 3–6 Nov. 2013, pp. 2089–2093, doi: 10.1109/ACSSC.2013.6810675.

[92] M. Andersson, R. F. Schaefer, T. J. Oechtering, and M. Skoglund, “Polar coding for bidirectional broadcast channels with common and confidential messages,” *IEEE Journal on Selected Areas in Communications,* vol. 31, no. 9, pp. 1901–1908, 2013, doi: 10.1109/JSAC.2013.130921.

[93] H. Mahdavifar and A. Vardy, “Achieving the secrecy capacity of wiretap channels using polar codes,” *IEEE Transactions on Information Theory,* vol. 57, no. 10, pp. 6428–6443, 2011, doi: 10.1109/TIT.2011.2162275.

[94] E. Hof and S. Shamai, “Secrecy-achieving polar-coding,” in *2010 IEEE Information Theory Workshop*, 30 Aug.–3 Sept. 2010, pp. 1–5, doi: 10.1109/CIG.2010.5592878.

[95] M. Andersson, V. Rathi, R. Thobaben, J. Kliewer, and M. Skoglund, "Nested polar codes for wiretap and relay channels," *IEEE Communications Letters,* vol. 14, no. 8, pp. 752–754, 2010, doi: 10.1109/LCOMM.2010.08.100875.

[96] O. O. Koyluoglu and H. E. Gamal, "Polar coding for secure transmission and key agreement," *IEEE Transactions on Information Forensics and Security,* vol. 7, no. 5, pp. 1472–1483, 2012, doi: 10.1109/TIFS.2012.2207382.

[97] E. Şaşoğlu and A. Vardy, "A new polar coding scheme for strong security on wiretap channels," in *2013 IEEE International Symposium on Information Theory*, Istanbul, Turkey, July 2013: IEEE, pp. 1117–1121, doi: 10.1109/ISIT.2013.6620400.

[98] T. C. Gulcu and A. Barg, "Achieving secrecy capacity of the wiretap channel and broadcast channel with a confidential component," *IEEE Transactions on Information Theory,* vol. 63, no. 2, pp. 1311–1324, 2017, doi: 10.1109/TIT.2016.2631223.

[99] E. Ekrem and S. Ulukus, "Capacity-equivocation region of the Gaussian MIMO wiretap channel," *IEEE Transactions on Information Theory,* vol. 58, no. 9, pp. 5699–5710, 2012, doi: 10.1109/TIT.2012.2204534.

[100] R. A. Chou and M. R. Bloch, "Polar coding for the broadcast channel with confidential messages: a random binning analogy," *IEEE Transactions on Information Theory,* vol. 62, no. 5, pp. 2410–2429, 2016, doi: 10.1109/TIT.2016.2539145.

[101] H. Wang, X. Tao, N. Li, and Z. Han, "Polar coding for the wiretap channel with shared key," *IEEE Transactions on Information Forensics and Security,* vol. 13, no. 6, pp. 1351–1360, 2018.

[102] Y. Zhao, S. Xu, and H. Chi, "Encrypted secure polar coding scheme for general two-way wiretap channel," *IET Information Security,* vol. 13, no. 4, pp. 393–403, 2019.

[103] R. Zamir, "Lattices are everywhere," in *2009 Information Theory and Applications Workshop*, 8–13 Feb. 2009, pp. 392–421, doi: 10.1109/ITA.2009.5044976.

[104] V. Forutan and R. F. H. Fischer, "On the security of lattice-based physical-layer network coding against wiretap attacks," in *SCC 2015; 10th International ITG Conference on Systems, Communications and Coding*, 2–5 Feb. 2015, pp. 1–6.

[105] A. Karrila, "Number-theoretic and geometric lattice code design for secure and reliable wireless communications," M.Sc. Dissertation, School of Science, Aalto University, Espoo, Finland, 2015.

[106] I. Stewart and D. Tall, *Algebraic Number Theory and Fermat's Last Theorem*, 4th Ed. ed. Boca Raton, FL: CRC Press, Taylor & Francis Group, 2016.

[107] J. Belfiore and F. Oggier, "Secrecy gain: A wiretap lattice code design," in *2010 International Symposium On Information Theory & Its Applications*, 17–20 Oct. 2010, pp. 174–178, doi: 10.1109/ISITA.2010.5650095.

[108] F. Oggier, P. Solé, and J. Belfiore, "Lattice codes for the wiretap Gaussian channel: construction and analysis," *IEEE Transactions on Information Theory,* vol. 62, no. 10, pp. 5690–5708, 2016, doi: 10.1109/TIT.2015.2494594.

[109] A. Ernvall-Hytonen, "On a conjecture by Belfiore and Solé on some lattices," *IEEE Transactions on Information Theory,* vol. 58, no. 9, pp. 5950–5955, 2012, doi: 10.1109/TIT.2012.2201915.

[110] A. Ernvall-Hytonen, "Some results related to the conjecture by Belfiore and Solé," *IEEE Transactions on Information Theory,* vol. 60, no. 5, pp. 2805–2812, 2014, doi: 10.1109/TIT.2014.2311075.

[111] F. Lin and F. Oggier, "A classification of unimodular lattice wiretap codes in small dimensions," *IEEE Transactions on Information Theory,* vol. 59, no. 6, pp. 3295–3303, 2013, doi: 10.1109/TIT.2013.2246814.

[112] J. Belfiore and F. Oggier, "Lattice code design for the Rayleigh fading wiretap channel," in *2011 IEEE International Conference on Communications Workshops (ICC)*, 5–9 June 2011, pp. 1–5, doi: 10.1109/iccw.2011.59 63544.

[113] L. Choo, C. Ling, and K. Wong, "Achievable rates for lattice coded Gaussian wiretap channels," in *2011 IEEE International Conference on Communications Workshops (ICC)*, 5–9 June 2011, pp. 1–5, doi: 10.1109/ iccw.2011.5963541.

[114] C. Ling, L. Luzzi, J. Belfiore, and D. Stehlé, "Semantically secure lattice codes for the Gaussian wiretap channel," *IEEE Transactions on Information Theory,* vol. 60, no. 10, pp. 6399–6416, 2014, doi: 10.1109/TIT.2014.23 43226.

[115] L. Choo and C. Ling, "Superposition lattice coding for Gaussian broadcast channel with confidential message," in *2014 IEEE Information Theory Workshop (ITW 2014)*, 2–5 Nov. 2014, pp. 311–315, doi: 10.1109/ITW.2014. 6970844.

[116] A. Shokrollahi, "LDPC codes: an introduction," *Progress in Computer Science and Applied Logic,* vol. 23, pp. 85–110, 2004.

[117] M. H. Taieb and J. Chouinard, "Reliable and secure communications over Gaussian wiretap channel using HARQ LDPC codes and error contamination," in *2015 IEEE Conference on Communications and Network Security (CNS)*, 28–30 Sept. 2015, pp. 158–163, doi: 10.1109/CNS.2015.7346824.

[118] D. Klinc, J. Ha, S. W. McLaughlin, J. Barros, and B. Kwak, "LDPC for physical layer security," in *GLOBECOM 2009–2009 IEEE Global Telecommunications Conference*, 30 Nov.–4 Dec. 2009, pp. 1–6, doi: 10.1109/ GLOCOM.2009.5426065.

[119] D. Klinc, J. Ha, S. W. McLaughlin, J. Barros, and B. Kwak, "LDPC codes for the Gaussian wiretap channel," *IEEE Transactions on Information Forensics and Security,* vol. 6, no. 3, pp. 532–540, 2011.

[120] A. Thangaraj, S. Dihidar, A. R. Calderbank, S. W. McLaughlin, and J. Merolla, "Applications of LDPC codes to the wiretap channel," *IEEE Transactions on Information Theory,* vol. 53, no. 8, pp. 2933–2945, 2007.

[121] A. Subramanian, A. Thangaraj, M. Bloch, and S. W. McLaughlin, "Strong secrecy on the binary erasure wiretap channel using large-girth LDPC codes," *IEEE Transactions on Information Forensics and Security,* vol. 6, no. 3, pp. 585–594, 2011.

[122] V. Rathi, M. Andersson, R. Thobaben, J. Kliewer, and M. Skoglund, "Performance analysis and design of two edge-type LDPC codes for the BEC wiretap channel," *IEEE Transactions on Information Theory,* vol. 59, no. 2, pp. 1048–1064, 2013.

[123] S. Shahbaz, B. Akhbari, and R. Asvadi, "LDPC codes over Gaussian multiple access wiretap channel," *IET Communications,* vol. 12, no. 8, pp. 962–969, 2018.

[124] M. Baldi, M. Bianchi, and F. Chiaraluce, "Coding with scrambling, concatenation, and HARQ for the AWGN wire-tap channel: a security gap analysis," *IEEE Transactions on Information Forensics and Security,* vol. 7, no. 3, pp. 883–894, 2012.

[125] M. Baldi, F. Chiaraluce, N. Laurenti, S. Tomasin, and F. Renna, "Secrecy transmission on parallel channels: theoretical limits and performance of practical codes," *IEEE Transactions on Information Forensics and Security,* vol. 9, no. 11, pp. 1765–1779, 2014.

[126] M. Baldi, G. Ricciutelli, N. Maturo, and F. Chiaraluce, "Performance assessment and design of finite length LDPC codes for the Gaussian wiretap channel," in *2015 IEEE International Conference on Communication Workshop (ICCW)*, 8–12 June 2015, pp. 435–440.

[127] Y. Huo, Y. Tian, L. Ma, X. Cheng, and T. Jing, "Jamming strategies for physical layer security," *IEEE Wireless Communications,* vol. 25, no. 1, pp. 148–153, 2017.

[128] T. T. Do, E. Björnson, E. G. Larsson, and S. M. Razavizadeh, "Jamming-resistant receivers for the massive MIMO uplink," *IEEE Transactions on Information Forensics and Security,* vol. 13, no. 1, pp. 210–223, 2017.

Chapter 8
Cybersecurity and privacy issues

Dennis A. Vincenzi[1], Yupeng Yang[2], and Dahai Liu[3]

A variety of challenges to cybersecurity and privacy currently exist. Technical issues and human factor-related hurdles have brought forth a range of research efforts to help mitigate or resolve these challenges. Legislation plays a significant role in controlling the use of networks all around the United States. Currently, as the definition of cybersecurity is relatively new, the explicit reference of the concept of cybersecurity is scant. The first time that a published U.S. court even used the word "cybersecurity" was in a 2007 Seventh Circuit court opinion. The issues associated with the concept of cybersecurity such as data security, cyber risk, data breach, etc., have been taken very seriously by a number of U.S. state and federal statutes, regulations, and courts.

Public outcry concerning unwarranted or unknown observation is nothing new. The privacy debate is inherent to American society so much so that the issue is embedded in the Fourth Amendment of the U.S. Constitution. With the advent of new surveillance technologies and techniques, concern that they may be used in violation of personal rights and protections has grown. Examples include wiretapping, electronic surveillance, video monitoring, and other types of law enforcement and related agency activities. In this chapter, we have identified themes that overlap with technologies related to cybersecurity. Furthermore, commonalities and occurrences in previous privacy-related confrontations were characterized in order to serve as a guide for efforts to resolve the cybersecurity privacy quandary.

8.1 Introduction

In the past, network security and security protocols have been implemented as an afterthought as a reactionary response to potential security breaches. With advances in technology and software, and the sharp increase in unauthorized use and access of various data networks, enhanced security protocols must be part of network design to effectively combat cyber threats. It is foreseeable that the next generation of cyber

[1]Department of Graduate Studies, College of Aeronautics, Embry–Riddle Aeronautical University, Daytona Beach, FL, USA
[2]Department of Electrical Engineering and Computer Science, Embry–Riddle Aeronautical University, Daytona Beach, FL, USA
[3]Department of Graduate Studies, College of Aviation, Embry-Riddle Aeronautical University, Daytona Beach, FL, USA

networks will contain a great emphasis on the security issues and protocols as opposed to being an afterthought built into a system or network to compensate for an unanticipated weakness. In today's environment, cybersecurity technology has been applied to every corner of our daily lives, society, industry, and military. In the big data era, with the development of sophisticated information technology, applications have been greatly extended to include a wide variety of civilian services and networks. At an individual level, a cybersecurity attack can come from anywhere, and this cybersecurity attack may result in identity theft, extortion attempts, financial fraud and theft, and the loss of crucial data.

Because everyone in today's society relies on data service companies and networks, it is essential for those companies to maintain effective security protocols that will maintain data security and safety for the individual and society as a whole. It makes sense to state that everyone will benefit from the work of cybersecurity researchers, professionals, and network designers. These efforts will keep our lives and society secure and protected from new threats and vulnerabilities. Cybersecurity training, design, and education will also help raise public awareness of cybersecurity threats and issues to enhance protection efforts.

8.1.1 Congressional mandate for cybersecurity integration and operation

The term "cyber risk" refers to any risk to information infrastructure, including physical or personal risks and security vulnerabilities, that, if exploited or not mitigated, could pose a significant risk of disruption to the operation of information infrastructure essential to the reliable operation of covered critical infrastructure [Cybersecurity Act of 2012 Title I Sec. 101]. Early back in 2012, the 112th Congress had already presented a bill whose goal was to enhance the security and resiliency of cyber networks and communications infrastructure of the United States.

Cyberattacks and security breaches play a crucial role in our daily life. With the development of sophisticated cyber techniques, the creation of government guidelines, rules, and regulations is also necessary. However, to write and pass a new set of rules or regulations that culminate in the passage of new laws and Acts of Congress is not an easy task. During the past decade, Congress failed to resolve the mandatory regulation versus voluntary cooperation debate. In 2014, the Cyber Intelligence Sharing and Protection Act (CISPA) was finally passed in the U.S. House of Representatives, and the bill was eventually passed by the U.S. Senate in 2015. This bill requires government agencies and businesses to share information between each other to help fight and combat cybercrime. In some states, there are currently more than 280 bills introduced or considered by state legislatures in session to deal with significant threats involving cybersecurity. These bills cover a wide array of cyber-related areas that include: [1].

- Requiring government agencies to implement training or specific types of security policies and practices and improving incidence, response and preparedness.
- Increasing penalties for computer crime or addressing specific crimes, e.g., ransomware.

- Regulating cybersecurity within the insurance industry or addressing cybersecurity insurance.
- Creating task forces, councils, or commissions to study or advise on cybersecurity issues.
- Supporting programs or incentives for cybersecurity training and education.

8.1.2 Public concerns and outcries for protection of privacy

As Steve Morgan, the founder of Cybersecurity Ventures, points out that, in the next two decades, cybersecurity activity will be one of the biggest challenges that society will face. In August 2016, Cybersecurity Ventures predicted that cybercrime will cost the world 6 trillion dollars annually by 2021, up from 3 trillion dollars in 2015. This prediction indicated that cybercrime, in general, will result in one of the greatest transfers of economic wealth in recorded history. At the same time, according to the 2018 PwC's Global Economic Crime and Fraud Survey, U.S. organizations faced more financial losses in every category (from under $50,000 to over $100 million) because of fraud compared to their global counterparts [2]. All the above-published data have shown that cybersecurity will play an ever increasingly important role in current human society.

According to recent statistics, the most common cybercrimes listed are hacking devices (53%), debit or credit card fraud (38%), compromised account passwords (34%), hacking email or social media accounts (34%), fraudulent online purchases (33%), and phishing scams (32%) [3]. Based on those statistics, another fact that is worth mentioning is that the adopters of newer security techniques, such as a personal Virtual Private Network (VPN), will be overconfident in their abilities to avoid becoming a cybercrime victim compared to those who are not familiar with such network and security techniques. Therefore, it is important to raise public concern and awareness of cybersecurity threats to keep the public informed of potential vulnerabilities.

While the Computer Fraud and Abuse Act (CFAA) is a U.S. cybersecurity bill that was enacted in 1986 as an amendment to existing computer fraud law (18 U.S.C.& 1030). The CFAA is part of the Comprehensive Crime Control Act of 1984. The law states that accessing a computer without authorization, or in excess of authorization and illegal misappropriation of customer privacy, is prohibited. This bill is recognized by the federal government of the United States as protecting a general right to privacy. There is growing public concern about the increasing use of the Internet of Things (IOT), despite the fact that most of the civil networks and data privacy networks are supervised and protected by the government. In January 2015, then President Obama proposed expanding the CFAA in his Modernizing Law Enforcement Authorities to Combat Cyber Crime proposal. President Obama claimed that in this interconnected and digital world, the phenomenon of crime through attacks on the Internet has gradually increased. These illegal intrusions not only endanger public safety but also cause great harm in terms of invasion of privacy.

These attacks are often complex and unpredictable, posing a great threat to the maintenance of people's daily lives. From one perspective, proponents of legislation such as the Modernizing Law Enforcement Authorities to Combat Cyber Crime argue that law enforcement must have the right tools to investigate, disrupt, and prosecute cybercrime, and argue that it is crucial for Congress to get bills such as this passed into law. From another perspective, opponents of the bill argue that the bill could dramatically expand law enforcement powers to the point where innocent Internet activities executed by ordinary law-abiding citizens could be construed as cybercrimes. In many states, legislators have voiced their concerns about the potential for privacy invasion posed by the development of the IOT. These and other examples have illustrated the growing concerns of the U.S. public as the use and potential abuse of technology and cyber networks within the United States becomes increasingly probable.

Civil rights surrounding privacy debates have been a major cause of concern for almost every government and commercial system put into use, and it is a crucial issue that cybersecurity entities such as Information Systems Security Association (ISSA) must face and ethically resolve before considering further development of information technology systems. As Jeff Kosseff [4] points out, a specific cybersecurity law is needed that promotes the confidentiality, integrity, and availability of public and private information, systems, and networks, through the use of forward-looking regulations and incentives, with the goal of protecting individual rights and privacy, economic interests, and national security. Without government support, protection, and regulation of the cybersecurity industry to help safeguard the privacy of American companies and individuals, public acceptance of information technology within the borders of the United States will be difficult to achieve.

8.2 Privacy law

Generally speaking, privacy is considered a human right. This can be defined as the ability of a single person or group to protect themselves from being observed by others, or the ability to keep information about themselves away from others, and thereby are free from other observation or distribution. For instance, if a person considers the information confidential or private, it usually means this information has a special meaning, value, or importance; something that should not be viewed by others for specific reasons. In the United States, there is no single comprehensive federal law that governs data privacy. However, there exists a batch of sector-specific and medium-specific laws, which contain guidance, rules, and regulations that address telecommunications, personal information, or other potentially sensitive information. In the following section, we will introduce and discuss some of the laws put forth to help protect personal and public privacy and help address cybersecurity and privacy issues.

8.2.1 The Fourth Amendment of the U.S. Constitution

For centuries, U.S. citizens have enjoyed the protection of specific rights provided by the U.S. Constitution. In today's society, the potential erosion of those protections afforded by the Fourth Amendment concerning the right of privacy and protection from unwarranted search and seizure is of great concern. In the United States, the Fourth Amendment to the U.S. Constitution states "The right of people to be secure in their persons, houses, papers, and effects, against unreasonable searches and seizures, shall not be violated." The concept of protection of an individual's right to privacy is not outdated or obsolete by any means. The Fourth Amendment is as significant and valid today as the day it was conceived and written into the Bill of Rights that was drafted by the first Congress on September 25, 1789, over 230 years ago. The current problem is that the technology being employed today did not exist 230 years ago, and the frame of the Bill of Rights never envisioned this technology or how its capabilities could be used against an individual's right. The Fourth Amendment was designed and worded within the context of the technology and capabilities in existence in late 1789, not 2021. The issues at the center of the cybersecurity privacy debate revolve around (1) the preservation of human rights, (2) the demand for greater security, (3) how to frame the policy discussion, (4) how to differentiate it from cybersecurity surveillance concerns, and (5) how to access the individual's personal data while protecting themselves from privacy invasions and cyber threats.

8.2.2 Legislation

The most notable event in recent history that has contributed to the expansion of government powers has been the coordinated terrorist attack against the United States on September 11, 2001. Citing the need to increase the ability of the U.S. Government law enforcement and intelligence communities to be able to collect information that may help prevent future terrorist attacks, and thereby better protect the public, the USA PATRIOT Act of 2001 was introduced and signed into law by Congress and then President Bush. Many people favored this expansion of power by the Federal Government, whereas some categorized it as a necessary evil that had great potential to erode the protections afforded law-abiding citizens under the Fourth Amendment of the U.S. Constitution.

However, the resistance to the potential erosion of the Fourth Amendment came years later with the advent of cybersecurity technology and the obvious surveillance capabilities this technology possessed. For instance, the focus of the Fourth Amendment of the U.S. Constitution is to protect the people of the United States from unreasonable searches and seizures by the government. Another main aspect of the Fourth Amendment is to protect peoples' privacy. However, in most instances, this protection conflicts with the techniques used by law enforcement to fight cybercrime. Some examples of these conflicts include the ways in which individual electronic devices and personal data are accessed by law enforcement. Information contained on a personal laptop, desktop, cell phone, monitoring devices such as Amazon, Echo, and Alexa, or many other personal data-recording or storage technologies, can be accessed and collected by law enforcement, but the methods or pretense under which these devices are obtained and accessed are questionable in nature and may not always be considered legal. Taking those questionable methods and techniques, and combining them with advanced technology to combat cyber offenders, only compounds the effect and perception that an individual's privacy is being violated in a questionable manner not consistent with the intent of the Fourth Amendment of the U.S. Constitution. This can result in unnecessary and illegal violation of people's privacy, access to personal information without legal authorization, unwarranted and unjustifiable prosecution of criminal offenses, and illegal confiscation and misappropriation of individual property.

In today's world of digital networks, cloud computing and storage, wireless connectivity, and advanced computer technology and software, distinct and definite boundaries for privacy and private information are ambiguous and difficult to define. As a comparison, in the real physical world, these boundaries of privacy and personal property are better defined and much more distinct and easier to identify. Nevertheless, despite this distinction between physical and digital information and property, law enforcement has the obligation to comply with legal standards and work within the boundaries set by documents such as the Bill of Rights and the U.S. Constitution that are constitutionally permissible in terms of privacy, search, and seizure protocols when investigating any potential crime, including cybercrimes. If law enforcement fails to adhere to these established and legal protocols, their actions may result in many undesirable consequences including, but not limited to, (1) suppression of evidence, (2) dismissal of charges, and (3) social panic and protest revolving around privacy violations.

8.2.3 USA PATRIOT Act of 2001

The Uniting and Strengthening America by Providing Appropriate Tools Required to Intercept and Obstruct Terrorism (USA PATRIOT) Act of 2001 was signed into law by President George W. Bush on October 26, 2001. The PATRIOT Act of 2001, which was a response to the terrorist attacks of September 11, 2001, significantly weakened the protection of U.S. citizens and removed many restrictions previously placed on law enforcement agencies that required law enforcement to abide by certain protocols and procedures when gathering intelligence within the United States against suspected terrorists. The immediate reaction to the USA PATRIOT Act of 2001 was overwhelmingly positive and supportive. It passed in the Senate by a vote of 98-1 and in the Hours of Representatives by a vote of 357-66.

Clearly, the American people and government of the United States were focused on taking steps to ensure that events similar to the September 11 attacks would not happen again. Additionally, the U.S. Government needed to reassure the American people that everything was under control and their government was taking positive steps to ensure their safety. However, the potential for abuse is also clear. It also clearly undermines the privacy protections afforded to American citizens through the Fourth Amendment by blurring the lines between the act and necessity to gather evidence for a criminal offense, and the act of gathering intelligence, two different things with two different purposes. It also provides the tools to allow the government to easily allow surveillance of American citizens, access to personal records, and "secret" searches of homes and offices under the guise of anti-terrorism.

Embedded within the USA PATRIOT Act were 10 sections, Title I–X, which greatly enhanced the power and authority of law enforcement and intelligence agencies throughout the country. Without going into extreme detail on each section, the 10 sections are listed below:

- Title I: Enhancing Domestic Security Against Terrorism.
- Title II: Enhanced Surveillance Procedures.
- Title III: International Money Laundering Abatement and Anti-Terrorist Financing Act of 2001.
- Title IV: Protecting the Border.
- Title V: Removing Obstacles to Investigating Terrorism.
- Title VI: Providing for Victims of Terrorism, Public Safety Officers, and Their Families.
- Title VII: Increased Information Sharing for Critical Infrastructure Protection.
- Title VIII: Strengthening the Criminal Laws Against Terrorism.
- Title IX: Improved Intelligence.
- Title X: Miscellaneous.

All of the ten titles included in the USA PATRIOT Act of 2001 are administrative and/or supportive of enhancing anti-terrorist initiatives to one degree or another. Title III, for example, deals with providing the resources and tools needed to identify, track, and freeze money laundering and financial activities that are linked to the funding of terrorist organizations and/or terrorist activities. Title IV deals with the steps necessary to protect and secure the borders of the United States. Title IV does not address actual securing of open borders to prevent illegal crossing of the border into the United States but rather deals with the denial of admission, verification of immigrant status, custody of aliens involved in terrorism, and eventual deportation or removal of the individual from the United States. Title VI deals with provisions for providing donations, payment, and support to victims of terrorism including the general public, public safety officers, and their families. Titles I, VII, VIII, IX, and X involve similar administrative and/or supportive guidance, but do not necessarily directly impact privacy or have any direct impact on the protections afforded by the Fourth Amendment.

Title II, however, potentially expands the powers of the Federal Government in ways that can easily undermine the Fourth Amendment protections if interpreted or implemented improperly or over-zealously by law enforcement agencies and the intelligence community of the United States. Title II discusses enhanced surveillance procedures and specifically enumerates powers listed below:

- Section 206: Allows for roving wiretaps under the Foreign Intelligence Surveillance Act, which allows the issuance of a court order enabling the government to employ electronic surveillance of a foreign power or an agent of a foreign power.
- Section 209: Allows law enforcement to seize voice mail messages pursuant to a warrant.

- Section 210: Allows law enforcement to subpoena additional subscriber records from service providers such as "records of sessions and duration" and "means and source of payment."
- Section 215: Allows the Director of the Federal Bureau of Investigation "access to certain business records for foreign intelligence and international terrorism investigations" where such an investigation is to e "conducted under guidelines approved by the Attorney General."
- Section 216: Allows a Pen Trap, a device that records the numbers dialed but not the content of the conversation, to be applied to internet dialing and email.
- Section 220: Allows for "Nation Wide Service of Search Warrant for Electronic Evidence."

On the surface, no one would object to "enhanced surveillance procedures" to help ensure the safety of the American people. However, recent news events have shown that these "enhanced surveillance procedures" can be interpreted very broadly and used to research and investigate data pertaining to all Americans, not just "foreign powers or agents of a foreign power." For example, under the USA PATRIOT Act of 2001, the FBI can secretly conduct a physical search or wiretap on American citizens to obtain evidence of a crime without providing probable cause to a court of law as is specifically required under the Fourth Amendment. The recent National Security Agency (NSA) phone records scandal used portions of the USA PATRIOT Act of 2001, which references the Foreign Intelligence Surveillance Act to obtain a court order allowing collection of data from phone records from all Verizon customers (foreign and domestic) for the purpose of metadata analysis of information pertaining to those records including what phone number was called, what time the call was made, and duration of the call. In theory, the court order was obtained to research and investigate potential links between foreign entities, known terrorist entities and individuals, and suspected terrorist entities and individuals. In reality, this revelation has led to a renewed debate over the legality and policy merits of broad and indiscriminate government surveillance of Americans under the guise of national security.

It is difficult to be certain as to the legality of this action because both the details of the program and the legal rulings on it are classified as secret. But civil liberties groups argue that the program exceeds the powers Congress has granted to the executive branch, and that such a broad surveillance program is inconsistent with the Fourth Amendment.

The program appears to be partly based on Section 215 of the Patriot Act, which allows the government to obtain business records that are relevant to an ongoing terrorism investigation. That's a pretty permissive standard, but the Electronic Frontier Foundation argues that Congress intended to authorize information requests relevant to a specific terrorism investigation. Demanding the phone records of every person in the United States seems inconsistent with that requirement because it is highly unlikely that ALL Americans are terrorists.

8.2.4 Computer Fraud and Abuse Act

The Computer Fraud and Abuse Act (CFAA) was enacted in 1986. It is an amendment to the first federal computer law and has been included in the Comprehensive Crime Control Act of 1984. The CFAA aims to prohibit the intentional access of a computer without authorization or in excess of authorization [Jarrett, H. Marshall; Bailie, Michael W. (2010). "Prosecution of Computer" (PDF). justice.gov. Office of Legal Education Executive Office for United States Attorneys. Retrieved June 3, 2013].

This law defines the "protected computer" as:

(A) exclusively for the use of a financial institution or the U.S. Government, or, in the case of a computer not exclusively for such use, used by or for a financial institution or the U.S. Government and the conduct constituting the offense affects that use by or for the financial institution or the Government; or
(B) which is used in interstate or foreign commerce or communication, including a computer located outside the United States that is used in a manner that affects interstate or foreign commerce or communication of the United States [18 U.S.C. §1030(e)(2) (2006)].

In practice, this law protects computers, cellphones, and any device that has the interstate nature of most Internet communications [Varma, Corey (2015-01-03). "What is the Computer Fraud and Abuse Act." CoreyCarma.com. Retrieved 10 June 2015]. Before this law, computer crimes were prosecuted as mail and wire fraud, which is often insufficient when applying the law.

However, this law made a number of researchers wary of revealing bugs, which could result in complications with law enforcement agencies or with special companies. As Tor Ekeland, an attorney who specializes in defending people accused of violating the CFAA, said, "Computer researchers are constantly afraid that a security test they run is going to run them afoul of the law. This law makes the Internet less safe because it chills legitimate information security research and it's bad for the economy because it chills innovation." In this law, the only way to prosecute a person is to find a direct link toward hacking into computers. But in our modern Internet, varieties of fraud and other crimes were ignored by this law.

8.2.5 Cyber Intelligence Sharing and Protection Act

The Cyber Intelligence Sharing and Protection Act (CISPA) was written by Rep. Mike Rogers and Dutch Ruppersberger. The goal of the CISPA is to protect and safeguard network and Internet security. This bill allows private companies, organizations, and the federal government to share information to defend and protect against potential cyber threats and other Internet threats. This highly controversial bill took nearly 4 years from drafting to final passage. In this bill, the biggest controversy is that this bill grants broad new powers, which allows companies and the government to go through customers' information and data when they are investigating (tracking or identifying) a cyber threat or information related to a cyber threat. The power granted by this bill is so broad that it allows the investigating company or agency access to an individual's personal information and data, and allows the investigating company or agency to share any information obtained with the government without judicial oversight. Many people are deeply concerned about the power this bill grants, and that it creates a potential threat to privacy and existing law. Others consider this bill the second attempt to protect digital information after the failure of the SOPA and PIPA (the Stop Online Piracy Act and the Protect IP Act).

The bill is in effect an amendment to the National Security Act of 1947, which includes no mention or reference to cybercrimes or cybersecurity, because such things did not exist at that time. There are no provisions specifically aimed at cybercrime or cybersecurity in the National Security Act of 1947. In today's environment and the exponential leaps in the development of new technologies to be used in that environment, the establishment of a new bill is necessary that takes into account new technology, systems, data networks, and future trends. According to the official explanation provided by the House Intelligence Committee, there are specific aspects that need to be addressed; the first includes cybersecurity threats from other countries, and the theft of personal information and intellectual property theft are becoming more and more prevalent and serious. Second, the sharing of information and intelligence is needed. Prior to the passage of CISPA, there was no way to share information on cybersecurity threats with the private sector. Sharing of information and warning of potential threats will help prevent commercial espionage and protect the legal rights of enterprises.

Looking back at history, we find that the CISPA is indeed a political derivative brought by technological evolution and changing tactics revolving around cybersecurity and privacy. It is necessary to have a specific bill that protects individuals, companies, and government data from cybersecurity breaches.

However, the core of the dispute with the CISPA is related to the Fourth Amendment of the Constitution. As previously discussed, The Fourth Amendment stipulates that the person, residence, documents, and property of any person shall not be unreasonably searched and seized.

As for the definition of "cyber security threats," the statement in the bill is that as long as the information is "pertaining to a vulnerability" of the system or network, it is defined as a cybersecurity threat. But opponents believe that such a definition is too broad. Faced with strong doubts, the drafters of the bill also agreed to make amendments that would clarify the definition and scope of "cyber threats."

As new bills are introduced, the new sections of these bills will conflict with previous bills to some extent, and also with the Fourth Amendment. With regard to the CISPA bill, the main issue of concern is how to resolve the conflict between the advent of new technologies and capabilities (that are obviously capable of violating an individual's right to privacy) and traditional ideas and concepts revolving around individual privacy and property, which is still problematic.

In this section, we discuss some of the federal laws related to personal privacy and cybersecurity. As the development of techniques designed to exploit weaknesses in networks and data storage systems continues, there will be more issues related to cybersecurity and privacy. Whether the definition of "without authorization" in the

CFAA is better defined and modified, or the definition of "cyber threats" in CISPA is modified to better integrate the concepts of cybersecurity and individual privacy, there will always be conflicts and controversies of one form or another. The existence of these disputes has enhanced people's awareness of new technologies as well as capabilities and threats, and as the years progress, these disputes will also result in amendments of bills related to privacy and cybersecurity privacy issues.

8.3 Privacy and cybersecurity privacy case law

It is not uncommon for there to be confusion or uncertainty the when actions or procedures conducted by individuals or government entities occur in untested circumstances. This is particularly an issue when new technologies are utilized in the conduct of criminal enforcement proceedings. Further, when the public learns of new and different ways in which they are subjected to observation, apprehension about general privacy typically arises. Because of the inherent nature of the Internet, which can facilitate communication and share information, it is not surprising that privacy has become a considerable topic of concern among the public. Because cybersecurity is a relatively new field dealing with relatively new technologies, it would be erroneous to assume that existing laws or interpretations thereof are adequate to manage the use, abuse, and misuse of these technologies in observation of the public and even more importantly, the ability to use the evidence collected by Internet or personal information obtained online in criminal proceedings. At the same time, it is not fair for Internet companies such as Twitter, Facebook, Google, and so on to claim that their applications be prohibited from use in any way related to the collection of evidence to the government to track criminals and criminal activity. In the past, when new technologies or procedures have been introduced, namely, in the collection of evidence by law enforcement or other government agencies, resolutions were only provided upon the testing of such specific incidents through legal challenges and court cases.

When these types of challenges occur, legal precedents are generated from the decisions. Stanford University defines a legal precedent as "the decision of a court (or other adjudicative body) that has a special legal significance. That significance lies in the court's decision being regarded as having practical, and not merely theoretical, authority over the content of the law" [18 Boyd v. United States, 116 U.S. 616 (1886)]. These decisions lay the ground for future cases heard on similar topics, as "if there are good reasons to believe that decisions made on earlier cases are valid, then there are good reasons for believing that the same decision would be correct in a later case" [18 Boyd v. United States, 116 U.S. 616 (1886)]. Once precedent (case law) has been established, it begins to "have practical authority because they are regarded as valid and in accordance with constitutional law." Simplifying this concept somewhat, "the law is what the court stated it to be" [18 Boyd v. United States, 116 U.S. 616 (1886)]. Precedents are established by courts at various levels, both state and federal, but often pivotal cases end up in the U.S. Supreme Court. Such instances have the most wide-ranging influence as they essentially supersede lower court decisions.

Many cases related to privacy as well as search and seizure have ended up in the Supreme Court. The resultant decisions have provided lower courts further guidance on how to handle the collection of evidence and to evaluate the legality thereof. In the past, any time law enforcement agencies have used novel techniques or technologies, these cases have seemingly ended up in a variety of levels of appeal for legal consideration. Even if a precedent is established, it is not guaranteed that even minor changes in the way evidence may be collected will be considered legal under the same case law. In order for a precedent to apply, a court must decide if the present case is "identical" or "relatively the same" as the precedent case [18 Boyd v. United States, 116 U.S. 616 (1886)]. This cycle of case trials, evaluations, and reevaluations has become widespread among the introduction of neophyte surveillance technologies and therefore should present no surprise that such is likely to occur in cases surrounding cybersecurity. To understand how current cybersecurity precedents have been created, the road to present-day case law must be examined.

General surveillance. The seminal document advocating for the privacy protection of citizens of the United States resides within the Fourth Amendment, which states that *"the right of the people to be secure in their persons, houses, papers, and effects, against unreasonable searches and seizures, shall not be violated, and no warrants shall issue, but upon probable cause, supported by oath or affirmation, and particularly describing the place to be searched, and persons or things to be seized".* Much interpretation has been made by law enforcement, citizens, and most importantly, by courts, as to what is considered to be "unreasonable searches and seizures." Also, debate has centered on the term "probable cause" as well as just about every other statement within the Fourth Amendment [19Weeks v. United States, 232 U.S. 383 (1914)]. One of the earliest tests to the attachment itself was in the 1886 case of *Boyd v. United States*. In this case, the precedent to what is today considered reasonable in terms of search and seizure was partially developed. Being forced to produce business records was determined by the Supreme Court to be unreasonable, as other means of remedy were available [18 Boyd v. United States, 116 U.S. 616 (1886)].

One of the principal guiding cases in terms of admissibility of evidence can be found in *Weeks v. United States*. In this case, search and seizure of the plaintiff were deemed to be unwarranted and thus illegal. The court decided that the evidence that had been obtained illegally cannot be admitted. This case has provided individuals protection from questionable evidence-collection techniques and technologies over the years and is still cited in cases concerning illegal gathering of evidence [19Weeks v. United States, 232 U.S. 383 (1914)]. A formative case involving probable cause, *Brinegar v. United States*, established that police, given obvious indications that an individual is conducting a crime – in this case a heavily loaded vehicle being driven by a known and previously convicted liquor smuggler being operated in a suspicious location – allows for the conduct of certain types of inquiry and search [Brinegar vs. United States]. The courts have also advocated for the ability of police to search even without a warrant specific to such a search. In *United States vs. Rabinowitz*, the court found that an arrest warrant in itself deemed probable cause and subsequent, reasonable search of premises within control of the accused could be conducted [United States vs. Rabinowitz].

Further focusing and defining the concept of what is deemed reasonable and probable cause, *Augilar v. Texas* created the precedent that warrant requests must be based upon credible informants and/or reliable information. Although definitions of these standards may vary, there must be more than hearsay or circumstantial evidence to provide the basis for search and seizure [Augilar v. Texas]. Police have also been limited in what they can provide as evidence-based upon how and where such proof is collected. Without proper probable cause, evidence gathered incidental to an arrest may not be, in fact, admissible, as was the case *Beck v. Ohio* [Beck v. Ohio]. Most subsequent case law, however, generally accepts evidence gathered post-arrest, such as in "pat downs" and personal possession items. Furthermore, if individuals consent to a search, in most cases, the evidence found thereafter has been found to be admissible [Schneckloth v. Bustamonte, 412 U.S. 218 (1973)].

Electronic monitoring and observation. With the advent of more sophisticated communications and the ability to "tap" or monitor them, the issue of privacy truly took on a different meaning. For instance, could talking on a telephone be considered to be a private transaction between speakers or is personal information collected from the social media company such as Facebook considered legal evidence? In one of the first cases involving evidence collected using a wiretap of phone lines, *Olmstead v. United States*, the Supreme Court found that tapping was not considered a search or seizure. The justices noted what was lacking was a "material ingredient," meaning there was no physical removal or confiscation [Olmstead v. United States]. Probably one of the most important precedent cases related to privacy was *Katz v. United States*, which challenged the aforementioned *Olmstead* case. In this dispute, the plaintiff stated that a conversation within a telephone booth should be considered private and immune from monitoring. The court favored the plaintiff, yet they noted that places are not protected – instead personal privacy is what is to be considered. Thus, even if an individual is visible in public, such as in a telephone booth, this does not mean that they lose all expectations of privacy. In fact, that is the nature of a telephone, so an individual can enter it to have a conversation that they reasonably expect to be private. Thus, this case created the Harlan two-part test for determining privacy protection: (1) A person must have an actual expectation for privacy and (2) that this expectation is reasonable [Katz v. United States, 389 U.S. 347 (1967)].

Since Katz, reasonable expectation for privacy has been interpreted in a variety of ways. In general, when someone is in their residence, they are afforded the presumption of privacy. This apparently even applies to use of technologies that "pierce" into the privacy of one's home. In the case of *Kyllo v. United States*, Federal agents used thermal imaging devices to detect heat from growing lamps used to produce marijuana within the plaintiff's home. The Supreme Court determined that such evidence could not normally have been detected without a warranted search of the physical interior of the home. This case essentially placed a limit on the use of advanced technologies and monitoring devices on the ability to infiltrate deemed to be reasonably private. However, as the justices noted in this case, citing Katz, police are able to use "plain sight" and other reasonable "sense" to procure evidence [Kyllo v. United States, 533 U.S. 27 (2001)]. A case that precedes *Kyllo* exemplifies this extra "loophole" available to police – *United States v. Cusumano*. Similar to *kyllo*, the defendant's home was "searched" by a thermal device. The difference in this case was substantial other evidence used to provide probable cause, such as power usage, admission of the defendant as to owning growing lamps, gardening supplies, payment of rent in cash, and the procurement of additional electrical supplies to the basement through the services of a professional electrician. The courts found that aside from the thermal evidence, other indications existed that were suspicious enough to merit closer scrutiny. Whilst the case law indicates that individuals should be immune from the use of advanced technologies to infiltrate a home, the precedent does not support the notion that all parts of an individual's home or adjacent property are immune from observation.

As law enforcement duties overlap into the realm of cybersecurity, several cases have been heard at the State and Federal Supreme Court levels to argue the admissibility of evidence from things such as phone records. Cellphones perform their wide and growing variety of functions by continuously connecting to a set of radio antennas called "cell sites." Each time a phone connects to a cell cite, it generates a time-stamped record known as cell-cited location information (CSLI) [Carpenter v. United States]. *Carpenter v. United States* states that to obtain historical CSLI from the third-party providers the law enforcement must obtain a warrant supported by probable cause. In this case, the government obtained 12,898 location points cataloging Carpenter's movements over 127 days an average of 101 data points per day. Because CSLI was provided by third-party wireless carriers, Timothy Carpenter argued that those records were collected without obtaining a warrant supported by probable cause violated the Fourth Amendment. However, the government argued that in this case it could access historical CSLI through a court order alone under the Stored Communications Act (the "SCA").

Several important precedents were established relating to cybersecurity, privacy, recording techniques, data collection tactics, and technology in *U.S. v. Miller and Smith v. Maryland*. In the case *U.S. v. Miller*, the Bureau of Alcohol, Tobacco, and Firearms (ATF) issued subpoenas to two of Miller's banks, The Citizens & Southern National Bank of Warner Robins, and the Bank of Byron requesting records of Miller's accounts. Miller was convicted based on the information collected, and appealed his conviction alleging that his Fourth Amendment rights were violated. The Court held that a defendant had no right to privacy in his banking records, as they were business records belonging to the bank. In another case *Smith v. Maryland*, the Court held that police did not require a warrant to use a pen register to monitor a suspect's outgoing call data. Those two cases established a concept called the third-party doctrine. The third-party doctrine is a U.S. legal doctrine that holds that people who voluntarily give information to third parties – such as banks, phone companies, Internet service providers (ISPs), and e-mail servers – have "no reasonable expectation of privacy." Looking back at the case of Carpenter mentioned earlier, the government argued that the historical CSLI constituted routinely collected business records protected by the Supreme Court's third-party doctrine, which means that the public cannot have an expectation of privacy for certain records that were held by third-party service providers [Carpenter v. United States].

In general, precedent has also prohibited the use of "dragnet"-type observation, i.e., prolonged, constant surveillance, except under the confines of a warrant. In both

United States v. Knotts and *United States v. Karo* [United States v. Knotts, 460 U.S. 276 (1983)] [United States v. Karo, 468 U.S. 276 (1984)], the allowance for tracking of individuals was deemed permissible. These cases have also been cited to support GPS location of individuals as long as such is conducted with a warrant. In *Knotts*, Justice Rehnquist stated, "twenty-four hour surveillance of any citizen of this country will be possible, without judicial knowledge or supervision. However, the fact is that the reality hardly suggests abuse; if such dragnet-type law enforcement practices are allowed as envisioned, there should be time enough to determine whether different constitutional principles may be applicable." Thus, simply because a technology exists does not mean that there should be laws to control their use—instead the resultant data that may be collected should be the subject of potential control or limitation [United States v. Knotts, 460 U.S. 276 (1983)].

Although this description of case law is not exhaustive, it does highlight the key cases related to current standards for privacy and surveillance. It is evident that there is a long and often challenged history of cases that have been considered in order to form a current day precedent. And because of the ever-changing legal landscape and challenges to the law, it is likely that cybersecurity will provide fodder for further legal disputes.

8.4 Cybersecurity-related application

With the development of cybersecurity-related technologies, some applications related to these technologies have permeated every aspect of our daily life. Cybersecurity privacy technology ensures the safety of the Internet and guarantees the privacy of information and property. Generally speaking, these technologies related to cybersecurity privacy provide a protection mechanism. This mechanism ensures that information cannot be obtained by unauthorized third parties from any terminal of the communication network. In this section, some specific applications with regard to cybersecurity privacy will be discussed.

8.4.1 Cybersecurity and drones

An unmanned aerial vehicle (UAV), also known as a drone, is an aircraft with no human pilot on board. Recently, with the development of advanced technology and open-source hardware/software initiatives [4], the UAV industry has developed several new applications that are essential toward the creation of a strong robotics industry and essential ingredients toward maintaining economic competitiveness [5]. These include development of vital tools to support industries such as agriculture, firefighters, scientists, news agencies, filmmaking, and law enforcement, to name a few [6]. Those applications make the ideal idea become practical. These functions are currently satisfied using other, typically more expensive, technologies, but these same functions can now be performed using much less expensive, much more efficient, UAV technologies and applications. The widespread interest in the UAS arena is quickly increasing within the aviation community; the task of integrating UAS into the NAS has resulted in much attention as well as the creation of many issues and dilemmas for the various different groups of stakeholders and researchers. Although UAVs can be used for the benefit of society, this technology can also be used by malicious entities to conduct physical and cyberattacks that threaten the stability of society [7].

UAV Systems: A Brief Introduction. Depending on the size, distance, and capabilities of the specific UAV, control of the UAV can involve a number of different technologies. Larger, more capable UAV systems such as those employed by the military require powerful control stations and satellite links, and the distance from the control station to the UAV may be beyond visual line of sight (BVLOS) operations spanning hundreds or thousands of miles. Smaller, less-capable UAV systems commonly owned and operated by business entities for commercial purposes typically (but not always) involve visual line of sight (VLOS) operations. These units are typically controlled by some handheld unit that can generate a radio signal, or they may be operated by a smartphone or tablet using WiFi signals. Drones controlled by WiFi typically use IEEE 802.11 standards. In traditional drone-control systems, the communication between the drone and ground station controllers that use WiFi networks can be easily compromised and attacked. As Figure 8.1 shows, the complete UAV system consists of UAV, a ground station, and communication interconnections for transmitting information. Among them, the UAV part includes the power system, the main controller, the communication link module, the sensor, and the task execution unit. The ground station part includes remote control, intelligent terminal, and communication link module. Control commands are transmitted to the UAV through the ground station, and the data collected by the UAV and its operating data are also transmitted to the ground station. The communication link part includes the communication link between the UAV and the ground station, the satellite communication link, and the UAV to UAV communication link.

As the security threats faced by UAV systems target different components, the types of threats are also different. Threats to ground stations include software vulnerabilities, viruses, malware, Trojan horses, and key loggers. Security threats to the communication links between system components include password cracking, identity spoofing, cross-layer attacks, and multi-protocol attacks. Attacks from the four aspects of UAV sensors, communications, software, and networks are more common and two specific attacks called De-authentication Attack and GPS Spoofing Attack have attracted widespread research interests recently.

De-authentication Attack. The de-authentication attack in drones refers to a type of denial-of-service attack that targets the communication between a drone and a Wi-Fi wireless access point. For example, various types of software such as Skyjack can be used to find and control other drones without any permission. This software involves a drone that flies around, seeks the wireless signal of any other drone in the area, forcefully disconnects the wireless connection of the true owner of the target drone, then authenticates with the target drone pretending to be its owner, then feeds commands to it and all other possessed zombie drones at the will of the hacker [8].

GPS Spoofing Attack. GPS spoofing attack attempts to deceive a GPS receiver by broadcasting fake GPS signals, structured to resemble a set of normal GPS signals, or by rebroadcasting genuine signals captured elsewhere or at a different time.

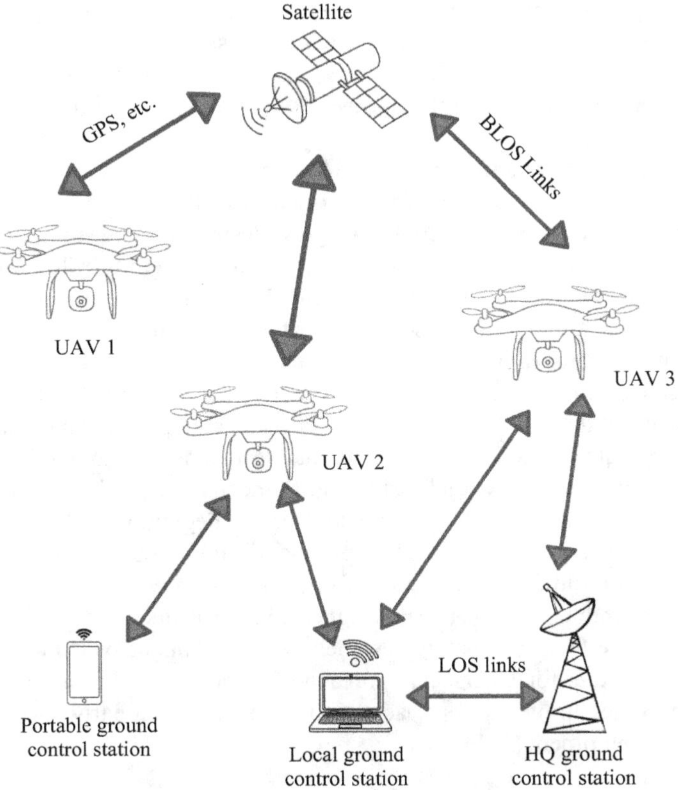

Figure 8.1 Drone communication architecture

High reliability and robustness are important requirements of safety-critical systems, especially the autopilot function of the UAV, which is usually aided by the GPS sensor. However, the civilian GPS signal itself is easily attacked because of its public nature [9]. In December 2011, Iranian forces claimed to have captured a Lockheed Martin RQ-170 Sentinel drone, operated by United States Air Force (USAF). The President of the United States asked for the return of the drone; this request was initially rejected by Iranian officials [10]. According to an Iranian engineer statement, the USAF drone was captured and controlled via a cyberattack on the GPS system, which could be classified as a GPS spoofing attack.

Cybersecurity Privacy and Drones. With the aforementioned exploration of the potential danger of drone intrusion, there are more and more public concerns about the increasing use of UAS in the open airspace, despite the fact that most of the operations are government-related and legal. For example, a protest led by the American Civil Liberty Union (ACLU) in Seattle, Washington, resulted in the Seattle Police Department ending UAV operations in the city as concerns were raised involving invasion of privacy issues. In many states, legislators have voiced their concerns about the potential for privacy invasion posed by the wide use of UASs, worrying about the personal information that could be collected by these small size drones. These and other examples have illustrated the growing concerns of the U.S. public as the use and potential abuse of UAS within the borders of the United States becomes more and more of a reality.

As privacy has been a major cause of concern for almost every government and commercial system put into use, and it is a critical issue that the UAS community must face and ethically resolve before considering a widespread agenda of integrating UAS into NAS. As the ACLU has pointed out, "we need a system of rules to ensure that Americans can enjoy the benefits of this technology without bringing our country a large step closer to a "surveillance society" in which every move is monitored, tracked, recorded, and scrutinized by the authorities."

To overcome those concerns, the Center for Democracy and Technology (CDT) has requested that the FAA issue rules on privacy and create a set of recommendations on the use of data collection. The most suitable means of maintaining privacy is regarded as "Privacy by Design" (PbD). PbD means the product, whether it is a network, Web browser, or UAV, is designed from the beginning with privacy as a top priority, not a lingering afterthought. A product designed with a specific quality or characteristic in mind (such as privacy) from conception will be much more secure compared to a product that must be redesigned and retrofitted with modifications that try to compensate for inadequately designed systems. This helps maintain standards and provide remedies for cybersecurity breaches. By adopting PbD principles, privacy intrusion becomes limited and privacy can be ensured at an early stage of product design and development [11].

8.4.2 Cybersecurity on smart grid systems

A smart grid is an electrical grid that includes a variety of operations and energy measures including smart meters, smart appliances, renewable energy resources, and energy-efficient resources [12]. It combines advanced technology including communication and information that manages an intelligent electrical network. In the United Kingdom, more than 7 million smart meters will be installed during the next years in households under a new project called Smart Meters—Project Spark funded by the European Investment Bank and six other commercial banks [13]. Currently, it is the largest existing smart meter program in Europe. This large project enables both utilities and customers to manage utility power and improve power quality. This promising project helps the local government and residents save energy, promotes economic development, and provides a variety of potential social benefits.

Design principle. In 2007, the U.S. National Energy Technology Laboratory (NETL) firstly proposed seven principal characteristics for modern smart grids [14]. Later, in 2009, the U.S. Department of Energy (DOE) merged two of them (self-heals and resist attack) and restated the design features and benefits for smart grid as follows [15]:

(1) Enabling Informed Participation by Customers.
(2) Accommodating All Generation and Storage Options.

(3) Enabling New Products, Services, and Markets.
(4) Providing the Power Quality for the Range of Needs.
(5) Optimizing Asset Utilization and Operating Efficiently.
(6) Operating Resiliently to Disturbances, Attacks, and Natural Disasters.

Cybersecurity Challenge on Smart Grid. Cyberattacks on smart grids, generally speaking, involve the attacker taking an illegal unauthorized remote control on the smart grid control unit, with a potential impact on the power system. Because of the fact that Information and Communications Technology (ICT) has been widely adopted in the smart grid environment, power communication and cybersecurity become more and more essential. Private communication networks for power systems can be vulnerable with respect to cyberattacks [16]. One of the most known and widely used malware in cyberattacks is the Stuxnet worm. This malware was discovered in 2017 [17]. The objective of this malware is to search for a specific type of programmable logic controller and reprogram parts of the code needed to operate the controllers and system. The malware will wait for certain conditions to occur and then take control of the system [18].

Broadly speaking, cybersecurity in the power industry combines all IT and communications issues that may affect the customer's privacy, and the management of the power infrastructure [19]. This includes securing the power grid in case of potential cyber events or natural disasters [19]. Wei *et al.* [20] pointed out that the development of a secure smart grid would encounter the following four challenges:

(1) Because the smart grid is completely different than the traditional power grid in terms of communication, the automation system for data communication in terms of bandwidth, delay, and new communication protocols requires a completely new design. Early obsolescence when developing a new security solution should be considered.
(2) Because many of the previous or existing automation components for traditional power grids are of an old design only used for control functionality, these components need to be replaced and re-designed with security and cybersecurity as a main objective and concern. These components were never designed with security or cybersecurity in mind, and are extremely vulnerable to breaches of security and cyberattacks.
(3) Many heterogeneous network technologies, such as ModBus [21], ModBus+ [21], ProfiBus (Process Field Bus) [22], ICCP(Inter-control Center Communication Protocol), DNP3 [23], etc., are used in current power grid. Nevertheless, most of them were designed for connectivity, without any consideration of privacy or cybersecurity.
(4) Many legacy devices that have been adopted in power automation systems are more than 20 years old. Many legacy devices were tailored to control functionality and may not have reserved computational power or memory space to perform security functionalities. Therefore, the task of combining the existing legacy devices into smart grids without weakening their control performance is a challenge, not to mention that they are totally inadequate for the purpose of combating cybersecurity threats.

Cybersecurity Privacy on Smart Grid. In smart grids, personal information is required to manage power consumption. Therefore, it is essential to protect the privacy between the customer and the service provider. Otherwise, any smart grid technology may be susceptible to security breaches and cyberattacks and is not acceptable. In August 2010 [19], the National Institute of Standards and Technology (NIST) proposed a guideline that contains a list of personal information that may be acquired through the smart grid as follows [24]:

(1) Name: the person who is responsible for the current account.
(2) Address: the location of the current service account.
(3) Account Number: unique identifier for the account.
(4) Meter Reading: power consumption recorded at 15–60 min intervals during the current billing cycle.
(5) Current Bill: current amount due on the account.
(6) Billing History: past meter readings and bills, including history of late payments/failure to pay, if any.
(7) HAN: in-home electrical appliances.
(8) Lifestyle: when the home is occupied and when it is unoccupied; when occupants are awake and when they are asleep, how many various appliances are used, etc.
(9) DER: the presence of on-site generation and/or storage devices, operational status, net supply to or consumption from the grid, usage patterns.
(10) Meter IP: the IP address for the meter, if applicable.
(11) Service Provider: identity of the party supplying this account, relevant only in retail access markets.

All the above information could be acquired by a third party without the customer's permission in a smart grid. Hence, to protect customer's privacy information, it is essential to abide by the corresponding laws and regulations designed for smart grids. Also, specific technologies, such as [25,26], were designed to secure data to prevent privacy information from being copied. Based on the study of NIST [19], three specific privacy concerns were discussed as follows:

(1) The gird system should prevent fraud. Changing the attributed location and modifying the collected data should be considered.
(2) Any collected data used to reveal customers' electricity consumption habits and living habits can be obtained by third parties and vendors.
(3) By obtaining near real-time data on energy consumption, it is possible to infer whether a house or facility is in use, people's specific activities in private houses, and so on.

Some research has already focused on those public concerns. Doohwang Chang *et al.* [27] used time and spectral domain analysis techniques to explore the transmit spectrum. They identified unauthorized transmissions and prevented potential cyberattacks. Costas *et al.* [28] provided a third-party escrow mechanism for preventing personal information from being tracked by unauthorized third parties. Additionally,

NIST [19] has already proposed a design principle that could address the privacy issues in smart grids.

The aforementioned applications are all part of a concept called smart cities, which are of interest to researchers today. A smart city incorporates ICT to reduce costs, optimize resource consumption, improve interactivity, and enhance the quality of life for its citizens [29]. When new technologies are applied to people's daily lives, many controversies and debates follow, especially with regard to new ICT technologies. The documentation of cybersecurity and privacy issues related to smart city infrastructure will increase for decades as vulnerabilities become apparent. At the same time, researchers are also working to address these public concerns. How to apply these new technologies on a large scale while ensuring public privacy will become a major topic in the coming decades. Failure to apply those technologies under the premise of ensuring public privacy will lead to unimaginable disasters. With the help of cybersecurity researchers, the government has promulgated related technology management laws and regulations, and this guidance will be one of the most effective solutions to these problems.

8.4.3 Cybersecurity and industrial environments

Day-to-day business operations depend on connectivity, technology, networks, and access to information and databases. Even the simplest transaction often depends on individual computer connectivity with larger networks and databases for successful execution. Any business, large or small, depends on communication and information exchange through a variety of networks for sales, solicitations, and service, and this includes individual customers, service personnel, and vendors to complete a successful business cycle. The aviation industry, for example, provides a classic example of a business infrastructure that involves thousands of manufacturers and service providers whose business dealings are inextricably interconnected through technology in every way. From the building and manufacture of products and components for actual aircraft, to building of airports and the many services they provide, to dealing ultimately with millions of individual customers, the aviation industry provides a perfect example of how vulnerable to cyberattack any industry is, and why cybersecurity safeguards need to be designed into the system from the beginning and not just an afterthought in response to a major industry cyberattack.

Cyberattacks, hacking, and fraud are common in any industry or business today. On a larger scale, things like cyber theft, cyber espionage, and cyber spying are typically associated with world governments looking for a way to exploit weaknesses for a variety of political and ideological reasons, but companies and industries also fall prey to the same type of exploits in attempts to steal industry secrets or disruption of competitor company operations.

On the manufacturing end of the industry, companies like Boeing and Airbus, two of the world's largest aircraft manufacturers, are potential targets for cyberattacks as their business dealings involve literally thousands of parts manufacturers and vendors who supply parts and services associated with the manufacture and sales of the products these companies manufacture, assemble, and sell. And, along with the actual transactions and contracts associated with the business operations, the individual people who work for these companies are also targets (and often the source or gateway) for cyberattacks against the companies themselves or individuals employed by the companies involved in the business cycle.

On the sales end of the aviation industry, companies such as Delta Airlines, United Airlines, and American Airlines purchase the products from Airbus and Boeing for use in consumer operations. Along with these obvious product purchases, there must also be service and maintenance contracts for the various equipment used in the transportation of cargo and people, but also thousands of supply and support services provided so cargo and people can smoothly travel on a daily basis.

Focal points of weakness and potential targets for cyber criminals are the points at which individuals, service personnel, and vendors must access a company's network in order to communicate with company personnel responsible for sales, maintenance, and service. Both internal and external threats must be considered when designing a system that is resistant to cyberattacks. Internal threats include personnel working for the company who may be careless when accessing company networks, or personnel within the company who may wish to cause harm and create havoc because of some personal motivation. External threats include any outside entity that may want access to sensitive information with regard to the company, its finances, employees, or any business practices or partners.

Airbus Exact figures are difficult to obtain, but according to various seminars and company websites (Airbus.com, 2020), it is estimated that the Airbus Company, for example, employs roughly 134,000 people made up of over 140 different nationalities, operates in over 35 countries worldwide, maintains 180 sites worldwide, and deals with over 12,000 vendors and suppliers. The company is a major supplier of commercial aircraft, helicopters, military equipment, and space-related equipment and telecommunications satellites. Cybersecurity concerns include anyone who accesses or tries to access any components of the Airbus system for any reason.

Boeing The Boeing Company is similar to Airbus in that it is estimated that Boeing employs over 140,000 people in over 65 different countries, and deals with over 12,000 vendors and suppliers worldwide (Boeing.com, 2020). The company is a major supplier of commercial aircraft, helicopters, military equipment, and space-related equipment and solutions. Similar to Airbus, cybersecurity concerns include a vast array of individuals and companies who may access or try to access any component of the Boeing system for any reason.

8.4.4 Industrial environments and cybersecurity intervention

Large companies in any sector are inviting targets simply because of the great potential for exploitation of a large and complex network, and for the great potential for disruption or operations, theft of industrial secrets, and extortion of financial resources. The aviation industry is only one example of an industry that is widespread throughout the world and potentially vulnerable from a number of exploitable access points from employees to customers to vendors and more. The objectives and techniques of an industrial hacker or cybercriminal are the same as the small-time hacker; to exploit some vulnerable point of access, infiltrate wider throughout the network/system, and obtain increased privileges within the system until critical access and control in achieved where information, resources, and/or money can be obtained or extorted.

With the extent of a large company's network in today's interconnected, worldwide business environment, a new term has emerged known as "Hyperconnectivity." Hyperconnectivity is defined as "the use of multiple systems and devices to remain constantly connected to social networks, networks, and other systems" that provide constant communication and connectivity to other sources of information (www.collinsdictionary.com, 2020). Hyperconnectivity is good in that it allows access to information and resources 24 h a day, but the negative side is that it dissolves or dilutes the concept of any type of network perimeter security as access is always available. Network perimeter security is typically the first line of defense against external threats. One analogy for network perimeter security describes it as a fence that defines the boundary between a company's internal network and business network system, and its external counterpart. And, like any fence, it can have gates and breaches that are potential external access points through which threats may enter and wreak havoc. A good perimeter security system has multiple layers of defense tactics working together to prevent unwanted cyberattacks, which could put the privacy and resources of anyone who uses the network at risk. This includes company employees, vendor and contractor employees, or company resources.

8.5 Discussion

8.5.1 Disputes about cybersecurity and privacy

The concept of privacy and the legality of various types of surveillance have provided significant precedent for cases to reference in future proceedings. Yet even in light of current case law, it is apparent from historical trends in such legal actions that a precedent specific to cybersecurity will likely need to be set. Some existing disputes are discussed as follows:

What Is a Cyberattack? Unlike other physical attacks, it is a challenge to evaluate how the law might be used to address cyberattacks and related issues. In cyberspace, many activities are different from, and even defy, the traditional physical world's principles. It is necessary to be specific in the definition of cyberattacks as this will probably be used in legal analysis of future events. A current widely cited definition of a cyberattack is "an action or actions by a nation-state to penetrate another nation's computers or networks for the purpose of causing damage or disruption." However, this definition only states that cyberattacks originate as a national-level action, thereby variations of non-national-level attacks will not be included in this definition. Another challenge is that the existing definition cannot distinguish the concept among cyber-crime, cyberattack, and cyber-war. In [30], the relationship among those three concepts is as shown in Figure 8.2

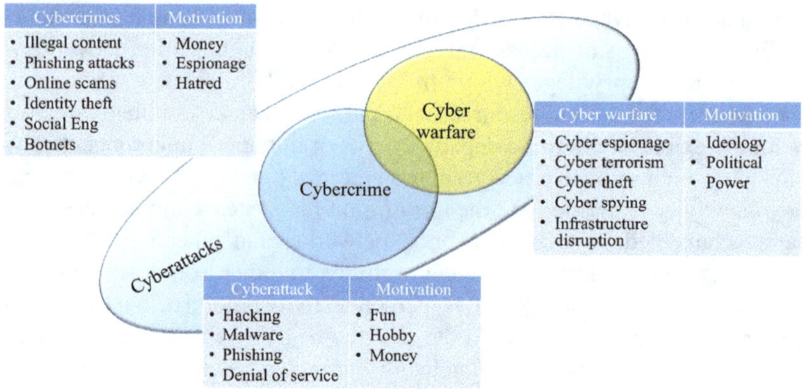

Figure 8.2 Relationship in cyberspace

In [26], the authors also adopt a narrow definition of cyberattack, one meant to focus on the unique threat posed by cyber-technologies.

Do "Ancient Bills" Still Apply to Today's Technological Environment The "ancient bill" or legacy bills such as the CFAA were enacted in the previous years. For those previous decades, the technological environment was vastly different than in recent years, and today's environment was unimaginable at that time. One well-known case is *the United States v. Swartz*. The prosecutors used the CFAA to prosecute Aaron Swartz for removing academic papers from the JSTOR database. The Federal prosecutors eventually charged him with two counts of wire fraud and eleven violations of the Computer Fraud and Abuse Act, charges carrying a cumulative maximum penalty of $1 million in fines plus 35 years in prison [3]. This case shows that CFAA gives law enforcement agencies great powers, and law enforcement agencies have the power to choose whether to arrest a person. The biggest problem with the CFAA is that under this bill, most people are more or less suspected of violating this bill. Therefore, law enforcement agencies can arrest people they want to arrest in accordance with this broad bill, even if a person like Swartz just downloaded some papers. This has the effect of causing great public concern, which creates great opposition from people resulting in demonstrations against these actions. Some good news is that recently the Supreme Court has agreed to hear and rule on the controversial US hacking law [31]. To be specific, this decision will be used to be considered how broadly to interpret the Computer Fraud and Abuse Act, America's main anti-hacking statute, and restrict the abuse of this bill.

8.5.2 Challenges to the enactment of laws relating to cybersecurity

Because of an increasingly popular Internet trend, the need for regulatory for cyber-security management has become urgent. Similarly, when referring to legislation, we are referring to the large-scale application of standards to enforce cybersecurity regulation at the national level. Generally, legislation is very effective in regulating behavior. However, when the law is not effectively or uniformly applied, it will face many challenges. Many precedents suggest that, when considering the application of new technologies and the impact of new technologies, the implementation of law is bound to affect a variety of commercial companies and ordinary individuals. These effects and conflicts will undoubtedly bring a variety of challenges to a legislation of this type. Some examples are discussed as follows:

Delay in promulgation of the law: Creating laws governing the regulation of new technologies requires consideration of various factors, such as related policies, impact on people's lives, or impact on business activities. It usually consumes a lot of time, effort, and resources beginning at the introduction of the bill through the final passage of the bill. Even if the bill is passed, there may still be various resistances to the final promulgation of the bill.

Laws falling behind in time and context: As Miguel, a security researcher, said, "Technology is advancing at a rapid rate; the development of standards may, therefore, fall far behind technological advances." When responding to present issues, the law must be at the forefront of these challenges, and must be willing to be flexible in interpretation, and implementation. However, this concept is dangerous and extremely open to abuse. Flexibility in interpretation and implementation of a law, while good from a law enforcement perspective, may easily lead to violations of an individual's freedom and privacy. Flexibility in interpretation and implementation must be carefully considered. As seen in the past, technology can advance so quickly that laws enacted today are fundamentally obsolete tomorrow because the laws of today do not anticipate what may be technologically possible on the horizon. Threats to freedom and privacy posed by rapid advances in technology must be dealt with using new laws and interpretations, not legacy laws and regulations.

Conflict between law and basic principles: Good legislation is one of the most effective ways to resolve this challenge. However, as the debates between privacy and security continue, it is evident that the law can always be improved and enhanced, particularly when specific entities and individuals remain determined to violate not only the ethical principles of the Internet, but even the foundations of constitutional human rights. In cybersecurity, the main concerns of human rights revolve around the concepts of privacy and freedom of expression. At this time, the debate surrounding the balance between technological capabilities, ethical principles, and constitutional rights may play an important role in mitigating or resolving these future challenges.

8.6 Conclusion

Because the development of the technology was difficult to predict for the framers of the U.S. Constitution, sections of their document might become or may be considered obsolete or meaningless. However, the concepts and principles of privacy and freedom of speech are solid and intact. It is virtually impossible to envision recent technology that may cause a serious debate on privacy with technology related to cybersecurity.

Some people feel that the technology used makes no difference and that existing laws, rules, and regulations that guard an individual's privacy are sufficient to handle any issue or challenges that may arise. Others believe that new technology and creative uses of that technology offer an opportunity or a gray area where the rules may be temporarily reinterpreted until they are challenged in court. This kind of debate will continue for decades to come. From one perspective, this phenomenon and debate may inevitably hinder the development of information technology and capabilities. From another perspective, this phenomenon and debate are necessary to improve and safeguard the development of information technology that is concerned with preserving privacy and human rights, thereby reducing public concern about the damage new technology may harbor against public and private interests. The delicate balance between combating cybersecurity intrusions and an individual's freedom and privacy must be maintained.

Cybersecurity technology could be used for a variety of purposes and in a variety of ways. This technology is a sharp "double-edged sword" that can not only protect people's privacy and security, but also expose people's privacy to the public. As is often the case, technology may not be a problem, but what people plan to do with that technology is what must be regulated. The development of technology is one issue; the development of carefully crafted laws and regulations is another issue. Careful and thoughtful development of new laws that consider the advanced capabilities and creative implementation of this new technology is one way to mitigate abuse and prevent erosion of Fourth Amendment rights and protections. Laws can be developed that consider a wide variety of situations and circumstances, but laws cannot be developed that cover every possible scenario.

In this chapter, we discuss the legislation related to personal privacy and cybersecurity. Through the discussion of some specific cases and legal challenges, guidance for the formulation of related rules, regulations, and legislation surrounding the development and use of this technology has been provided. Through this chapter, we hope to provide researchers and developers of this technology a window of opportunity to consider the consequences of this new technology. This window of opportunity should allow researchers and developers to calmly explore the benefits of this technology to people, while avoiding public privacy concerns by designing new technology with appropriate safeguards to ensure individual rights are not violated.

References

[1] [Online]. Available: https://www.ncsl.org/research/telecommunications-and-information-technology/cybersecurity-legislation-2020.aspx.

[2] [Online]. Available: https://www.pwc.com/us/en/services/consulting/cybersecurity-privacy-forensics/library/global-economic-fraud-survey.html.

[3] K. Farahbod, C. Shayo, and J. Varzandeh, "Cybersecurity indices and cybercrime annual loss and economic impacts," *Journal of Business and Behavioral Sciences*, 32(1), 63–71, 2020.

[4] M. Asadpour, B. Van den Bergh, D. Giustiniano, K. A. Hummel, S. Pollin, and B. Plattner, "Micro aerial vehicle networks: an experimental analysis of challenges and opportunities," *IEEE Communications Magazine*, vol. 52, no. 7, pp. 141–149, 2014.

[5] J. Villasenor, "'Drones' and the future of domestic aviation [Point of View]," *Proceedings of the IEEE*, vol. 102, no. 3, pp. 235–238, 2014.

[6] C. Barrado, R. Messeguer, J. Lopez, E. Pastor, E. Santamaria, and P. Royo, "Wildfire monitoring using a mixed air-ground mobile network," *IEEE Pervasive Computing*, vol. 9, no. 4, pp. 24–32, 2010.

[7] E. Vattapparamban, İ. Güvenç, A. İ. Yurekli, K. Akkaya and S. Uluağaç, "Drones for smart cities: Issues in cybersecurity, privacy, and public safety," *2016 International Wireless Communications and Mobile Computing Conference (IWCMC)*, Paphos, 2016, pp. 216–221, doi: 10.1109/IWCMC.2016.7577060.

[8] [Online]. Available:http://samy.pl/skyjack/

[9] J. Su, J. He, P. Cheng, and J. Chen, "A stealthy GPS spoofing strategy for manipulating the trajectory of an unmanned aerial vehicle," *IFAC-PapersOnLine*, vol. 49, no. 22, 291–296, 2016.

[10] CNN Wire Staff, "Obama says U.S. has asked Iran to return drone aircaft," Dec. 2011. [Online]. Available: http://www.cnn.com/2011/12/12/world/meast/iran-us-drone/

[11] A. Cavoukian, "Privacy and drones: Unmanned aerial vehicles," Technical Report, Ontario, Canada, Aug. 2012

[12] M. S. Saleh, A. Althaibani, Y. Esa, Y. Mhandi and A. A. Mohamed, "Impact of clustering microgrids on their stability and resilience during blackouts," *2015 International Conference on Smart Grid and Clean Energy Technologies (ICSGCE)*, 2015, pp. 195–200. doi:10.1109/ICSGCE.2015.7454295

[13] R. Khatoun and S. Zeadally, "Cybersecurity and Privacy Solutions in Smart Cities," *IEEE Communications Magazine*, vol. 55, no. 3, pp. 51–59, 2017, doi: 10.1109/MCOM.2017.1600297CM.

[14] U.S. NETL, "A systems view of the modern grid," White Paper, Jan. 2007, available at: http://www.smartgrid.gov/white papers

[15] U.S. DOE, "Smart grid system report," White Paper, Jul. 2009, available at: http://www.oe.energy.gov/SGSRMain 090707 lowres.pdf.

[16] G. N. Ericsson, "Cyber security and power system communication—essential parts of a smart grid infrastructure," *IEEE Transactions on Power Delivery*, vol. 25, no. 3, pp. 1501–1507, 2010.

[17] T.M. Chen and S. Abu-Nimeh, "Lessons from Stuxnet," *Computer*, vol. 44, no. 4, pp. 91–93, 2011.

[18] J. Xie, A. Stefanov, and C.-C. Liu, Physical and cyber security in a smart grid environment. WIREs Energy Environ, vol. 5, pp. 519–542, 2016. https://doi-org.ezproxy.libproxy.db.erau.edu/10.1002/wene.202

[19] U.S. NIST, "Guidelines for smart grid cyber security (vol. 1–3)," NIST IR-7628, Aug. 2010, available at: http://csrc.nist.gov/publications/PubsNISTIRs.htmlNIST-IR-7628.

[20] D. Wei, Y. Lu, M. Jafari, P. Skare, and K. Rohde, "An integrated security system of protecting smart grid against cyber attacks," *Innovative Smart Grid Technologies (ISGT 2010)*, Gaithersburg, MD, Jan. 2010, pp. 1–7.

[21] ModBus, "Modbus specifications and implimentation gudes," ModBus Protocol Specification v1.1b.

[22] ProfiBus, "Profibus standard," Profibus Specifications & Standards, available at: http://www.profibus.com/downloads/specifications-standards/.

[23] National Communications System, "Supervisory control and data acquisition (SCADA) systems," Technical Report, Oct. 2004, available at: http://www.ncs.gov/library/tech bulletins/2004/tib 04-1.pdf.

[24] J. Liu, Y. Xiao, S. Li, W. Liang, and C. L. P. Chen, "Cyber Security and Privacy Issues in Smart Grids," *IEEE Communications Surveys & Tutorials*, vol. 14, no. 4, pp. 981–997, 2012, doi: 10.1109/SURV.2011.122111.00145.

[25] J. Y. Liu, W. Guo, C. Fan, L. Chang, and C. Cheng, "A Practical Privacy-Preserving Data Aggregation (3PDA) Scheme for Smart Grid," *IEEE Transactions on Industrial Informatics*, vol. 15, no. 3, pp. 1767–1774, 2019, doi: 10.1109/TII.2018.2809672.

[26] H. Fan, Y. Liu, and Z. Zeng, "Decentralized privacy-preserving data aggregation scheme for smart grid based on blockchain," *Sensors*, vol. 20, no. 18, 5282, 2020.

[27] D. Chang, G. Bhat, U. Ogras, B. Bakkaloglu, and S. Ozev, "Detection mechanisms for unauthorized wireless transmissions," *ACM Transactions on Design Automation of Electronic Systems*, vol. 23, no. 6, pp. 1–21, 2018.

[28] C. Efthymiou and G. Kalogridis, "Smart Grid Privacy via Anonymization of Smart Metering Data," *2010 First IEEE International Conference on Smart Grid Communications*, Gaithersburg, MD, 2010, pp. 238–243, doi: 10.1109/SMARTGRID.2010.5622050.

[29] A. Giyenko and Y. I. Cho. Intelligent UAV in smart cities using IoT. *Proceedings of the 2016 16th International Conference on Control, Automation and Systems (ICCAS)*, Gyeongju, Korea, 16–19 October 2016; pp. 207–210.

[30] O. A. Hathaway, R. Crootof, P. Levitz, *et al.* "The Law of Cyber-Attack," *California Law Review*, 100(4), 817–885, 2012.

[31] [Online]. Available: https://arstechnica.com/tech-policy/2020/11/the-supreme-court-will-finally-rule-on-controversial-us-hacking-law.

Chapter 9
Workforce training and development in cybersecurity

Jiahao Yu[1], Sang-A Lee[1], and Dahai Liu[1]

9.1 Introduction

With the rapid expansion and increasing application of information technologies and digital solutions among enterprises and the general public, the commitment to a safer cyber environment and qualified cybersecurity professionals are surging. According to a study conducted by the International Information System Security Certification Consortium ((ISC)^2) in 2019, about 805,000 cybersecurity professionals were estimated to be working in the United States (U.S.) with a nearly-500,000 gap existing, which requires 62% growth in the U.S. cybersecurity workforce to meet the demand. From a broader perspective, the global cybersecurity gap is estimated to be 4.07 million, which means the global cybersecurity workforce needs to grow by 145% to meet the global businesses' demand [1].

The underlying reason for this phenomenon is the lack of a qualified cybersecurity workforce. Various factors hinder people from joining the cybersecurity community, including stereotypes, salary, work-life balance, and experiences. Gender bias, for example, is one of the major stereotypes for cybersecurity professionals. Cybersecurity personnel is more than twice as likely to be male, and hiring companies are likely to take male as an ideal type for the personnel they are looking for [1]. It will make a difference if hiring companies can clear the prerequisites. As more people become interested in the cybersecurity industry, another fundamental question for the cybersecurity reserve force is how to be trained and appropriately educated to fit the position. The main question for training institutions, colleges, and universities that offer cybersecurity programs, vice versa, would be how to design the course curriculum and structure properly to provide students adequate and appropriate training. Due to the fast-shifting and advancement of information technology (IT), the required cybersecurity knowledge is exploding as well. Training institutions, colleges, and universities are altering their programs correspondingly to adhere to this fast-shifting environment. Many institutions provide easy-access online classes to cultivate more cybersecurity professionals [2,3]. Some colleges opened high-end

[1]College of Aviation, Embry–Riddle Aeronautical University, Daytona Beach, FL, USA

cybersecurity graduate programs specifically to meet the demand for high-quality talent in the cybersecurity domain [4,5].

Even though these organizations have done various practices, the information covering the psychology perspective, program content, and other facets like certificates and training approaches are still understudied. In this chapter, the person-vocational fit, person-motivational fit, and person-organization fit for cybersecurity professionals were analyzed. Cybersecurity training programs, certificates, and strategies were also discussed to help both students and educators find their positions and approaches to cybersecurity training.

The remainder of this chapter is structured as follows. The rest of this section provides an overview of workforce training and education. Section 2 analyzed the cybersecurity training from theory, existing programs, and certificate perspectives. Section 3 introduced common training approaches. This chapter ends with an overall summary.

9.1.1 Workforce training overview

At the beginning of the training and education for cybersecurity students, educators should consider the topics to be covered and the gaps to be addressed when designing the course/program structure. In other words, educators should consider what Knowledge, Skills, and Abilities (KSAs) are imperative for cybersecurity training [2]. In 2017, the National Institute of Standards and Technology (NIST) published the third version of the National Initiative for Cybersecurity Education (NICE) Framework that included seven cybersecurity function categories, 33 specialty areas, and 52 work roles [6]. The NICE framework systematically structured the knowledge in cybersecurity and connected the cybersecurity work roles with specific KSAs. The NICE framework could be used to efficiently analyze the courses/programs and easily find the existing gaps. In a more recent NICE framework published in November 2020, the NIST provided a new streamlined approach of connecting KSAs with task statements [7].

González-Manzano and de Fuentes [2] analyzed 35 online classes using the NICE framework and found a massive difference in classes' coverages on NICE categories and specialty areas. It is noted that some specialty areas were not covered in the online classes, and there were some level gaps in the specialty areas. For example, they offered advanced-level knowledge for systems architecture, but there are no beginner or intermediate levels available for this specialty area. Other than the levels provided by the institutions, the required knowledge varied depending on the specific application environment. According to Yang [5], the 5 most frequently required specialty areas among 37 cybersecurity graduate programs in the business school were (1) IT security, (2) network security, (3) cybersecurity planning and management, (4) IT infrastructure, and (5) policy, legal, ethics, and compliance. Cabaj *et al.* [4], however, found that there was a surging interest in other less-technological areas, including human, organizational, and societal security, while reviewing a broad-spectrum of cybersecurity graduate programs. Besides the knowledge/level gaps in the current curriculum, educators should also consider the teaching format and related content. Learning management systems such as WebCT/Blackboard and WebTycho were incorporated to improve the learning performance in support of the online programs. However, it was also found that for technology-based courses, like cybersecurity, the traditional instruction-based method was not enough [3]. Several institutions used various platforms and online virtual labs to increase hand-on skills [3,8–15].

Several platforms and virtual labs, such as Cybersecurity Training and Operation Network Environment (CyTrONE), Leaf, Platoon, Security Requirement Education Game (SREG), SEED Labs, were designed to improve the students' hands-on learning experiences [8,11,15,16]. Various platforms or virtual labs have different approaches to cybersecurity training. It could be challenging to design a cybersecurity training platform; Son *et al.* [3] described their experiences designing a virtual lab to train thousands of students. They proposed six attributes regarding designing a new virtual lab: cost, performance, software license, network connectivity, virtual lab management, and support. Besides the six attributes, educators should also consider an appropriate deployment model as applied to their unique educational purposes and backgrounds [3]. Wahsheh and Mekonnen [14] proposed a virtual lab that offers a wide range of cybersecurity topics and scenarios. The hand-on exercises enable the application of these topics and scenarios in a controlled virtual environment. Teachers and students can set the topic and scenario according to their current or desired levels. However, the proposed training laboratory exercises require manual setups, which may be tedious and time-consuming. Beuran *et al.* [16] introduced an integrated cybersecurity framework that did not require manual setups. This integrated framework, called CyTrONE, used open-source tools that allowed the automated generation of required scenarios and contents. CyTrONE had several advantages, including the improved reliability of the task setups, less time and cost, and repeatable scenario for multiple students without redundant manual setups.

Except for the personal training, group cybersecurity training is also imperative since cyber threats were usually handled by a cybersecurity team in the enterprises [17]. When cybersecurity professionals cooperate in dealing with cyber threats, they should understand how to work with other professionals. Team members usually can learn how to work together through long-term projects. However, sometimes teams are formed in a short period for quick responses. For example, the military sometimes has short-term team-based training. Deckard and Camp [18] described the military training events in 2015, in which ad hoc teams practiced a team-based approach to cybersecurity defense. Group training exercises can help cyberteam members to know each other quicker, which is more suitable for short-term team building. For example, Platoon is a team-oriented cybersecurity training platform that has a relatively low cost. The Platoon allowed users to create one or multiple virtual networks similar to the enterprise's real network [11]. The generation of virtual networks could be completed quickly and automatically without using specially designed computers. It is much cheaper to use a team-oriented platform like Platoon to conduct cybersecurity team exercises compared to individual-oriented platforms. Ficco and Palmieri [8] also gave an open-source approach called Leaf, which could be used for cooperative and competitive cybersecurity training for various cybersecurity domains. The Leaf platform could simulate the cyberinfrastructure and reproduce realistic scenarios. Leaf was designed explicitly for cybersecurity exercises that were customized for Edge-Internet of Thing-oriented scenarios and security profiles.

Besides the training for current and future cybersecurity professionals, cybersecurity training for other occupations is also necessary. The human breach is a significant proportion of cybersecurity breaches, and humans are always the weakest link in the security chain [19]. Evans *et al.* [20] mentioned a framework for human breaches that could be used for human reliability assessment when preparing for cyber threats. Instead of giving them assessments, employees like clinical workers, who do not have a cybersecurity-related background but need to deal with the secret, personal, or classified data, could benefit from certain cybersecurity training. The training would improve their awareness of the importance of a safe cyber environment and decrease the possibility of cybersecurity breaches. Dameff *et al.* [21] detailed novel high-fidelity clinical simulations for clinical workers to recognize, identify and prevent possible threats. Furthermore, other enterprises provide cybersecurity training and awareness programs to employees to improve their information security behaviours [22]. For the general public, cybersecurity knowledge could also help them protect themselves from cyberattacks and identity theft [19,23–25]. Some cybersecurity concepts could also be taught to kids through exciting and educational games to give them early exposure to these security issues [9].

As a developing discipline, there is no direct answer for which content or structure is more suitable to accommodate the cybersecurity training. It is an open issue that educators need to address with their experiences and understandings. This chapter includes workforce training and education overview, cybersecurity workforce training from psychology, program, and certificate perspectives, and common training and education approach and strategies intended to provide educators with guidance to find better cybersecurity training solutions.

9.2 Cybersecurity workforce training

9.2.1 Theory and foundation of the cybersecurity training

Before the workforce training, desired candidates should be chosen as the reserve force. The selection of a cybersecurity workforce is a bidirectional process. People choose this field according to their interests, career goals, and market expectations, among other factors. Based on the students' academic performance, extracurriculars, and other factors, institutions or organizations choose cybersecurity professions to show their promises and potentials in this field. The organizations need to find the right program and the right person for the job. For this purpose, the theory and psychology that underlies this training and selection process should be investigated first.

Furthermore, identifying which kind of work force, their knowledge, and skillsets are suitable for certain areas and organizations is beneficial to both the institution and the students. Even though the psychological theory does not work as a criterion to assess the qualification of cybersecurity students or professionals, it can provide a foundation for better understanding between individuals and the cyber industry. In this section, the person-vocational (P-V) fit, person-motivational (P-M) fit, and person-organizational (P-O) fit needed for cybersecurity personnel are investigated.

Person-vocational fit. People usually think that technical skills are the only trait that matters in the cyber domain, but various studies also supported the need for social skills [23,26–28]. As mentioned earlier, the NICE framework connects the KSAs with specific work roles. The NICE framework identified over 1000 types of KSAs, but fewer than 10 KSAs are related to social skills or teamwork [27]. Thus, the NICE framework alone is not enough to describe cybersecurity professionals' aptitude, so other widely accepted social models should be considered. This section described the five-factor model of personality and Holland's vocational model.

The five-factor personality characteristics model includes the five most important factors: openness, conscientiousness, extraversion, agreeableness, and neuroticism, as shown in Figure 9.1 [29]. The five-factor model is generally used to organize personal traits, and each factor is a continuum variable instead of binary categories. Thus, everyone can be posited somewhere between two ends of each factor, and the values remain stable for a long time. Personality characteristics can be combined with vocational interests to show vocational preferences and possibly work performance [27].

The first characteristic of the five-factor model is agreeableness. Agreeableness is the willingness to help and care for others, and two ends refer to egocentric and altruism. People who score high in agreeableness are usually trusting and well-liked [30], and it is also a characteristic for cybersecurity professionals to build trust. Besides the trust between cybersecurity professionals and employers, cybersecurity

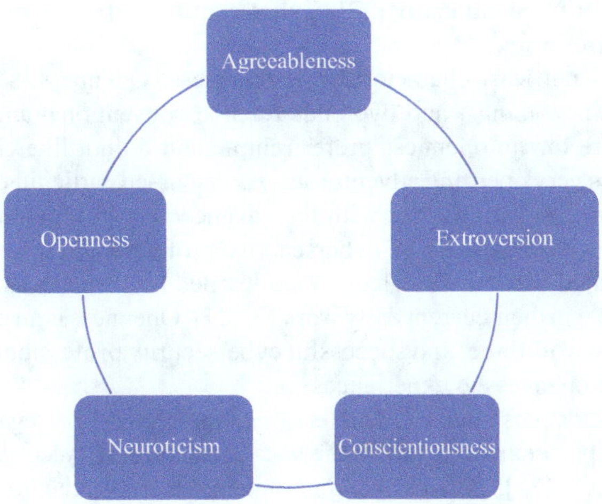

Figure 9.1 Five-factor model (adapted from [29])

professionals sometimes need to work as a team, especially in the military, in which that trust is also essential [18]. People should be comfortable with each other in the modern workforce teamwork [27]. Team players should be cooperative, which means certain agreeableness is required for cybersecurity professionals.

The second characteristic of the five-factor model is extroversion. Extroversion is the willingness to interact with the people and environment, and two ends refer to quiet and sociable [30]. As mentioned at the beginning of this section, the cyber domain requires not only technical skills but also social skills [23,26–28]. Cybersecurity professionals should also communicate with people who are not in the same field with fewer technology backgrounds, so they need to perceive their questions and concerns and explain in a way that others can understand. It is not just the communication between cybersecurity professionals with other colleagues in the company. This is also a way for them to build trust as others understand more about their work. Extroversion is required for a cybersecurity professional, but they do not have to score high in this factor as only certain social skills are needed.

The third characteristic is conscientiousness. Conscientiousness is the willingness to fulfil the responsibility thoroughly, and two ends refer to carelessness and goal-oriented [31]. It is self-evident that it is considered important to describe a person as responsible, conscientious than lax and negligent across most cultures and domains [29], especially for the cyber domain which a high level of trust is required. The cyber domain includes jargon, so the management may not understand cybersecurity professionals' actual work. More details about the relationship between the employer and employee will be discussed in the P-O fit, but the cybersecurity professionals need to be reliable and trustworthy, so they are expected to score high in conscientiousness.

The fourth characteristic is neuroticism. Neuroticism is emotional stability, and two ends refer to calm and dramatic. People who score high in neuroticism are usually emotionally unstable. It is important for cybersecurity professionals to be mentally stable at all times, or else it will create opportunities for potential human breaches, either viciously or unintentionally [32]. Cybersecurity professionals are expected to score low in neuroticism.

The last but not least characteristic is openness. Openness is the willingness to learn and try new things, and two ends refer to conventional and extraordinary. People who score low in openness prefer routine and do not like changes, while a high score means very open and adventurous. As discussed earlier in the introduction, the cyber domain content changes with the advancement in IT. Continued learning is a fundamental characteristic of cybersecurity professionals. Most cybersecurity professionals stated that the knowledge they learned by themselves after education was more valuable in their current daily work [23,27]. Openness is an essential trait for the cybersecurity workforce, and successful cybersecurity professionals are expected to score high for openness to experiences.

The five-factor personality characteristics model could be regarded as a basis that matches the personal interest with the vocational interest. According to Holland's vocational interests [33], how the personal interest matches the vocational interest could depend on how long a person will stay in this company and lacks accordance between personality and the environment resulting in dissatisfaction, unstable career paths, and impaired performance. Dawson and Robert [27] stated six key features for the future cybersecurity workforce: systemic thinker, team player, technical and social skills, civic duty, continued learning, and communications. These features are similar to the five-factor model of personality, as discussed above. The P-V fit can be shown at the beginning of the program that the program teaches and the specific vocational role the program is designed for. For example, a special graduate cybersecurity program in a business school is designed for cybersecurity professionals interested in preserving professional business confidentiality [5]. It is noted that the five-factor model is loosely correlated to the achievements of the career, but its main benefit is to be used to understand which types of students should select the cyber domain field generally [27].

Person-motivational fit. To cultivate successful cybersecurity professionals, it is crucial to improve their motivations as it is a basis to improve the learning outcome [34]. Educators can design cybersecurity classes that can spur students' motivation which makes them want to get involved. With the proper motivation, students usually will do better in the field. In Keller's Attention, Relevance, Confidence, and Satisfaction (ARCS) model of motivation, four components include the specific guidance to design appropriate classes that can promote and sustain students' motivation [9]. Four components of the ACRS model are shown in Figure 9.2 [35].

According to Keller [35], attention can be attained by perceptual arousal or inquiry arousal. Perceptual arousal is to gain students' attention by surprise, and inquiry arousal is to reach that point by asking challenging questions. In the ACRS

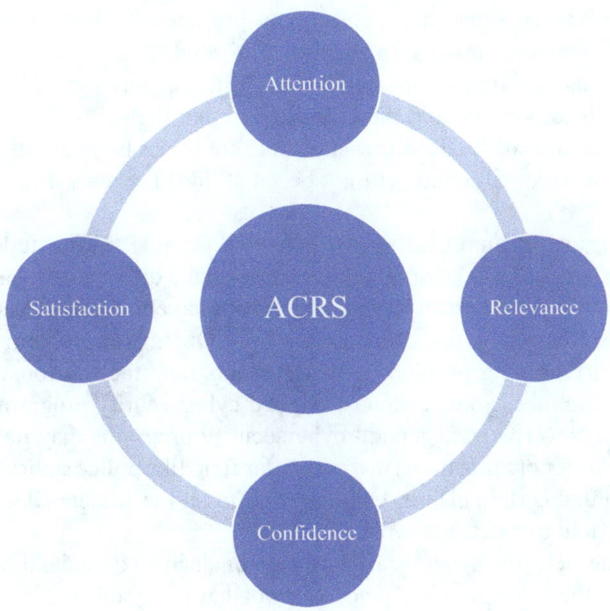

Figure 9.2 ACRS model (adapted from [35])

model of motivation, a few methods were given to stimulate students' attention, including active participation, variability, humour, incongruity and conflict, specific examples, and inquiry. They are suitable methods that educators can consider adding to their classes according to their circumstances. In cybersecurity training, these methods can be understood as fun games, different teaching formats, questions, and so on [15,36].

Relevance is the second factor in the ARCS model that could stimulate students' motivation. The course content should be converted to something students are familiar with or something they can relate to. Keller [35] gave several strategies to improve relevance, including previous experience, present worth, future usefulness, needs matching, modelling, and choice. Educators can use these strategies to make them know more about the course.

Confidence is the third factor in the ARCS model of motivation. Students should be confident about themselves in this program. Strategies to improve confidence include self-growth, objectives and prerequisites, feedback, learners' control. Educators can set objectives for them, allowing them to understand the required performance and evaluation criteria [35]. Educators can also set up some positive figures in the same field or even the same program for students to compare and understand the requirements. Other than the program's possible success, students also need to have a certain level of control about their learning objectives and future expectations [37]. They will be more confident if they have the control because they can relate the success to their effort directly [35]. The more confident the students are, the more motivated they might be.

Satisfaction is the last factor. As mentioned in the confidence factor, students should feel confident about future success in the program. Students should also be satisfied during the learning process to be more motivated. Satisfaction can be improved by praise, rewards, and immediate application [35]. Students should be praised when they accomplish something, and they should feel that the skills they learned can be useful in their future work. However, students should not be overpraised, especially on small, simple tasks. Students will not be satisfied if the rewards or praise are too frequent.

Person-organizational fit. Even with the P-V fit and P-M fit, students could not be suitable for positions as various organizations have different values or organizational climates. How the person fits the organizational environment can predict how long they may remain with this organization [33]. One major problem for designing the cybersecurity training problem is that students may face various organizations in different disciplines. Some special-designed cybersecurity programs may aim at specific fields. Nevertheless, general cybersecurity programs may not differentiate special programs for organizations in a particular field like police stations or hospitals. It is suggested that certain classes can be set up to let the students decide according to the desired field they want to work in.

Besides the field, the cybersecurity professionals in an organization should communicate with the management, who may not have a good understanding of the technical skills. This is because the organization could use the cyber domain for non-cyber-related purposes. Unlike other domains, the cyber domain is difficult to understand for people outside of the IT industry. In this situation, it is more likely to have exploitation either by unethical or accident behaviors which undermine the internet's security [32]. Therefore, a high level of trust is needed between employers and employees. Cybersecurity professionals should be reliable and trustworthy [27].

According to Bagchi-Sen *et al.* [23], different hard and soft skills are needed for cybersecurity professionals, depending on their cybersecurity career stage. In the beginning, it is essential for cybersecurity professionals to have the required technical skill. However, it is more important to relate the knowledge to the business goals of the company. Later on, as they are promoted, other skills like cooperating, trustworthiness, loyalty to the company, and relationships with the clients become critical. The cybersecurity professionals should solve the problem to reduce the risk for the organization and have a positive relationship with the clients while keeping them informed [23]. Career advancement in cybersecurity is related to the understanding between the employee and the organization.

The P-V fit, P-M fit, and P-O fit can be regarded as a reference when selecting the cybersecurity reserve force. It is noted that there will always be some outliers that can be successful in this field without perfect fits. The next section provides an overview of the programs.

9.2.2 Program overview

Some researchers reviewed cybersecurity programs in the U.S. and across the world [2,4,5]. As mentioned in the introduction, González-Manzano and de Fuentes [2] analyzed 35 online cybersecurity classes to find the training gaps. They used the NICE framework to compare the coverage of the current online cybersecurity programs that were mainly taught on Coursera, Cybrary.it, edX, and Udacity. Most of the courses were developed and released by U.S. educational institutions. The online classes all covered the seven cybersecurity function categories in the NICE framework, including Security Provision (SP), Operate and Maintain (OM), Oversee and Govern (OG), Protect and Defend (PD), Analyze (AZ), Collect and Operate (CO), Investigate (IV) [2]. One class usually covered two or more categories, but the frequency of the categories varied a lot, as shown in Table 9.1 [2].

Besides the cybersecurity function categories, 33 specialty areas were also investigated and compared. Educators decided the coverage for specialty areas according to their understanding of the course content and level. Among 33 specialty areas, some were only covered by one class, including systems architecture, data administration, training, education, and awareness, program/project management and acquisition,

Table 9.1 Frequency of category covered (adapted from [2])

Category	**SP**	**OG**	**OM**	**AZ**	**PD**	**CO**	**IV**
Frequency	19	18	13	13	9	3	3

all-source analysis, language analysis, and some were not covered, including technology R&D, systems requirements planning, knowledge management, customer service and technical support, systems analysis, incident response, collection operations, and cyber operational planning [2]. These specialty areas identified by the NICE framework are not fully covered by the training courses. Moreover, even the online classes covered some areas, but level gaps existed for the covered areas. As mentioned in the introduction, courses were taught at various levels, including beginner, intermediate, and advanced. Some areas were only covered by beginner classes, and some were only covered by advanced classes. The level gaps need to be addressed in the later program design. It is suggested that the cybersecurity program cover different levels for the specialty areas that the program includes to have a holistic and systematic approach. González-Manzano and de Fuentes [2] gave further advice based on the general analysis of the 35 online classes and detailed analysis of students' performance in Cybersecurity basics: A hands-on approach. Before the course design, educators need to consider the topic, level, teachers' commitment, gender adaption, regional adaption, interaction features, and age groups. For modification of the courses, educators should consider the suitability of content design, activities review, and whether the video is watched to the end for online classes [2].

For cybersecurity graduate programs, more advanced content is required for an in-depth learning outcome. Cabaj *et al.* [4] reviewed 21 cybersecurity graduate programs across the top 700 universities from eight countries. They first reviewed the evolution of the cybersecurity discipline by analyzing the coverage of various cybersecurity knowledge frameworks across time. These programs usually last one to two years, with 6–14 courses per year, with 8–10 courses on average. The courses needed for graduation varied from university to university. Some universities like the University of York, the Queen's University Belfast, the Lancaster University, the George Washington University, and the University of South Australia only required core courses.

In contrast, some other universities like Washington University in St. Louis and Boston University required no core courses that students can choose courses from predefined sets according to their own interests [4]. A combination of these two models is recommended with both core and elective classes. Students can have certain controls and flexibility, as discussed in the P-M fit, while ensuring the program's characters.

On the other hand, the program also changes frequently. Two kinds of revisions, including content revision and structure revision, were identified in the classes' evolution [4]. As mentioned in the introduction, required cybersecurity knowledge changes rapidly, and the course content changes according to the recent trend. The change of program includes the change of core and elective classes. The cybersecurity program's content and structure should be continuously modified to keep it up to date. In the analysis, Cabaj *et al.* [4] also noticed the increasing emphasis on the social areas, which is less technology-focused.

Cybersecurity programs are not unique programs for colleges with computer science or engineering department. Businesses also started to offer cybersecurity programs to satisfy the vast gap of cybersecurity workforce requirements in the business market. Yang [5] surveyed 529 Association to Advance Collegiate Schools

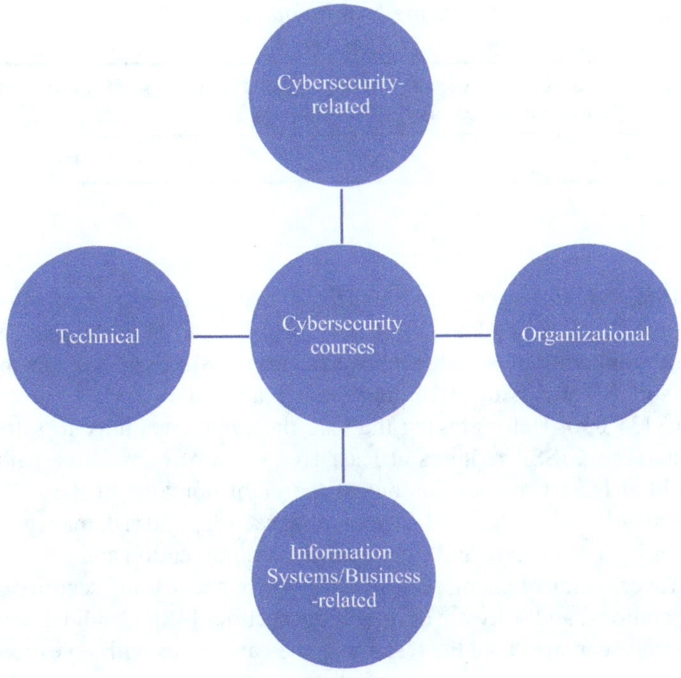

Figure 9.3 Four dimensions for cybersecurity curriculum (adapted from [5])

of Business–accredited schools and identified 37 cybersecurity graduate programs. Cybersecurity graduate programs in business schools usually include IT security, network security, cybersecurity planning and management, IT infrastructure and policy, legal, ethics, and compliance [5]. Yang [5] also developed a new cybersecurity curriculum framework for business schools which includes four dimensions, as shown in Figure 9.3. The cybersecurity program in the business school can be easily customized with this framework.

9.2.3 Cybersecurity certificate program

Besides the university degree programs, there are also a lot of cybersecurity certificates available. The top three most popular security certifications based on the $(ISC)^2$ cybersecurity workforce study 2019 will be introduced in this section. The top ten most popular security certificates are listed in Table 9.2, reproduced from the $(ISC)^2$ cybersecurity workforce study 2019 [1].

The top three certifications held by cybersecurity professionals were Certified Information Systems Security Professional (CISSP), CISSP with concentrations, and Cisco Certified Network Associate (CCNA) security. Each certification is unique in terms of the content.

Table 9.2 Top 10 security certificate held (adapted from [1])

Certificate	CISSP	CISSP with concentrations	CCNA security	CCSP	CCNP security	CIW	GSAE	CCSK	CISA	CASP+
Percentage	36%	26%	24%	18%	17%	14%	14%	14%	13%	13%

The CISSP is a well-recognized certificate that cybersecurity professionals use to demonstrate the required knowledge and technical skills to design, develop and manage an organization's overall security posture [38]. Over 142,000 professionals hold this certificate as of 2020, and the average salary for a CISSP certificate holder was $131,030. Before taking the test, the candidates have to satisfy certain requirements. The CISSP requires at least five years of cumulative paid full-time experience in at least two domains out of the eight domains of the (ISC)^2 CISSP Common Body of Knowledge (CBK), including security and risk management, asset security, security architecture and engineering, communication and network security, identity and access management, security assessment and testing, security operations, software operations, and software development security [38]. Candidates with no relevant experience can also take the test. Once the candidates with no experience pass the test, they will become an ISC associate and have six years to earn the required experience in cybersecurity.

The test of CISSP evaluates candidates' knowledge and skills across the eight CBK domains mentioned above. Candidates must score 700 out of 1000 to pass the test. Those who pass the test will have nine months to complete the endorsement process and be certified. Recertification is required every three years with continued learning and a $125 annual maintenance fee.

CISSP with concentrations is based on the CISSP, but it includes more specific architecture, engineering, and management expertise. To earn this certificate, the candidate needs to be a CISSP certificate holder. The concentration provides a way for CISSP certificate holders to improve themselves and grow focus on one of three specializations.

CCNA is another popular Information Technology (IT) certification from Cisco Systems. Unlike the CISSP, work experiences are not required for the CCNA certificate, but candidates should be familiar with the exam topics. CCNA candidates usually have the experience of implementing and administering Cisco solutions, basic knowledge of Internet Protocol (IP) addressing, a comprehensive understanding of network fundamentals [39]. The exam for the CCNA is a 120-minute exam that tests candidates' knowledge and skills in six areas. Cisco placed 25% of the exam on IP connectivity, 20% on network fundamentals, 20% on network access, 15% on security fundamentals, 10% on IP services, and 10% on automation and programmability. The CCNA certification, the same as the CISSP, is also valid for three years.

Having the certificate provides apparent benefits and advantages for employment. In the (ISC)^2 cybersecurity workforce study 2019, cybersecurity professionals make

$69,000 a year on average across the world. The cybersecurity professionals with the certificate annually earn around $71,000 on average, while those without certificates have an average income of $55,000 [1].

9.3 Common training approaches

As introduced in the first section, various platforms and virtual labs have been used for cybersecurity education and training. These platforms contained various topics and scenarios, and some could automatically generate scenarios according to students' levels. In this section, virtual labs and platforms mentioned in the introduction were reviewed in detail.

CyTrONE is an integrated cybersecurity framework proposed by Beuran *et al.* [16], which allowed the automatic generation of the content and environment. Before the development of the CyTrONE framework, three main tasks were identified for training cybersecurity professionals, including attack-oriented training, forensic analysis-oriented training, and defense-oriented training [16]. The training is needed to prepare cybersecurity students for real-life attacks. Two key elements to fulfil these training requirements are knowledge and practice. In the cybersecurity training program, educators need to provide students with adequate knowledge and a corresponding training curriculum. In terms of the cybersecurity training platform, educators should have the ability to add or modify the content and generate a training environment. As the text-based YAML was used in the CyTrONE, the database's training content could be easily added and modified through a text editor [16]. YAML is a human-friendly data serialization standard for all programming languages [40]. The training coordinator, therefore, will not have to be technically eligible to change the content. For the training environment, setting up customized environments requires high-tech expertise, which may be costly. Reusing the same environment may limit the quality due to the lack of flexibility or cause information leakage. CyTrONE added the cyber range instantiation function, which allows the automatic generation of the environment.

CyTrONE includes the user interface, training database, training management, content description processing, cyber range instantiation, and learning management system [16]. The training coordinator uses the user interface to decide the content for each training session. The training database includes the training content description and the corresponding environment. The training management can be regarded as the communication mechanism that connects all the components of the CyTrONE. The content description processing converts the content description into the learning management system to provide knowledge to end-users. The cyber range instantiation generates cyber ranges according to the content description.

The CyTrONE was evaluated regarding the functionality and performance. CyTrONE includes all the features covered in the NIST technical guide to information security testing and assessment and better cyber range instantiation performance, as shown in the evaluation experiments [16]. According to Beuran *et al.* [16], the CyTrONE made the cybersecurity training easier to be managed, reduced the overall complexity and cost in designing the training, provided accessible training content in different formats, and made future customization and development possible as open-source software.

The Leaf is another cybersecurity training platform that could simulate cyber-infrastructure and reproduce realistic Internet of Things (IoT) scenarios, including the giant network that connects devices and shares data among them [8]. Ficco and Palmieri [8] also described how they designed the platform from scratch. They suggested using a top-down approach to identify requirements, outline platform organization, and decide involved entities. The platform should be flexible and robust to fulfil different training needs.

Platoon is a team-oriented cybersecurity training platform that allows an acceptable amount of cost for team training. Virtual networks can be automatically generated with a regular computer, and its superior network design makes it the perfect platform for team training. Before developing the Platoon, Li and Xie [11] suggested cybersecurity training requirements, including special support for teamwork, cost-effectiveness, functionality, and deployability. Platoon includes five components: Platoon blue team server network, Platoon edge router, Platoon central switch, Platoon scoring engine, and Platoon perimeter firewall [11]. It is a good team training platform for cybersecurity courses or competitions.

The selection of the platform should be based on the curriculum, and a mix of the individual-oriented and team-oriented platform can be used to help students fit different working environments. Manual setup can be used at the beginning to let students learn the process of configuration and have a better understanding of the system. An automatic training platform can be used at advanced training to save time and cost.

9.4 Summary

As cybersecurity professionals' demands are surging, it is necessary to provide more learning opportunities for people interested in this field. In-depth education and training for the cybersecurity workforce are necessary. A large number of cybersecurity professionals are required, and high-quality talents are also wanted. This chapter raised two suggestions regarding the training of cybersecurity professionals. The first problem suggested that traditional colleges and universities open more online learning events and some free initiation classes to draw more attention to this field. The second problem recommended opening more high-end cybersecurity graduate programs to fulfil the requirements of high-end cybersecurity professionals. The graduate programs can relate to the undergraduate programs and contain similar specialty areas with intermediate and advanced level classes. For the graduate program, some specialty areas not covered in the undergraduate program can be introduced to give students a broader view, which may help them be a leader in a cybersecurity team with members from diverse backgrounds.

Before the cybersecurity training, the right person should be selected as the reserve force. The selection is a bidirectional process. The candidate should fit with the requirements of vocation, motivation, and organization. Six key traits suggested for the future cybersecurity workforce were systemic thinker, team player, technical and social skills, civic duty, and continued learning [27]. For the cybersecurity reserve force, educators need to design appropriate programs to provide proper training. The training content can take advantage of the NICE framework. The new revised 2020 version of the NICE framework can be applied as connecting the KSAs with tasks. Educators can select the required knowledge and skills according to the features of their programs. It is recommended that the level be consistent; in other words, beginner classes should be provided before the advanced classes. The program should also continuously modify the content due to the rapid advancement in information technology. It is also recommended that educators refer to the knowledge needed for certificates, even though some certificates require around five-year work experience. The contents of the certificates can reflect the trend of this field. During cybersecurity education and training, some virtual labs and platforms can improve students' hands-on experiences. Various platforms can be provided appropriately to students at different phases of learning to reach maximum performance.

References

[1] International Information System Security Certification Consortium ((ISC)^2). (2019). *Strategies for building and growing strong cybersecurity teams, (ISC)^2 cybersecurity workforce study, 2019.*

[2] González-Manzano, L., and de Fuentes, J. M. (2019). Design recommendations for online cybersecurity courses. *Computers & Security*, 80, 238–256. https://doi.org/10.1016/j.cose.2018.09.009.

[3] Son, J., Irrechukwu, C., and Fitzgibbons, P. (2012). A comparison of virtual lab solutions for online cyber security education. *Communications of the IIMA*, 12(4), 81. https://scholarworks.lib.csusb.edu/ciima/vol12/iss4/5.

[4] Cabaj, K., Domingos, D., Kotulski, Z., and Respício, A. (2018). Cybersecurity education: Evolution of the discipline and analysis of master programs. *Computers & Security*, 75(C), 24–35. https://doi.org/10.1016/j.cose.2018.01.015.

[5] Yang, S. C. (2019). A curriculum model for cybersecurity master's program: A survey of AACSB-accredited business schools in the united states. *Journal of Education for Business*, 94(8), 520–530. https://doi-org.ezproxy.libproxy.db.erau.edu/10.1080/08832323.2019.1590296.

[6] National Initiative for Cybersecurity Education (NIST). (2017). *National initiative for cybersecurity education (nice) cybersecurity workforce framework.* https://doi.org/10.6028/NIST.SP.800-181.

[7] Peterson, R., Santos, D., Smith, M. C., Wetzel, K. A., and Witte, G. (2020). *Workforce Framework for Cybersecurity (NICE Framework)*. Available from: https://nvlpubs.nist.gov/nistpubs/SpecialPublications/NIST.SP.800-181r1.pdf.

[8] Ficco, M., and Palmieri, F. (2019). Leaf: An open-source cybersecurity training platform for realistic edge-IoT scenarios. *Journal of Systems Architecture*, 97, 107–129. https://doi.org/10.1016/j.sysarc.2019.04.004.

[9] Giannakas, F., Papasalouros, A., Kambourakis, G., and Gritzalis, S. (2019). A comprehensive cybersecurity learning platform for elementary education. *Information Security Journal*, 28(3), 81–106. https://doi.org/10.1080/19393555.2019.1657527.

[10] Kim, B., Kim, K., Hong, S., Hong, S., and Oh, S. (2017). Development of cyber information security education and training system. *Multimedia Tools and Applications*, 76(4), 6051–6064. https://doi.org/10.1007/s11042-016-3495-y.

[11] Li, Y., and Xie, M. (2016). Platoon: A virtual platform for team-oriented cybersecurity training and exercises. *Proceedings of the 17th Annual Conference on Information Technology Education*, 20–25. https://doi.org/10.1145/2978192.2978230.

[12] Mao, J., Chua, Z. L., and Liang, Z. (2017). Enabling practical experimentation in cyber-security training. *2017 IEEE Conference on Dependable and Secure Computing*, 516–517. https://doi.org/10.1109/DESEC.2017.8073870.

[13] Tsai, P., and Yang, C. (2018). Testbed@TWISC: A network security experiment platform. *International Journal of Communication Systems*, 31(2). https://doi.org/10.1002/dac.3446.

[14] Wahsheh, L. A., and Mekonnen, B. (2019). Practical cyber security training exercises. In *2019 International Conference on Computational Science and Computational Intelligence (CSCI)*, 48–53; IEEE.

[15] Yasin, A., Liu, L., Li, T., Wang, J., and Zowghi, D. (2018). Design and preliminary evaluation of a cyber security requirements education game (SREG). *Information and Software Technology*, 95, 179–200. https://doi.org/10.1016/j.infsof.2017.12.002.

[16] Beuran, R., Tang, D., Pham, C., Chinen, K., Tan, Y., and Shinoda, Y. (2018). Integrated framework for hands-on cybersecurity training: CyTrONE. *Computers & Security*, 78, 43–59. https://doi.org/10.1016/j.cose.2018.06.001.

[17] Østby, G., Lovell, K. N., and Katt, B. (2019). EXCON teams in cyber security training. 2019 International Conference on Computational Science and Computational Intelligence (CSCI), 14–19. https://doi.org/10.1109/CSCI49370.2019.00010.

[18] Deckard, G. M., and Camp, L. J. (2016). Measuring efficacy of a classroom training week for a cybersecurity training exercise. *2016 IEEE Symposium on Technologies for Homeland Security,* 1–6. https://doi.org/10.1109/THS.2016.7568940.

[19] Ricci, J., Breitinger, F., and Baggili, I. (2019). Survey results on adults and cybersecurity education. *Education and Information Technologies*, 24(1), 231–249. https://doi.org/10.1007/s10639-018-9765-8.

[20] Evans, M., Maglaras, L. A., He, Y., and Janicke, H. (2016). Human behaviour as an aspect of cybersecurity assurance. *Security and Communication Networks*, 9(17), 4667–4679. https://doi.org/10.1002/sec.1657.

[21] Dameff, C. J., Selzer, J. A., Fisher, J., Killeen, J. P., and Tully, J. L. (2019). Clinical cybersecurity training through novel high-fidelity simulations. *The Journal of Emergency Medicine*, 56(2), 233–238. https://doi.org/10.1016/j.jemermed.2018.10.029.

[22] He, W., and Zhang, Z. (2019). Enterprise cybersecurity training and awareness programs: Recommendations for success. *Journal of Organizational Computing and Electronic Commerce*, 29(4), 249–257. https://doi.org/10.1080/10919392.2019.1611528.

[23] Bagchi-Sen, S., Rao, H. R., Upadhyaya, S. J., and Chai, S. (2010). Women in cybersecurity: A study of career advancement. *IT Professional*, 12(1), 24–31. https://doi.org/10.1109/MITP.2010.39.

[24] Knowles, W., Prince, D., Hutchison, D., Disso, J. F. P., and Jones, K. (2015). A survey of cyber security management in industrial control systems. *International Journal of Critical Infrastructure Protection*, 9, 52–80. https://doi.org/10.1016/j.ijcip.2015.02.002.

[25] McCrohan, K. F., Engel, K., and Harvey, J. W. (2010). Influence of awareness and training on cyber security. *Journal of Internet Commerce*, 9(1), 23–41. https://doi.org/10.1080/15332861.2010.487415.

[26] Anwar, M., He, W., Ash, I., Yuan, X., Li, L., and Xu, L. (2017). Gender difference and employees' cybersecurity behaviors. *Computers in Human Behavior*, 69, 437–443. https://doi.org/10.1016/j.chb.2016.12.040.

[27] Dawson, J., and Thomson, R. (2018). The future cybersecurity workforce: Going beyond technical skills for successful cyber performance. *Frontiers in Psychology*, 9, 744–744. https://doi.org/10.3389/fpsyg.2018.00744.

[28] Ghafir, I., Saleem, J., Hammoudeh, M., Faour, H., Prenosil, V., Jaf, S., Jabbar, S., Baker, T. (2018). Security threats to critical infrastructure: The human factor. *The Journal of Supercomputing*, 74(10), 4986–5002. https://doi.org/10.1007/s11227-018-2337-2.

[29] Widiger, T. A., and Costa, P. T. (2012). Integrating normal and abnormal personality structure: The Five-Factor model. *Journal of Personality*, 80(6), 1471–1506. https://doi.org/10.1111/j.1467-6494.2012.00776.x.

[30] Lim, A. G. Y. (2020). The big five personality traits. *Simply Psychology*. https://www.simplypsychology.org/big-five-personality.html.

[31] Bakker, A. B., Demerouti, E., and ten Brummelhuis, L. L. (2012). Work engagement, performance, and active learning: The role of conscientiousness. *Journal of Vocational Behavior*, 80(2), 555–564. https://doi.org/10.1016/j.jvb.2011.08.008.

[32] Umphress, E. E., and Bingham, J. B. (2011). When employees do bad things for good reasons: examining unethical pro-organisational behaviors. *Organization Science*, 22(3), 621–640. http://doi.org/10.1287/orsc.1100.0559.

[33] Holland, J. L. (1996). Exploring careers with a typology: What we have learned and some new directions. *The American Psychologist*, 51(4), 397–406. https://doi.org/10.1037/0003-066X.51.4.397.

[34] Hodges, C. B. (2004). Designing to motivate: Motivational techniques to incorporate in e-learning experiences. *The Journal of Interactive Online Learning*, 2(3), 1–7. https://www.ncolr.org/jiol/issues/pdf/2.3.1.pdf.

[35] Keller, J. M. (2010). *Motivational design for learning and performance: The ARCS model approach*. https://doi.org/10.1007/978-1-4419-1250-3.

[36] Jalali, M. S., Siegel, M., and Madnick, S. (2019). Decision-making and biases in cybersecurity capability development: Evidence from a simulation game experiment. *The Journal of Strategic Information Systems*, 28(1), 66–82. https://doi.org/10.1016/j.jsis.2018.09.003.

[37] Walker, C. O. (2011). Student perceptions of classroom achievement goals as predictors of belonging and content instrumentality. *Social Psychology of Education*, 15(1), 97–107. https://doi.org/10.1007/s11218-011-9165-z.

[38] International Information System Security Certification Consortium ((ISC)^2). (2020). *The ultimate guide to the CISSP, how to achieve the world's premier cybersecurity certification*.

[39] Cisco Systems. (2019). *CCNA Exam v1.0 (200–301)*. Available from: https://www.cisco.com/c/dam/en_us /training-events/le31/le46/cln/marketing/exam-topics/200-301-CCNA.pdf.

[40] Ben-Kiki, O., Evans, C., and Ingerson, B. (2009). Yaml ain't markup language (yaml™) version 1.1. Working Draft 2008-05, 11.

Chapter 10
Aviation cybersecurity: a cyber-physical systems perspective

Faisal Alrefaei[1], *Abdullah Alzahrani*[2], *Houbing Song*[1], *and Mohamed Zohdy*[2]

The aviation sector has been experiencing an increasing revolution in the last century. It becomes a vital transporter for different purposes such as transport, cargo, delivery, and military purposes. The importance of aviation attracts governments and researchers to focus intensely on the aviation sector to improve aviation modernity and be robust. Recently, cyber-physical systems (CPSs) have emerged as intelligent embedded systems to improve areas such as aviation, smart grid, power systems, and aerospace sectors. Therefore, the CPSs have become prominent and eligible for the aviation sector to improve performance, productivity, reliability, safety, and efficiency to meet the stakeholder's satisfaction. On the other hand, applying CPS to the aviation sector opens new challenges related to system security caused by its operation neutrality. This chapter innovatively introduces the aviation CPSs (ACPSs) framework in terms of creating secure systems. Moreover, it discusses what are the possible policies and solutions to make this system robust and secure.

10.1 Introduction

The aviation sector developed in the last century and played a crucial role in improving aviation's different aspects. Aviation is used for various purposes, such as transporting people, cargo, or military purposes to defend or attack [1]. These facts attracted governments, organizations, enterprises, and researchers to improve the aviation sector to meet their purposes, for example, productivity, efficiency, and controllability are the characteristics that are required for that purpose [2].

Recently, the emergence of cyber-physical systems (CPSs) has attracted government and enterprise organizations' intention of using this new intelligent embedded system in the aviation and aerospace sector. It is used as an advanced technology to

[1]Department of Electrical Engineering and Computer Science, Embry–Riddle Aeronautical University, Daytona Beach, FL, USA
[2]Department of Electrical and Computer Engineering, Oakland University, Rochester, MI, USA

navigate and for surveillance by integrating cyber with physical space such as Automatic Dependent Surveillance-Broadcast (ADS-B) [3]. The application of CPSs has met the requirements and has driven the aviation sector toward its goals. The nature of a CPS is the integration of the cyber world with the physical world. This integration provides some benefits and opens up new challenges represented by the integration's security issues and vulnerabilities. Aviation is used to transport passengers, cargo, and military personnel [4]. This revolution makes using aviation an integral part of future intelligent plans and modernization of organizations and the government in the upcoming century [5]. Besides, the aviation management system helps in this revolution and makes the aviation sector easier to be used with less cost. These days, aviation CPSs (ACPSs) integrate the cyber layer such as computation, network, digital storage with the physical layer such as sensors, actuators, cameras, etc. [6], as shown in Figure 10.1. This integration makes ACPSs become an intelligent embedded system. This system has led to an increased number of users and devices on the wireless networks, so cognitive radio is able to handle this considerable number of users [7]. The ACPS, hence, provides valuable characteristics such as automation, efficiency, controllability, and productivity, which must be presented in intelligent systems. For example, the ACPS can be applied as a passenger-centric system PCS with deployed sensors in the airport that gather data to improve airline services [5]. In contrast, integrating these systems and applying them in aviation opens up some

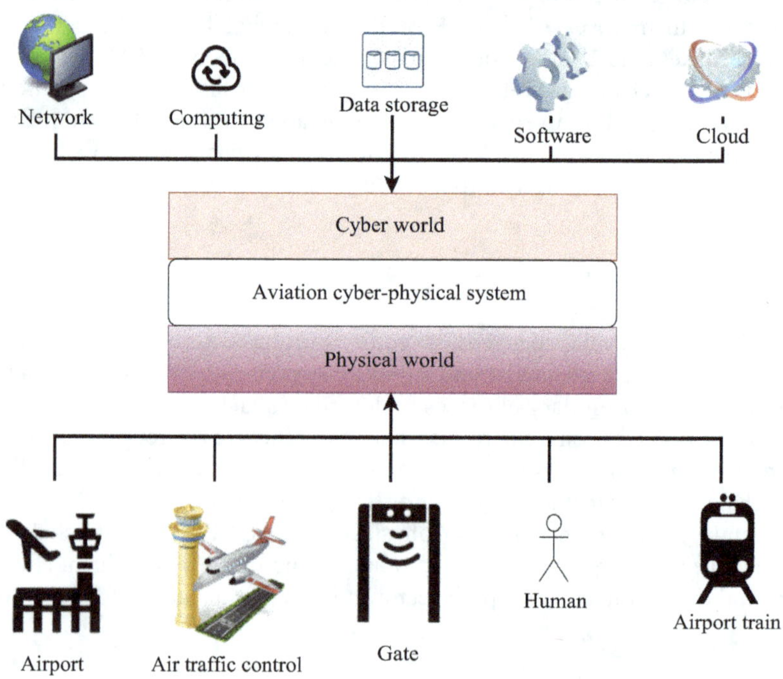

Figure 10.1 Integrating cyberspace with physical world

new challenges, such as security issues and vulnerabilities, which are caused by this integration [8]. Therefore, the security of the ACPSs has to be effectively addressed and investigated.

Security in ACPSs has to be well designed to be more effective and robust. The importance of security of the ACPSs is crucial, where the potential threat attacks are presented as threats to other organizations. This chapter focuses on the security issues raised in the physical layer to the application layer through the network layer [9]. Besides, it shows in detail the typical type of attacks that execute on the cyber and physical components with the impacts on their assets [10].

10.2 Importance of security in ACPSs

The urgent need and the nature of aviation's interoperation require security a high priority to be improved and developed. The ACPSs are integrated into two primary layers, leveraging cloud computing to achieve some goals such as increasing system performance, reduce the cost to retrieve, and use the data stored in the cloud [11]. Each one of them has a specific attack technique that is either active or passive. The cyber layer is represented by intangible components such as a network, software, data storage, etc [6]. During transmission of data, command signals, or measurement, the transmission is exposed to be damaged or data leak [12], as shown in Figure 10.2. On the other hand, the hackers intend to execute their malicious actions on the sensor or

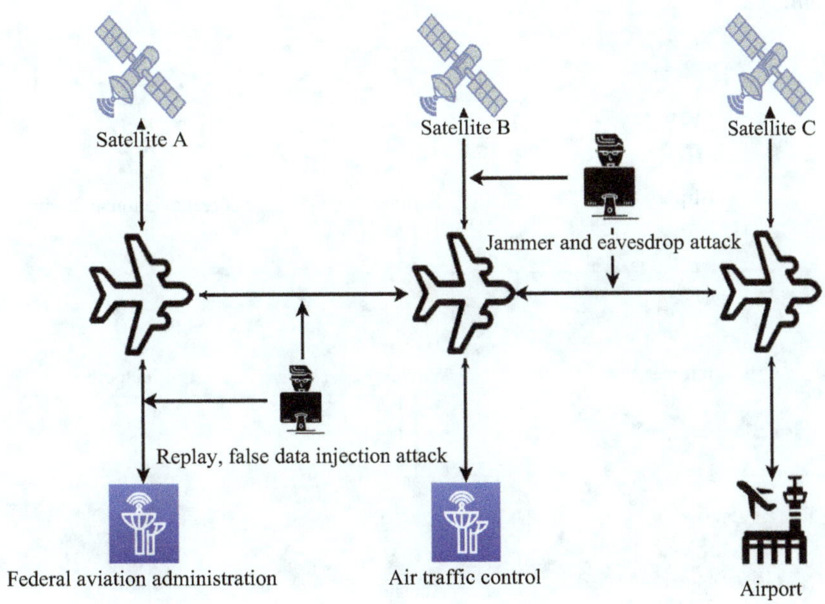

Figure 10.2 Several potential paths of cyberattacks

actuator nodes in the physical layer. Therefore, defense strategies and attack detections play a pivotal role in protecting commands and data transferred using the network.

The nature of the propagation of signals over the ACPS network is exposed to cyberattacks. The security mechanisms are applied to the sensitive part in aviation, such as cargo, passenger screen no-fly passengers, gates, aircraft, etc.; disclosing that sensitive information to hackers is a high-level risk for an industry or airlines [13]. The hackers utilize some intelligent techniques to compromise the system and drive it to catastrophic consequences.

10.3 ACPS architecture

The basic concept of the ACPS is built on integrating the cyber layer with the physical world. This integration allows logically combining two layers to constitute the ACPSs. The main three parts of the ACPSs are the application layer, network layer, and physical layer as shown in Figure 10.3. Every layer relies on the one below [14]. Therefore, any compromise on any one layer will affect all systems or at least one layer. Furthermore, the operating manner of ACPSs is constrained, so, real-time plays a vital role in this performance. Securing each one is not a proper technique: the technique has to cover every security issue in all the layers. The security technique must cover every single threat, bug, or vulnerability presented by the malicious attack. This section shows the main layers that constitute the ACPS with their functionality. Moreover, the proper protection technique must be used to present stability to the system:

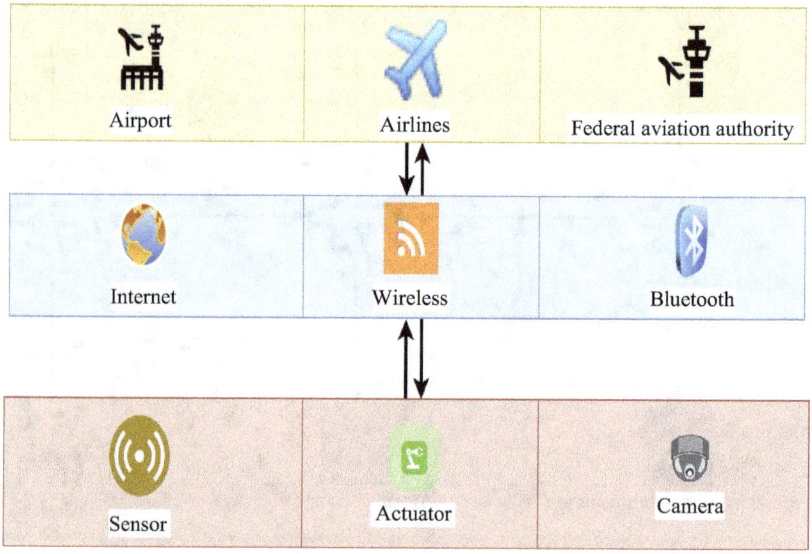

Figure 10.3 ACPS architecture

10.3.1 Application layer

The application layer is defined as the layer that stores, analyzes, and adjusts the physical world to the desired behavior [12]. It consists of the controller of the system that decides and releases the commands based on the measurements that come from the physical layer (source) [2]. The Application layer executes a complex algorithm on the aggregated data and transfers it back to the physical layer based on the system's status [15]. These layers have also sensitive data such as airlines, passengers, airports, etc. [1].

The attacks on the application layer have an enormous impact on the entire system. In this layer, attacks are intended to damage the data by violating privacy [1]. It happens by leaking the data unauthorized to this layer and manipulating the commands released to the physical layer [15]. Unauthorized access to this layer means that the system is driven by hackers who maliciously intend to damage the system for some purpose. The type of attacks executed in this layer is malicious code installed in the software deceptively or running buffer overflow on the resource to violate the availability and make the system unavailable [15]. Most of the stack holders, organizations, and governments focus on this layer because it controls the physical layer's operations and services [1].

10.3.2 Transmission layer

The transmission layer can be defined as a layer located between two layers, namely, the application and the physical layers [1]. The data transferred in this layer fragments into small pieces called packet [15]. This layer's data are transferred either via Bluetooth, local area network, wireless, or the Internet [1]. The Internet is most commonly used to protect the transmitted data [15].

During data transmission, security issues arose in this layer, especially over wireless communications [1]. The nature of the wireless communication propagating the data is suspicious to cyberattacks. In this layer, integrity confidentiality is violated. As a result, the hackers can intercept the traffic and modify the data and inject the commented data.

10.3.3 Physical layer

The physical layer is defined as a layer that has sensors to gather information and acting on the physical world to adjust the system to work in the expected behavior [16]. The aggregated data can be collected by the sensor, camera, light, etc. [15]. After that, the aggregated data are sent to the application layer for processing and releasing the commands to be executed by the actuator.

The security issues in this layer are various, where the hackers can exploit some weaknesses in this layer. The limitations of the sensors such as computation and memory are the main targets of this equipment [17]. The hackers can use malicious action to consume the power or compromise the equipment and send an inaccurate measurement [15]. This equipment's nature is preferable for the hackers where they can use the natural of wireless to damage the link between the sensor and the application layer [18].

10.4 Cyberattack on the ACPSs

The importance of using aviation across the world makes it a primary goal for hackers. The hackers always come up with an intelligent technique to disclose data and identify the vulnerabilities in the system components [19]. The hackers usually consider different types of attacks to get their damage tasks completed.

According to the architecture of the ACPSs, it is integrated between the cyber and physical worlds. Each layer must be applied by an appropriate mechanism to protect their assessment, which can be measurements, command signals, or equipment [1]. The cyber layer presents vulnerabilities, for example, malicious code, spoofing packets, and network protocols. In the physical world, the equipment's vulnerabilities or execution throw the jamming attack [20]. Therefore, the security engineers have to address the security in all ACPS aspects where the influence compromising the cyber layer affects the physical layer directly [21]. This section will describe the most common cyberattacks on the CPSs.

10.4.1 Eavesdropping attack

The eavesdropping attack is defined as the attack intercepting the transmitted data over the wireless medium to gather sensitive data [22], as shown in Figure 10.4. It can be classified into two forms in wireless networks: Active eavesdropping and passive eavesdropping [23]. Recently, a majority of researchers have been doing their researches on passive eavesdropping more than active eavesdropping. This is because

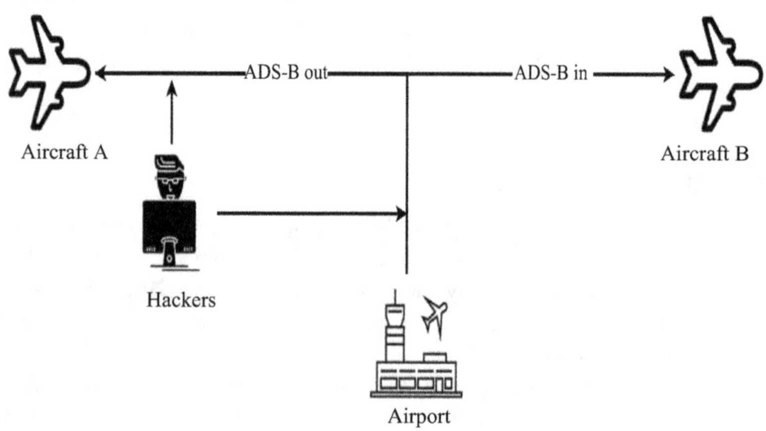

Figure 10.4 Eavesdropping attack

passive eavesdropping is more important than active eavesdropping. Moreover, active eavesdropping needs to understand the structure of the friendly node by executing passive eavesdropping to penetrate as a friendly node in the network [23]. Active eavesdrop has to do the first passive eavesdrop to complete its action.

The eavesdropping attack does not do much to the system, like damaging or catastrophic consequences [24]. It gathers valuable information about the aviation sector assets either in the physical or cyber layers [25]. In the United States, it is mandatory for all airplanes to be equipped with ADS-B to reduce cost and define the aircraft location accurately [26]. Its system is classified data into two parts ADS-B in and ADS-B out [27]. This system broadcasts unencrypted data of the airplane, such as location, altitude, speed, weather, etc. [27]. As a result, these data put aviation operations at a high level of risk. Therefore, the designer system must be aware of this issue.

10.4.2 Denial of service

Denial of service (DOS) attack, is a typical attack that targets the physical layer to make the system unavailable [15], as shown in Figure 10.5. The hackers disrupt both the network and service to make the system useless as an inaccessible service. The hackers execute this attack by using some techniques to achieve the mission [28]. For example, the hackers can target some airport resources, such as preventing the air traffic management system from sending their data to airlines, airplanes, or stakeholders [12]. Moreover, the hackers have another technique, which is corrupting the memory by using buffer overflow attacks on airport systems [29]. This type of malicious technique will degrade the system performance, in order to make the system interact slowly until being shut down.

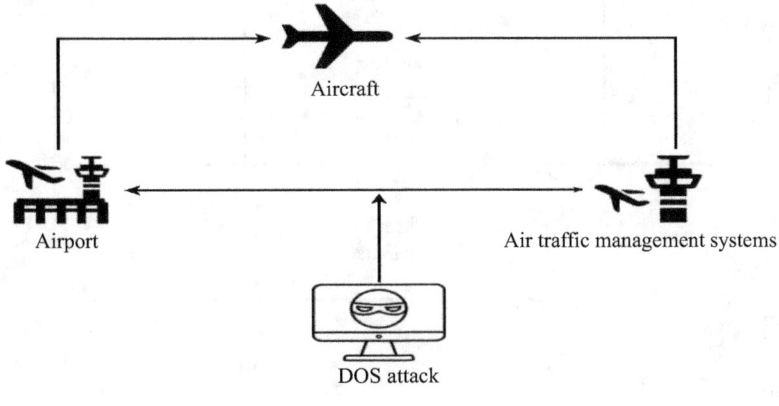

Figure 10.5 DOS attack

10.4.3 Jamming attack

A jamming attack is a form of DOS attack that is used to emit unwanted signals to interfere with transmission between the sender and the receiver. It is also used to block the receiver from receiving valid signals as shown in Figure 10.6. Therefore, the receiver fails to decode the signals when the interference rate in the received signals is high [30]. The hackers have two ways to execute this kind of attack by either corrupting the medium access control (MAC) protocol or sending signals to degrade the MAC protocol [31]. The main work of this attack is to send signals to the receiver or sender in terms of blocking them from sending and/or receiving valid signals. The hackers attempt to block the links between the sensor nodes or the link between sensors to control, in order to block the measurement from being transmitted to the controller and then to operate services. The jamming attacks can be divided into four main categories: constant jammer, deceptive jammer, random jammer, and reactive jammer [20]. Each one of these attacks has its own technique and procedure to damage the connection.

10.4.4 Constant jammer

A constant jammer is a jammer that continuously sends radio signals to the receiver or the legitimate transmitter for different purposes, for example, degrading signals or making congestion on the link connection as shown in Figure 10.7. There are several types of attacks the hackers can do, but there are two of them. One attempt is that the hackers can use an arbitrary signal to degrade the original signal

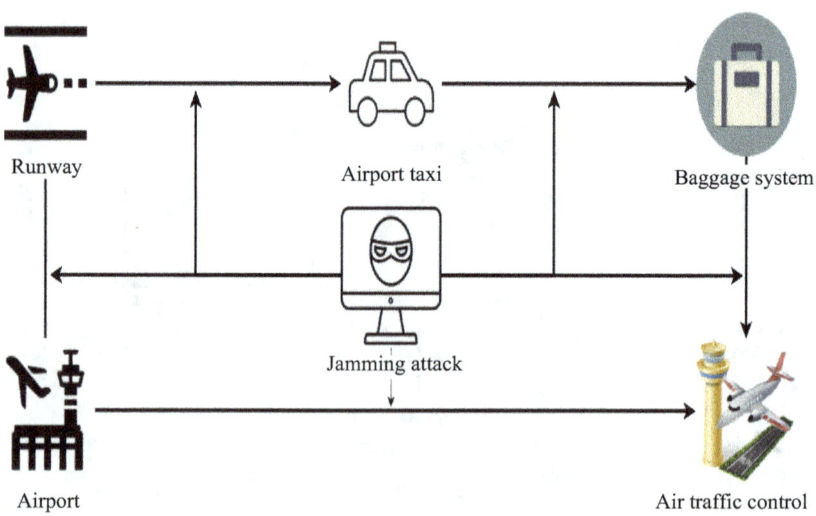

Figure 10.6 Jamming attack security model

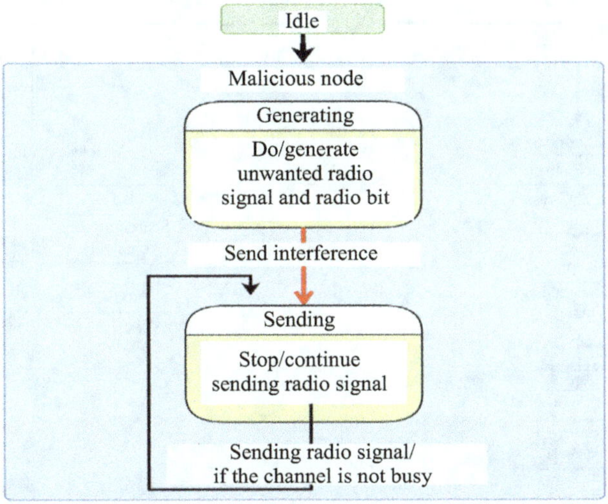

Figure 10.7 Constant jammer model

quality to make the signal at the receiver's channel undecoded. Another attempt is the hackers can transmit numerous signals on the network channel to make it busy [32]. These numerous signal transmissions cause a delay in fresh signals, where the legitimate transmitter cannot transmit the gain receiver channel as a result of this congestion on the network channel. Thus, this technique is the main reason for degrading the system performance and blocking the link between the sender and the receiver.

10.4.5 Intermittent jammer

It is a jamming that sends signals randomly from time to time as shown in Figure 10.8. This kind of jamming attack preserves its power by consuming rather than continue emitting signals. It is used to send malicious signals for the sake of destructive connection or transmit signals with malicious intent. It is similar to the constant jammer, but its approach is based on saving its energy where it is moving between active or sleeping mode [32,33].

10.4.6 Deceptive jammer

The deceptive jammer is the most innovative kind of jammer attack. It leads to injecting the data packet from the transmission while the sender sends data packets, there will be a gap, so the deceptive attack can exploit that gap to inject a valid data packet with a useless payload as shown in Figure 10.9.

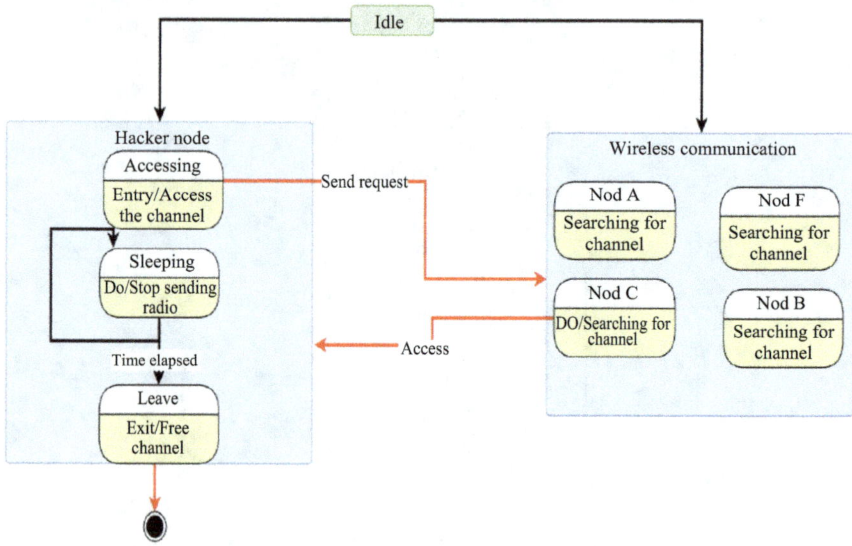

Figure 10.8 Intermittent jammer model

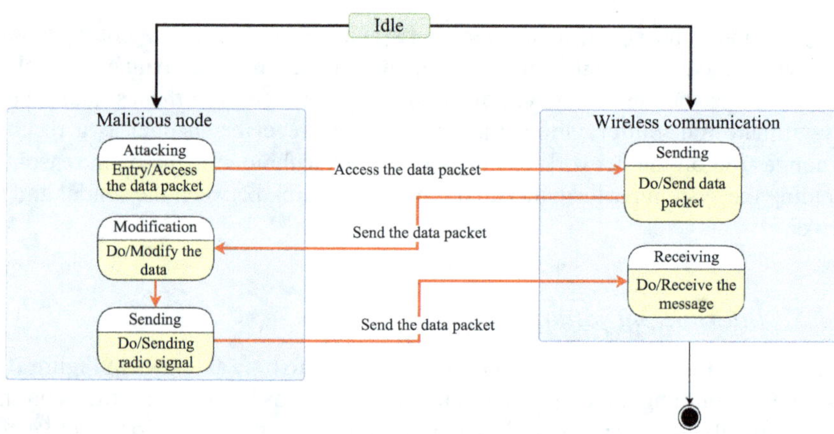

Figure 10.9 Deceptive jammer model

10.4.7 Reactive jammer

A reactive jammer is an attack that is in idle mode until the sense legitimate node is activated and then it starts sending a data packet as shown in Figure 10.10. Additionally, it senses whether the wireless channel is busy to start emitting its malicious signals to degrade the transmitting data at the receiver. The reactive jammer has the ability to distinguish whether the active legitimate node signal is weak or strong in terms of sending signals. Hence, the reactive jammer is moved to be ineffective if it

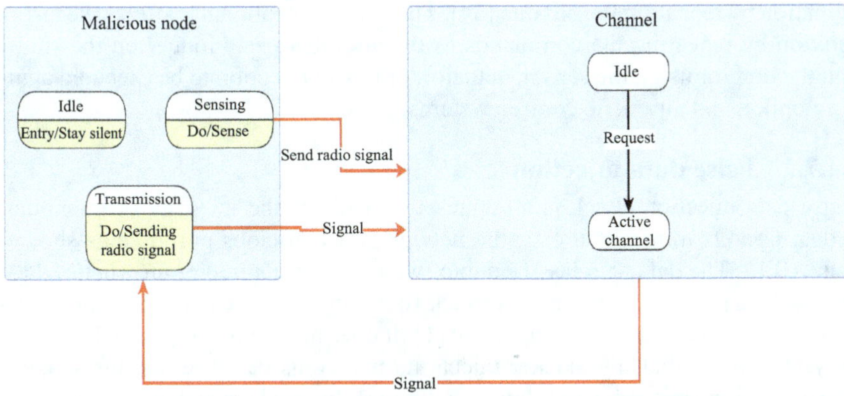

Figure 10.10 Reactive jammer model

detects weaknesses of the signals. Besides that, it continues sensing the channel until the sense that this channel is active. This type of attack consumes more energy on the sensing, so it uses a minimal amount of power for sensing.

10.4.7.1 Replay attack

A replay attack is an attack that retransmits the data packet maliciously to repeat the measurements or signal commands to degrade the system as shown in Figure 10.11. This action is a reason for creating an infinite loop and a way to increase end-to-end latency [34,35]. It, therefore, aims to delay the new data packet from source to

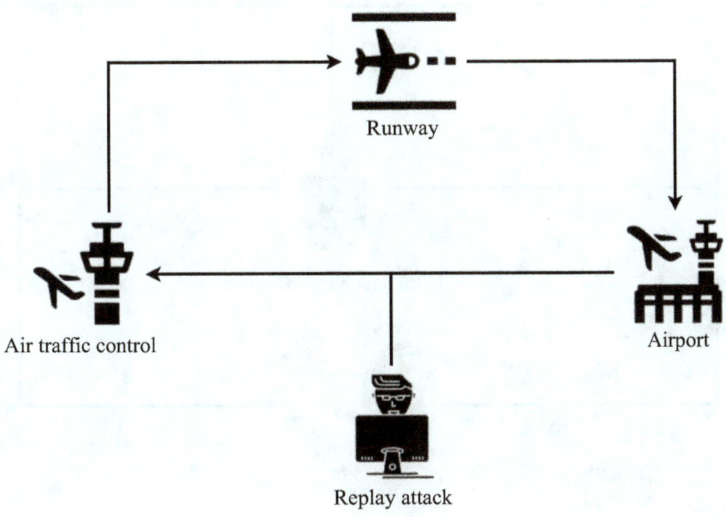

Figure 10.11 Replay attack

destination by repeating the old one [34]. This attack intentionally affects the airline's operation by repeating the commands to the aircraft to stay longer on the runway, through compromising the sensor, actuator, or link connection to be executed against the autopilot, and air traffic control systems.

10.4.7.2 False data injection

A false data injection attack is an attack that violates the integrity or manipulates the data packets transmitted over the network for malicious purposes as shown in Figure 10.12. The data are classified into two types: structured or unstructured [36]. What hackers need in this malicious attack first is unauthorized access to the network and then start listening to the data traffic [1]. It does not require any knowledge about the system. After that, the hacker injects its malicious data to drive the system to undesirable performance. False data injection can be used against the ground station or the aircraft [37]. The ground station packet injected directly affects the route and arrival time by misleading the pilot and aircraft control system. This injection data drives the aircraft to catastrophic consequences, either collision in the air or on the runway. For example, all aircraft are equipped to use ADS-B to navigate and broadcast some information to other aircraft in the surrounding area while flying. The hacker can inject a malicious data packet to the broadcast link as valid ADS-B signals to the aircraft around or to the ground station [38]. Moreover, the connection link between the surrounding aircraft and the ground station might be injected with invalid data. The destinations will not be able to define the valid data packet among the enormous received.

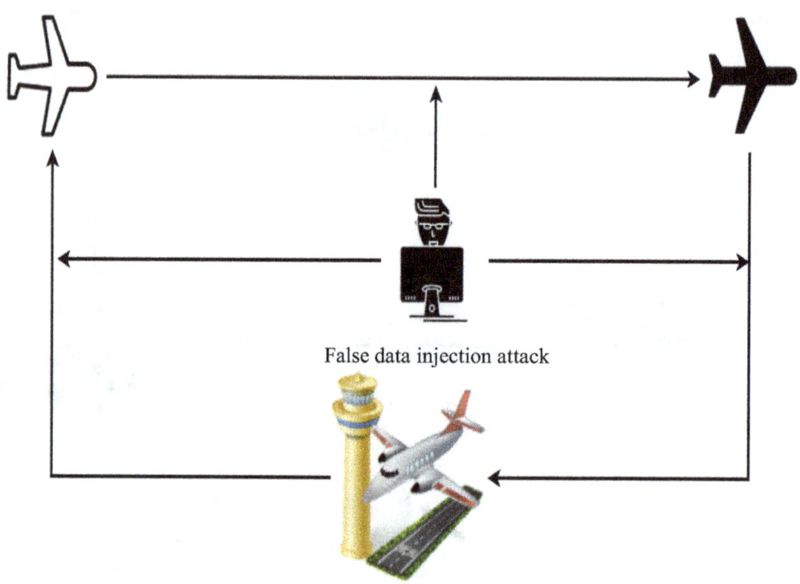

Figure 10.12 False data injection attack

10.5 Threats in the ACPS

Threats are vulnerabilities exploited by hackers to disrupt the system in terms of reaching undesirable behavior. The threats are classified into two forms: cyberattack threats on physical assets and physical threats on cyber assets. As the aviation sector is crucial, it is of great importance to define the threats and their reasons at the beginning before designing the system. This system design will make our system more trustable and robust. The threats in aviation are presented in various aspects [39]. It can be executed by installing malicious software or manipulating data transmitted over the network [40]. Therefore, the threats of aviation have been presented in the cyber world, the physical world, or the cyber-physical layers. Cyber threats affect the physical world directly and vice versa. Therefore, addressing all different scenarios that will lead to present threats at any layer in the system is needed to prevent any unauthorized behavior.

10.6 Security of wireless sensor networks in aviation

The wireless sensor network (WSN) is known as one of the ACPSs components. This technology plays a crucial role in enhancing the performance of the ACPSs. It is an intelligent, multifunction, and self-organizing technology [41]. For example, it can be deployed in the aircraft's wing to reduce fuel consumption while flying [42]. The sensors are the main components of the WSN and link each other in a short-range. The WSN can increase the remote level airport [43]. It can be used to connect all operations, for example, assigning gates, taxis, and increasing interoperation in the airport. The WSN is widely used in the aviation sector in terms of lowering cost and increasing efficiency [44]. It can also be used to sense and gather data from the physical world. Additionally, it is used to monitor and observe the behavior of the physical realm and adjust it based on erratic behavior, such as it is used in airplane monitoring system [45]. Also, the WSN is applied in terminals to sense the temperature in airports, lounges, stores, etc. [46]. Moreover, it is used to drive the aircraft on the runway to the assigned gate [1].

In contrast, the WSN has some security obstacles that can be exploited by hackers to degrade the performance of the system. For example, it can be used to limit the computation, battery, and memory resources [47]. Therefore, it is impossible to apply the traditional encryption techniques to protect transmission data over the wireless medium [48]. These weaknesses are exploited by hackers to execute their malicious actions. For instance, the hackers can damage or drop the data packet when congestion happens in the network [49]. Attacks can cause a delay in responding and could lead to dangerous events. Furthermore, the hackers contain the sensor channel to provide wrong measurements or data in order to attack the systems. Moreover, the importance of securing the WSN in the aviation sector is an urgent request [50]. The WSN is applied to constitute in the airport to efficiently and effectively sense the data to keep the controller ready for any expected or unexpected situations.

10.7 Conclusion

The aviation sector has been evolving and has become more efficient in the last century. The emerging of the CPS improves aviation operations and makes them more efficient and effective. The ACPS integrated a two-cyber and a physical layer. Therefore, the integration increases the performance and provides new services to airlines, stakeholders, governments, and enterprises. In contrast, it opened up new challenges to the system where the security issues expanded and some vulnerabilities presented. The security engineer and researchers started to focus on the security issues to innovate new techniques to enhance the security and reliability of the ACPSs.

References

[1] Alrefaei F, Alzahrani A, Song H, *et al.* Security of Cyber Physical Systems: Vulnerabilities, Attacks and Countermeasure. In: 2020 IEEE International IOT, Electronics and Mechatronics Conference (IEMTRONICS). IEEE; 2020. p. 1–6.

[2] Sun Y, Song H. *Secure and trustworthy transportation cyber-physical systems*. Springer; 2017.

[3] Sampigethaya K, Poovendran R. Cyber-physical system framework for future aircraft and air traffic control. In: 2012 IEEE Aerospace Conference. IEEE; 2012. p. 1–9.

[4] Song H, Fink G, Jeschke S. *Security and Privacy in Cyber-Physical Systems*. Wiley Online Library; 2017.

[5] Aboti CD. A Review: Forthcoming Commercial Aviation and Cyber Risks. –;.

[6] Sampigethaya K, Poovendran R. Aviation cyber–physical systems: Foundations for future aircraft and air transport. *Proceedings of the IEEE*. 2013;101(8):1834–1855.

[7] Aboti C. Cyber Security for Connected Commercial Aviation: Case Study. –;.

[8] Li W, Song H, Wei Y, *et al.* Toward more secure and trustworthy transportation cyber-physical systems. In: *Secure and Trustworthy Transportation Cyber-Physical Systems*. Springer; 2017. p. 87–97.

[9] Song H, Srinivasan R, Sookoor T, *et al. Smart Cities: Foundations, Principles, and Applications*. John Wiley & Sons; 2017.

[10] Liu Y, Weng X, Wan J, *et al.* Exploring data validity in transportation systems for smart cities. *IEEE Communications Magazine*. 2017;55(5):26–33.

[11] Ren L, Liao H, Castillo-Effen M, *et al.* Transformation of mission-critical applications in aviation to cyber-physical systems. In *Cyber-Physical Systems*. Elsevier; 2017. p. 339–362.

[12] Alrefaei F. The Importance of Security in Cyber-Physical System. In: 2020 IEEE 6th World Forum on Internet of Things (WF-IoT); 2020. p. 1–3.

[13] Xu C, Zhang K, Jiang Y, *et al.* Communication Aware UAV Swarm Surveillance Based on Hierarchical Architecture. *Drones*. 2021;5(2):33.

[14] Ekedebe N, Yu W, Lu C, *et al.* Securing transportation cyber-physical systems. In: *Securing Cyber-Physical Systems*. CRC Press; 2015. p. 163–196.

[15] Ashibani Y, Mahmoud QH. Cyber physical systems security: Analysis, challenges and solutions. *Computers & Security*. 2017;68:81–97.

[16] Hu P, Ning H, Qiu T, *et al.* Security and privacy preservation scheme of face identification and resolution framework using fog computing in internet of things. *IEEE Internet of Things Journal*. 2017;4(5):1143–1155.

[17] Dartmann G, Song H, Schmeink A. *Big Data Analytics for Cyber-Physical Systems: Machine Learning for the Internet of Things*. Elsevier; 2019.

[18] Kim D, Song H, Cano JC, *et al. Efficient Data Forwarding in Internet of Things and Sensor Networks. Hindawi*; 2018.

[19] Jiang Y, Wang M, Jiao X, *et al.* Uncertainty theory based reliability-centric cyber-physical system design. In: 2019 International Conference on Internet of Things (iThings) and IEEE Green Computing and Communications (GreenCom) and IEEE Cyber, Physical and Social Computing (CPSCom) and IEEE Smart Data (SmartData). IEEE; 2019. p. 208–215.

[20] Wang J, Liu Y, Song H. Counter-Unmanned Aircraft System (s)(C-UAS): State of the Art, Challenges, and Future Trends. *IEEE Aerospace and Electronic Systems Magazine*. 2021;36(3):4–29.

[21] Xu Q, Ren P, Song H, *et al.* Security-aware waveforms for enhancing wireless communications privacy in cyber-physical systems via multipath receptions. *IEEE Internet of Things Journal*. 2017;4(6):1924–1933.

[22] Zou Y, Zhu J, Zheng B. Defending against eavesdropping attack leveraging multiple antennas in wireless networks. In: 2013 8th International Conference on Communications and Networking in China (CHINACOM). IEEE; 2013. p. 699–703.

[23] Dai HN, Wang Q, Li D, *et al.* On eavesdropping attacks in wireless sensor networks with directional antennas. *International Journal of Distributed Sensor Networks*. 2013;9(8):760834.

[24] Xu Q, Ren P, Song H, *et al.* Security enhancement for IoT communications exposed to eavesdroppers with uncertain locations. *IEEE Access*. 2016;4:2840–2853.

[25] Sun Y, Song H, Jara AJ, *et al.* Internet of things and big data analytics for smart and connected communities. *IEEE Access*. 2016;4:766–773.

[26] Cretin A, Vernotte A, Chevrot A, *et al.* Test Data Generation for False Data Injection Attack Testing in Air Traffic Surveillance. In: 2020 IEEE International Conference on Software Testing, Verification and Validation Workshops (ICSTW). IEEE; 2020. p. 143–152.

[27] Pollack J, Ranganathan P. Aviation Navigation Systems Security: ADS-B, GPS, IFF. In: Proceedings of the International Conference on Security and Management (SAM). The Steering Committee of The World Congress in Computer Science, Computer Engineering, and Applied Computing; 2018. p. 129–135.

[28] Jeschke S, Brecher C, Song H, *et al. Erratum to: industrial internet of things. In: Industrial Internet of Things*. Springer; 2017. p. E1–E1.

[29] Wei W, Sun Z, Song H, *et al.* Energy balance-based steerable arguments coverage method in WSNs. *IEEE Access*. 2017;6:33766–33773.

[30] Niu J, Ming Z, Qiu M, *et al.* Defending jamming attack in wide-area monitoring system for smart grid. *Telecommunication Systems*. 2015;60(1):159–167.

[31] Wang L, Wyglinski AM. A combined approach for distinguishing different types of jamming attacks against wireless networks. In: Proceedings of 2011 IEEE Pacific Rim Conference on Communications, Computers and Signal Processing. IEEE; 2011. p. 809–814.

[32] Zou Y, Zhu J, Wang X, *et al.* A survey on wireless security: Technical challenges, recent advances, and future trends. *Proceedings of the IEEE*. 2016;104(9):1727–1765.

[33] Jiang Y, Song H, Yang Y, *et al.* Dependable model-driven development of CPS: From stateflow simulation to verified implementation. *ACM Transactions on Cyber-Physical Systems*. 2018;3(1):1–31.

[34] Ding D, Han QL, Xiang Y, *et al.* A survey on security control and attack detection for industrial cyber-physical systems. *Neurocomputing*. 2018;275: 1674–1683.

[35] Reji P, Natarajan K, Shobha K. Secured Wireless Network for Aircraft Intra Communication. In: 2020 IEEE International Conference on Electronics, Computing and Communication Technologies (CONECCT). IEEE; 2020. p. 1–6.

[36] Ahmed M, Pathan ASK. False data injection attack (FDIA): an overview and new metrics for fair evaluation of its countermeasure. *Complex Adaptive Systems Modeling*. 2020;8:1–14.

[37] Li W, Song H. ART: An attack-resistant trust management scheme for securing vehicular ad hoc networks. *IEEE transactions on intelligent transportation systems*. 2015;17(4):960–969.

[38] Song H, Rawat DB, Jeschke S, *et al.* *Cyber-physical systems: foundations, principles and applications*. Morgan Kaufmann; 2016.

[39] Lv Z, Song H, Basanta-Val P, *et al.* Next-generation big data analytics: State of the art, challenges, and future research topics. *IEEE Transactions on Industrial Informatics*. 2017;13(4):1891–1899.

[40] Liu Y, Wang J, Song H, *et al.* Blockchain-based secure routing strategy for airborne mesh networks. In: 2019 IEEE International Conference on Industrial Internet (ICII). IEEE; 2019. p. 56–61.

[41] Chelli K. Security issues in wireless sensor networks: Attacks and countermeasures. In: Proceedings of the world congress on engineering. vol. 1; 2015. p. 876–3423.

[42] Schmidt JF, Neuhold D, Bettstetter C, *et al.* Wireless Connectivity in Airplanes: Challenges and the Case for UWB. *IEEE Access* . 2021;9:52913–52925.

[43] Mehmood A, Mukherjee M, Ahmed SH, *et al.* NBC-MAIDS: Naïve Bayesian classification technique in multi-agent system-enriched IDS for securing IoT against DDoS attacks. *The Journal of Supercomputing*. 2018;74(10): 5156–5170.

[44] Butun I, Österberg P, Song H. Security of the Internet of Things: Vulnerabilities, attacks, and countermeasures. *IEEE Communications Surveys & Tutorials*. 2019;22(1):616–644.

[45] Gao S, Dai X, Hang Y, *et al.* Airborne wireless sensor networks for airplane monitoring system. *Wireless Communications and Mobile Computing*. 2018;2018.

[46] Lin C, Wang P, Song H, *et al.* A differential privacy protection scheme for sensitive big data in body sensor networks. *Annals of Telecommunications*. 2016;71(9):465–475.

[47] Zhu J, Zou Y, Zheng B. Physical-layer security and reliability challenges for industrial wireless sensor networks. *IEEE Access*. 2017;5:5313–5320.

[48] Sun X, Ng DWK, Ding Z, *et al.* Physical layer security in UAV systems: Challenges and opportunities. *IEEE Wireless Communications*. 2019;26(5):40–47.

[49] Yue X, Liu Y, Wang J, *et al.* Software defined radio and wireless acoustic networking for amateur drone surveillance. *IEEE Communications Magazine*. 2018;56(4):90–97.

[50] Albalawi M, Song H. Data Security and Privacy Issues in Swarms of Drones. In: 2019 Integrated Communications, Navigation and Surveillance Conference (ICNS). IEEE; 2019. p. 1–11.

Index

www.ingramcontent.com/pod-product-compliance
Lightning Source LLC
LaVergne TN
LVHW011848200926
848484LV00002B/420
* 9 7 8 1 8 3 9 5 3 3 2 1 1 *